THE DEVELOPING HEART

THE DEVELOPING HEART

CLINICAL IMPLICATIONS OF ITS MOLECULAR BIOLOGY AND PHYSIOLOGY

EDITED BY MARIANNE J. LEGATO, M.D.
Associate Professor of Clinical Medicine
Columbia University College of Physicians and Surgeons

MARTINUS NIJHOFF PUBLISHING
A MEMBER OF THE KLUWER ACADEMIC PUBLISHERS GROUP
BOSTON/THE HAGUE/DORDRECHT/LANCASTER

Distributors for North America:
Kluwer Academic Publishers
190 Old Derby Street
Hingham, MA 02043

Distributors for all other countries:
Kluwer Academic Publishers Group
Distribution Centre
P.O. Box 322
3300 AH Dordrecht
The Netherlands

Library of Congress Cataloging in Publication Data
Main entry under title:

The Developing heart.

Includes bibliographies and index.
1. Heart — Growth. 2. Molecular biology. 3. Developmental biology.
I. Legato, Marianne J., 1935– .
[DNLM: 1. Cardiovascular Diseases. 2. Cardiovascular System — embryology.
3. Cardiovascular System — growth & development. WG 100 D489]
QP111.4.D48 1984 612′.17 84-1118
ISBN 0-89838-672-1

To Dr. George Hamilton Wilkie

CONTENTS

CONTRIBUTING AUTHORS

Peter Danilo, Jr.
Assistant Professor of Pharmacology
Columbia University
College of Physicians and Surgeons
630 West 168th Street
New York, NY 10032, USA

M. Irené Ferrer
Professor Emeritus of Clinical Medicine
Department of Medicine
Columbia University
College of Physicians and Surgeons
630 West 168th Street
New York, NY 10032, USA

Sture Forsgren
Institute of Anatomy
University of Umea
Umea, SWEDEN

Margaret Ann Goldstein
Associate Professor of Medicine and Cell Biology
Baylor College of Medicine, Room 515B
Section of Cardiovascular Sciences
Texas Medical Center
Houston, TX 77030, USA

Luisa Gorza
Assistant Professor of Pathology
University of Padova
Institute of General Pathology
Via Loredan 16
Padova, ITALY

Kent Hermsmeyer
Professor of Pharmacology
University of Iowa
College of Medicine
Department of Pharmacology
2-370 Bowen Science Building
Iowa City, Iowa 52242, USA

Jay M. Jarmakani
Professor of Pediatrics
University of California at Los Angeles
College of Medicine
Center for the Health Sciences
Los Angeles, CA 90024, USA

Glenn A. Langer
Professor of Medicine and Physiology
University of California at Los Angeles
College of Medicine
Center for the Health Sciences
Los Angeles, CA 90024, USA

Jean-Jacques Mercadier
Unite de Recherches Sur Le Metabolisme
Du Coeur et des Vaisseaux
U 127
Batiment Inserm
Hospital Laribisiere
41 Boulevard de la Chapelle
Paris, FRANCE

Henry N. Neufeld
The Chaim Sheba Medical Center
Sackler School of Medicine
Tel-Hashomer 52621, ISRAEL

Suzanne Oparil
Professor of Medicine
University of Alabama
College of Medicine
Cardiovascular Research Training Center
Department of Medicine
University Station
Birmingham, AL 35294, USA

Z. J. Penefsky
Associate Professor of Physiology and Biophysics
Department of Physiology and Biophysics
Mt. Sinai School of Medicine
1 Gustave Levy Place
New York, NY 10029, USA

Richard B. Robinson
Assistant Professor of Pharmacology
Columbia University
College of Physicians and Surgeons
Department of Pharmacology
630 West 168th Street
New York, NY 10032, USA

Saverio Sartore
Associate in Pathology
University of Padova
Institute of General Pathology
Via Loredan 16
Padova, ITALY

Stefano Schiaffino
Professor of General Pathology
University of Padova
Institute of General Pathology
Via Loredan 16
Padova, ITALY

Adam Schneeweiss
The Chaim Sheba Medical Center
Sackler School of Medicine
Tel-Hashomer 52621, ISRAEL

Ketty Schwartz
Maître de Recherches CNRS
Unite de Recherches Sur Le Metabolisme
Du Coeur et des Vaisseaux
U 127
Batiment Inserm
Hospital Laribisiere
41 Boulevard de la Chapelle
Paris, FRANCE

Lars-Eric Thornell
Department of Anatomy
University of Umea
Umea, SWEDEN

Laurel Traeger
Instructor, Department of Neurology
Baylor College of Medicine, Room 515B
Section of Cardiovascular Sciences
Texas Medical Center
Houston, TX 77030, USA

Radovan Zak
Professor of Medicine and Pharmacological and Physiological Sciences
Department of Medicine
University of Chicago
1025 East 57th Street
Chicago, IL 60637, USA

FOREWORD

Heart disease, despite recent improvements, continues to be the single most important cause of death and disability in the United States. It is estimated that the direct cost of medical care for cardiovascular disease is $6 billion dollars per year. Data compiled by the National Center for Health Statistics reveal a dramatic decline in deaths from cardiovascular disease in the United States (greater than 20% since 1968). This phenomenon has been the subject of in-depth study. It is clear that the decline is real and not a statistical artifact. The decrease in mortality has been noted in all sections of the country, though the onset and rate of decline varies in different regions of the country. Both primary prevention, through changes in risk factors, and basic and applied research leading to earlier recognition and improved treatment have contributed to the decline. They do not fully explain the decline. Further research is needed to clarify this issue.

Clinical cardiologists have been exposed to a veritable explosion of new knowledge of mechanisms of cardiovascular disease, development of new improved non-invasive diagnostic techniques, and the pharmacodynamics of agents affecting the cardiovascular system. This new knowledge results from contributions made by individuals from diverse disciplines including cellular and molecular biologists, geneticists, hematologists, cardiologists, and cardiovascular surgeons. Epidemiologists have made observational contributions of tremendous importance in community settings leading to population patterns of disease and to development of programs aimed at prevention of these diseases. Clearly safe, effective, preventive programs offer more hope for continuing the decline in morbidity and mortality than does treatment of far-advanced disease. Both approaches, however, are valid.

Chronic cardiovascular disease in adults often has its roots in childhood. Heart disease at any age may be an expression of events occurring in embryonic or fetal life. Cellular responses frequently are controlled by genetic influences, while intrauterine environmental factors further influence and complicate developmental patterns in the fetus and developing infant and child. Maternal behavior and habits (smoking, drugs, etc.) may affect the fetus during critical growth periods, and the birth process itself may alter postnatal homeostatic processes. Developmental biology provides the fundamental basis for unraveling the pathogenesis of cardiovascular disorders in this sequence from embryo and fetus to child and adult. A clear understanding of the developmental biology of the cardiovascular system is a prerequisite for prevention and optimal management of heart disease at any age.

Dr. Legato has devoted her professional career to the study of the cardiac cell and developing myocardium. Her important contributions to knowledge of ultrastructural composition and function of the myocardium have been supplemented by her unique ability to stimulate and collaborate with investigators from diverse disciplines. The organization of this valuable book and the expertise of the contributors further illustrate her talents.

This book provides essential information for those research-oriented pediatric cardiologists as they seek to explore new frontiers in their specialty. With even greater significance, it provides the basis for understanding developmental biology and thus mechanisms of disease for physicians treating and preventing heart disease in individuals of all ages. It presents the frontier of science in cardiology in a well-organized and eminently readable manner. One looks forward with anticipation to the contemplated companion volume.

Sidney Blumenthal, M.D.

FOREWORD

The study of developmental changes in cardiopulmonary function is by no means new; for example, in 1670, Robert Boyle reported in *Philosophical Transactions* on "the Phenomena afforded by a newly kitten'd Kitling in the Exhausted Receiver." The animal model was the newborn cat, and the experimental method involved the removal of air from a container (i.e., the "Exhausted Receiver") and the accurate measurement of the passage of time, using a "Watch that marks Minutes and Quarter-Minutes." And now, over 300 years after Boyle, many of us still sit in our laboratories studying the "Kitling," still using the Watch (although measuring time in even smaller fractions of minutes), and — even if the exhausted receiver may no longer be immediately recognizable to some — still wondering about hypoxia and its effects.

Another factor which has not changed much through the years is the relative lack of attention that has been paid to developmental biology and to what it can teach us about normal cardiac function, and the pathogenesis and manifestations of various disease processes. The fact that this has occurred puzzles me. After all, any disease process whose manifestations develop and persist over a long period of time (such as cardiovascular disease) finds its expression subject to two basic determinants: one is the pathophysiology of the disease itself; the other is the overall condition of the individual, the so-called "raw material" on which the disease is superimposed. This raw material changes considerably with growth and development and, hence, might be anticipated to modify and influence the course of the disease. Applying this to the cardiovascular system, then, the diseased heart is changing in its function not only as a result of the disease process but as a result of the normal growth and development that would occur in any case. Hence, a developmental biological approach to cardiovascular disease should be a useful and important one to the investigator interested in understanding it better. By identifying what is normal for any age, and how and why normal values change with age, we can begin to appreciate and understand the inroads made by cardiac disease and why and how the expression of disease and the response to treatment may change with age.

To obtain this information we must learn how the heart develops from the fetus through the neonate, the young, and the adult. In studying these different populations, we must learn more about the maturation of the heart biochemically, electrophysiologically, mechanically, and structurally, as well as about those structures that strongly influence cardiac function — such as the autonomic nervous system and its mediators.

For the above reasons I was pleased to learn that Dr. Legato had decided to edit this book. Both Dr. Legato and her collaborators in this book have contributed consistently and importantly over the years to our understanding of the roles played by growth and development in the structure and function of the normal heart as well as the diseased heart.

The book opens with a review of the changing anatomy of the heart cell during postnatal development. Drs. Goldstein and Traeger lay the groundwork here for understanding the structural basis for cardiac function. Dr. Danilo then describes the electrophysiology of the fetal and neonatal hearts, emphasizing not only the changes seen in this age range but how they relate to the mature adult. Dr. Ferrer expands on both earlier chapters, drawing on structural and electrophysiological information to provide a description of developmental changes in the electrocardiogram. Dr. Forsgren's section considers Purkinje and ventricular cell differentiation and their functional correlates. His data indicate that the two groups of cells develop quite separately; one is not a more specialized version of the other but an equally specialized and fundamentally different entity. The chapter by Dr. Robinson reviews critically the various models that are now in use for research on the developing myocardium, pointing out the advantages and limitations of each. Drs. Langer and Jarmakani then show how one of those models, the myocyte grown in tissue culture, has yielded important information about the ionic control of contractility in the neonatal heart. Their data explain several clinical features peculiar to young myocardium, including its requirement for higher doses of digitalis than the adult and its relative resistance to the calcium paradox. Dr. Penefsky reviews the elements influencing functional characteristics of the neonatal and developing heart, emphasizing the relationships of function and of structure that continue into the postnatal period. The differentiation of the myosin filaments is the subject of the chapters by Drs. Schiaffino, Gorza, Sartore, and Thornell and by Drs. Schwartz and Mercadier. These emphasize the importance of a developmental approach to the study of cardiac disease, showing that a stress placed on the adult heart can evoke the production of myosin isoenzymes comparable to those which are predominant in fetal or neonatal myocardium. The overall role of the synthesis and degradation of protein during cardiac growth is reviewed by Dr. Zak. He stresses the role of extrinsic factors such as the type and duration of the load placed on the heart in modulating the genes which manufacture new proteins. A transition then occurs to a discussion of the vasculature. Dr. Hermsmeyer considers the influence of neural elements on the differentiation of the peripheral vasculature. This compliments Dr. Oparil's observations on hypertension in the young. Finally, bringing back the theme of the chronicity of cardiac disease and the importance of development in its consideration, Drs. Schneweiss and Neufeld trace the origins of coronary artery disease into childhood in their review of arteriosclerosis in the young.

In closing, I should mention that I approach most books with trepidation these days, given the extent of duplication of information that one often sees. I was pleased with this book; pleased because I learned so much from it, including much that was readily applicable to my understanding of cardiac disease. And — interestingly enough — not only did I find that many questions circulating today were considered and some answered but so were a few that Robert Boyle posed so many years ago.

Michael R. Rosen, M.D.

PREFACE

There are compelling reasons for the clinical cardiologist to be aware of the recent explosion of research by molecular biologists on the developing myocardium. Essentially, it appears that the enlarging cardiac cell uses the same mechanisms to increase its size during normal growth as it does in the compensatory hypertrophy of the adult heart: a study of either phenomenon provides important insights into the other. Certain facts to support this notion are now apparent: First, cardiac growth is adaptive. The stress to which the heart is put determines in an essential way the structural and functional responses of individual myocytes and of the heart as a whole. In the adult myocardium, for example, work against increased pressure produces a cardiac cell with entirely different ultrastructural composition and functional properties than the myocyte of the thyrotoxic heart. Pressure-overloaded myocardium contracts more slowly and efficiently than the normal heart, and its myocytes have less mitochondria and more sarcomeres by volume than do controls. In thyrotoxic myocardium, as in other volume-overloaded hearts, on the other hand, mitochondrial mass increases and pressure is generated more rapidly and with less efficiency than in the normal heart. As time goes on, however, it is apparent that the composition and even some of the functional parameters of the heart tend to return to normal; it is evident that the myocyte strives to maintain its extraordinary constancy of composition. If the stress placed on the heart is not too great or imposed too rapidly, the transient imbalances in cell composition that characterize early responses to a need for increased work disappear.

The same phenomena observed by students of compensatory hypertrophy are mirrored by the changes in myocyte composition which occur during the normal development of the heart. The left ventricular myocyte, for example, responds at the time of partuition to a need for work against the relatively higher pressures in the systemic circuit than it experienced *in utero* with a burst of sarcomerogenesis. Similarly, both right and left ventricular cells augment mitochondrial volume as they adapt to handle the increasing blood volume of the growing fetal and newborn organisms. Shortly after birth, however, the balanced and constant proportions of the subsets of organelles in the myocyte are attained, and that remarkably stable composition remains throughout the greater part of the life of the animal. We have begun to see that the aging process is also attended by important changes in the myocyte at the molecular level, but to date the behavior of the heart during this period of life has been less well characterized than has the other end of the spectrum.

As more and more data evolve, it is obvious that the phenomenon of congestive heart failure will have to be re-examined from an entirely new perspective. Indeed, it may ultimately be defined as the inability of the myocyte to restore the optimal and original balance of organelle structure and function within the cell to normal values after the initial accommodations to stress have been met. The enlarged myocyte of the hypertrophied heart in order to be considered compensated may have to be defined as having adapted to stress by manufacturing new organelles which exist in the same relative proportions and which have the same functional parameters as in the smaller, "normal" cell. "Failure" at the cellular level may be most accurately defined as the inability of the myocyte to restore that balance. This may precede by a significant space of time the overt evidences of clinical congestive failure, but be no less casual because of that. The clinician has known for a long time that congestive heart failure is not a homogeneous entity, either with regard to its presentation or in its response to therapy. The molecular biologist has now suggested that, depending on the nature of the stress imposed on the heart, quite different compensatory mechanisms are called upon. It follows, then, that the reasons for the eventual failure of the myocardium as a whole are specific and distinct. Therapy, obviously, can be much more intelligently focused if the precise nature of these disturbances is understood.

This book is an effort to show some of the areas in which important and clinically relevant advances have been made in our understanding of the normally developing myocyte. The changing anatomy of the heart cell as the myocardium enlarges, reviewed by Goldstein and Traeger, is amplified by Danilo's description of how sarcolemmal differentiation influences the evolving electrophysiology of the growing heart. Ferrer expands on these notions in her chapter on the developing electrocardiogram, reviewing the changes in the electrical activity of the whole heart from intrauterine life into the postnatal period. Forsgren discusses the separate lines of Purkinje and ventricular cell differentiation, addressing the older concept of the Purkinje cell as an "embryonic rest" which is a precursor of and less differentiated than the ventricular myocyte. His data indicate that, in fact, the two groups of cells develop quite separately; one is not at all a more specialized version of the other, but an equally sophisticated and fundamentally quite different entity. In his interesting discussion of the various models of the intact heart, Robinson reviews the preparations now available for research on the myocardium, pointing out the advantages and limitations of each. Langer and Jarmakani show how one of those models, the myocyte grown in tissue culture, has yielded crucial data about the ionic control of contractility in the neonatal heart. The data explain several clinical features peculiar to young myocardium: its requirement for higher doses of digitalis than the adult and its relative resistance to the calcium paradox. Penefsky reviews the elements influencing the unique functional characteristics of the neonatal and developing heart, emphasizing that the tailoring of function like that of structure, continues well into the neo- and postnatal periods. There are two chapters on the differentiation of the myosin filaments themselves; these are particularly exciting because a stress placed on the adult heart can evoke the production of myosin isoenzymes which were predominant in fetal or immediately postnatal myocardium. Zak's summary of some of the elements governing the production of protein during normal growth stresses the role of extrinsic factors like the type and duration of the load placed on the heart in switching the genes which manufacture new proteins on and off. He provides us with important insights applicable to the heart enlarging in compensatory hypertrophy. Finally, we have included recent work on the vasculature, both coronary and peripheral: Hermsmeyer's discussion of the influence of neural elements on the differentiation of the peripheral vasculature is discussed and paired with Oparil's observations on hypertension in the young. Schneeweiss and Neufeld trace the origins of coronary artery disease far back into childhood in their review of arteriosclarosis in the young.

Ideally, this book will be the first of a pair; the second does not exist yet because there is not a critical mass of data assembled about the adult hypertrophied heart which reflects all the transient but dramatic changes seen in the normally growing myocyte. The work on ultrastructure, isomyosins, and protein synthesis and degredation is the most advanced and shows the striking parallels between the normally developing cell and that enlarging in the adult in response to an increased load. What we have learned so far more than warrants a search for further congruity in the behavior of the two populations.

ACKNOWLEDGMENTS

Two scholars were particularly influential in shaping my ability to edit this book. Michael Rosen, whose creativity and intelligence have amplified the knowledge and interest of many investigators (some of whom are represented in these chapters) working in the field of developmental myocardiology, suggested that I create the book. He has encouraged, refined, and fostered my notions about the importance of ultrastructure in exploring the cellular mechanisms of myocardial growth and differentiation. Norman Alpert, whose work in compensatory hypertrophy has revolutionized the ways in which molecular biologists and clinical cardiologists think about the enlarged heart, has contributed in no small measure to my ideas about the parallels between normal growth and enlargement of the stressed adult myocardium. Both men are critical, demanding, and nourishing teachers and scientists of exceptional talent.

Jeff Smith's advice, counsel, and expertise helped shape and expedite the production of the book. Karen Caldwell's cheerful and capable assistance made the technical tasks of editing much easier for us all.

THE DEVELOPING HEART

1. ULTRASTRUCTURAL CHANGES IN POSTNATAL DEVELOPMENT OF THE CARDIAC MYOCYTE

MARGARET ANN GOLDSTEIN AND LAUREL TRAEGER

The appearance, distribution, and arrangement of cellular structures in the cardiac myocyte gradually change to give the overall architecture of a mature cell (figure 1–1). Postnatal development is part of a continuum of changes that begins in the embryonic stages. In general, the period is characterized by the progressive addition of cellular components with more precise orientation of each. The time sequence of organelle appearance relative to each other is remarkably similar, but the actual time of appearance varies with species. In some mammals, such as the rat and dog and perhaps human, the period of postnatal development is rather prolonged. As a result the sequence of changes is well defined and the events are well separated in time. Hyperplasia continues after birth but is gradually replaced by cell hypertrophy.

MAJOR COMPONENTS

In the newborn, myofibrils are randomly oriented with incomplete sarcomeres. Intact myofibrils are located in the subsarcolemmal region [1, 2, 3], but there is no lateral registration of myofibrils. Clusters of ribosomes, rough endoplasmic reticulum (ER), mitochondria, glycogen, and amorphous Z material are scattered throughout the cell. An extensive Golgi and a centriole are seen near the nucleus [4]. The nucleus is round and prominently located in the center of the cell. Nuclear volume, as a percentage of cell volume, decreases steadily after birth as the cell hypertrophies [5, 6, 7, 8]. Ribosome clusters, rough ER, and Golgi profiles all indicate an active protein synthesis compatible with a rapidly growing cell. The

M.J. Legato (ed.) THE DEVELOPING HEART, pp. 1–20. *Martinus Nijhoff Publishing. Boston/The Hague/Dordrecht/Lancaster.*

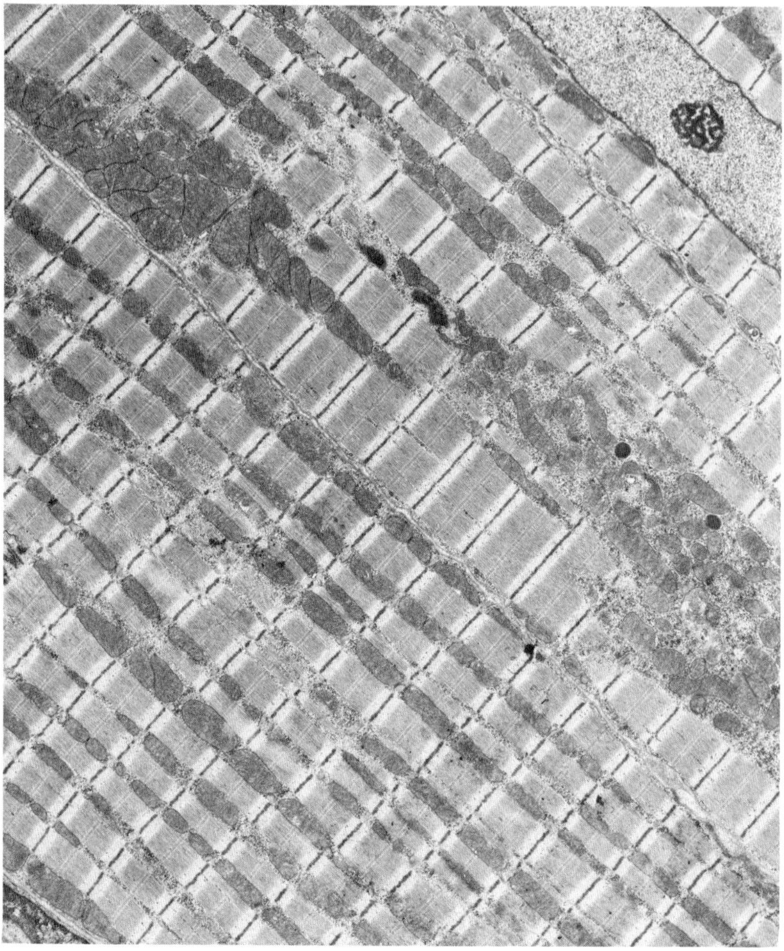

Figure 1–1. Left anterior papillary muscle cells from normal dog. The cells are closely apposed to each other and exhibit normal morphology. Characteristic features of ventricular myocardium, such as an adjacent capillary, a central nucleus, an intercalated disk, and the regular array of myofibrils and mitochondria are shown. X12,850. (From Goldstein, M.A. 1979. Ultrastructure of the ischemic myocardium. *Cardiovascular Research Center Bulletin* 18, 1–33.)

sparse and randomly oriented sarcomeres [9] are probably the major reason that very young hearts develop less force per unit area compared to adult hearts [10]. The morphometric studies of cat myocardium provide a structural basis for the age related changes in mechanical performance of the cat myocardium [2].

CELL SIZE

Myocyte volume increases during postnatal development, resulting in a decrease in surface-to-volume ratio. In rat ventricle, cell size increases 90% between 1 and 5 days, 30 to 40% between 5 and 11 days [11], and 500% by maturity [12]. The increase in cell volume is greater than the increase in cross-sectional area, indicating increased cell length [11]. In rat ventricle, cell length increases 100% by 11 days, and growth is faster in the first 5 days [11]. In rabbit papillary, cell length continues to increase until maturity [8]. The increase in cell length is linearly related to body weight changes in both rat and rabbit, and occurs by an increase in sarcomere number [13].

Cell length-to-width ratio remains constant after 30 days in rat ventricle, indicating concommitant increases in cell length and width, which are correlated to total body growth [12]. Cell diameter increases 2.5-fold between 1 and 16 days in cat papillary [2], and two-fold after 8 days in rabbit papillary [8]. There are no differences between ventricles in cell diameter increases in rats [14], guinea pigs [15], or cats [16]. In the dog, however, the increase in cell diameter is larger in the left than right ventricle up to five months, after which no differences are seen [9]. The right ventricle myocyte is wider at birth, but does not grow in width during the first five months as does the left ventricle myocyte. Overall structural changes in postnatal rat heart cells have been described by David and Marx [17] and indicate synthesis and degradation.

Cell size varies with left ventricle weight [18]. This observation is based on experiments in which heart weight was changed by altering body weight growth in rats from birth to 21 days. Cell volume, length, and diameter increased, and these were directly related to the increases in heart weight. Statistical analysis showed a dependence of cell length and width on left ventricle growth.

Myofibril volume density (Vv), as a percentage of cell volume, increases from 37% to 45% between one and five days in rat left ventricle. In right ventricle, myofibril Vv increases from 40% to 47% between 5 and 11 days [19]. After 11 days myofibril Vv does not change, indicating proportional growth of myofibrils and cell volume [6, 14]. In rabbit heart, myofibril Vv increases 15% during the first eight days of life [8, 20], then continues to increase slightly until maturity [8]. Myofibril Vv increases slightly, but significantly, between 1 and 30 days in hamster left ventricle [7]. However, in cat atria and ventricles, and dog ventricle, myofibril Vv does not change during postnatal development, remaining constant at approximately 60% of cell volume [9, 16]. At two months, myofibril Vv is slightly greater in dog left compared to right ventricle, but the ventricles are similar by five months.

Cytoplasm Vv decreases early in postnatal life. In rat this is due to increased myofibril and mitochondria Vv [19]. Glycogen decreases after one day in rat atrium and ventricle [15]. Rough ER and ribosomes are still abundant at five months in dog ventricle [4, 5].

Cell number changes in certain species during postnatal development. Overy and Priest [21] reported mitotic cell division in postnatal rats from 2 to 9 weeks and in postnatal lambs from 3 to 28 days. Anversa et al. [11] report that myocyte number increases 100% by 11 days in rat left ventricle but only 24% (nonsignificant) in right ventricle. These findings are in agreement with those of Martin and Mayer who studied mass development using histologic-morphometric techniques in rat ventricles from 1 to 14 days after birth. They concluded that the increase in myocardial area was due primarily to an increase in the left ventricle. They concluded that growth was by hyperplasia in the first week and by hypertrophy in the second week. Myocyte proliferation has also been suggested in rabbit papillary [8]. Zak [22] has questioned whether the frequency of mitosis is significant enough to alter cell number and whether the resulting daughter cells are viable. He suggests that the weight gain of the human heart after birth can be accounted for by cell hypertrophy alone, whereas only 75% of the growth in rat heart results from hypertrophy. The remaining 25% presumably is due to cell proliferation. There is evidence for cellular hyperplasia in developing dog heart. Bishop and Hine [23] measured single nucleated cell size in isolated myocytes. They found no increase until four to six weeks after birth although left ventricular weight increased linearly from birth.

There is evidence that mitosis occurs in the postnatal cardiac myocyte (figure 1–2). Well differentiated cells incorporate ^{3}H-thymidine [24] and mitotic cells can be seen with electron microscopy. Cell proliferation stops after two to three weeks [25]. Rumyantsev [26] has shown by EM autoradiography that rat ventricular myocytes proliferate during the first postnatal week more actively than do atrial ones. By the 17th to 18th day postnatally almost all myocytes stop dividing, with 0.1% to 0.5% of atrial or conducting cells continuing to enter the mitotic cycle. The number of binucleate cells also increases during postnatal development, indicating nuclear but not cellular proliferation [11]. In the dog heart at birth 96% to 100% of myocytes contain a single nucleus. Multiple nuclei are seen at 4 weeks in 55% of myocytes, at 6 to 10 weeks in 85%, and beyond 15 weeks in 55% to 60%. It has been suggested that both hyperplasia and hypertrophy occur in the left ventricle in response to increasing pressure load during postnatal life, while only hypertrophy occurs in the right ventricle due to the higher volume load [11].

Sarcomere

In the neonate, sarcomeres are incomplete with irregular I and A bands and AI junction [1, 4, 5, 9]. The Z band is irregular, of variable width, and wider than in the adult [4]. The M band does not appear until 5 to 11 days in rat myocardium [27].

Contractile filaments are randomly arranged in the neonate [4, 5]. Alignment of thin and thick filaments into distinct I and A bands is incomplete, resulting in an irregular AI junction. During development, the sarcomere becomes more organized, with filaments more in register (figures 1–3, 1–4). No studies have been reported on changes in lengths or arrangements of thin and thick filaments in postnatal cardiac muscle. Studies on postnatal rat skeletal muscle show changes in both length and arrangement of thin filaments between three and nine days [28, 29]. Thin filaments become longer, more variable in length, and more regularly arranged during postnatal development. It is possible that these changes also occur in developing cardiac muscle. Actin content in rat cardiac muscle increases to adult

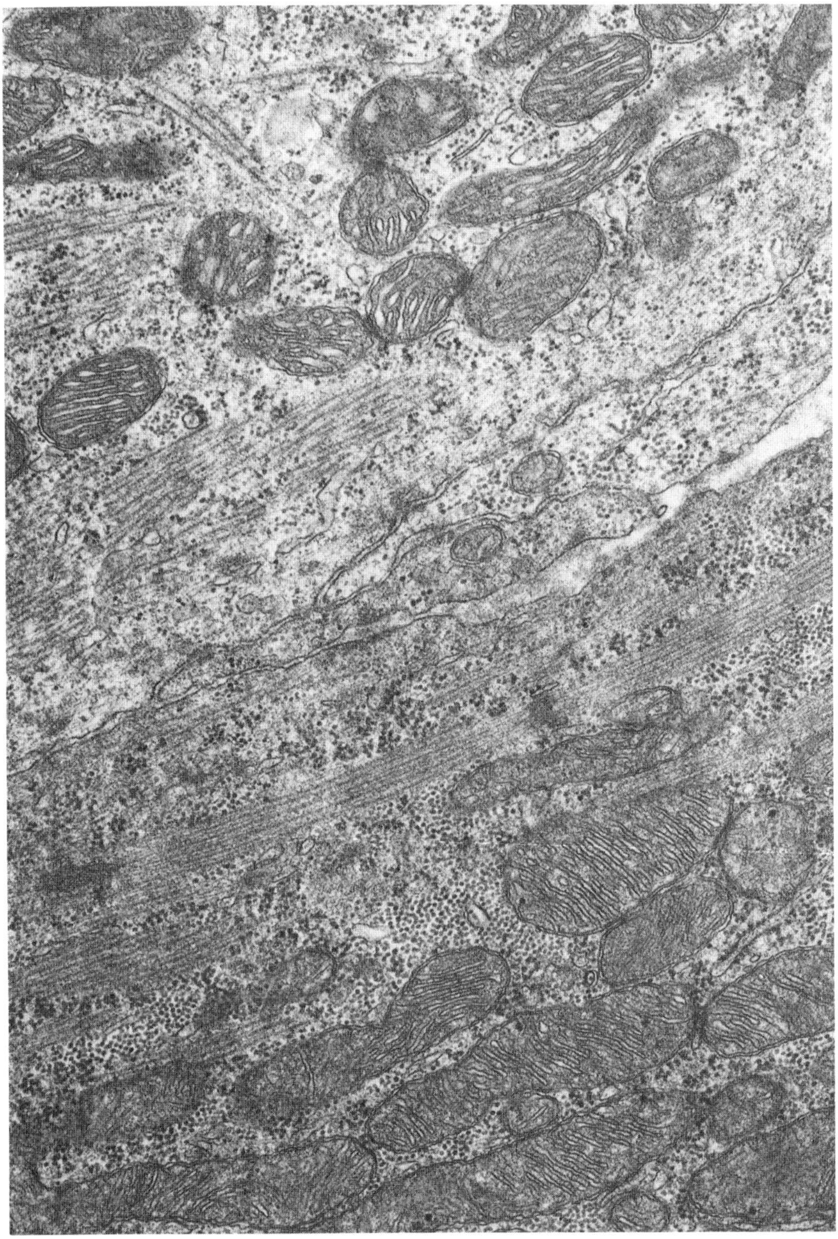

Figure 1–2. Longitudinal section of neonatal rat heart showing portions of a dividing cell and an apposing non-dividing cell. Small but organized myofibrils are present in the non-dividing cell, but this sarcomere organization is disrupted when mitosis is initiated. Numerous mitochondria, glycogen granules, and ribosomes are present in both, but the cytoplasm is less dense in the dividing cell. X42,000.

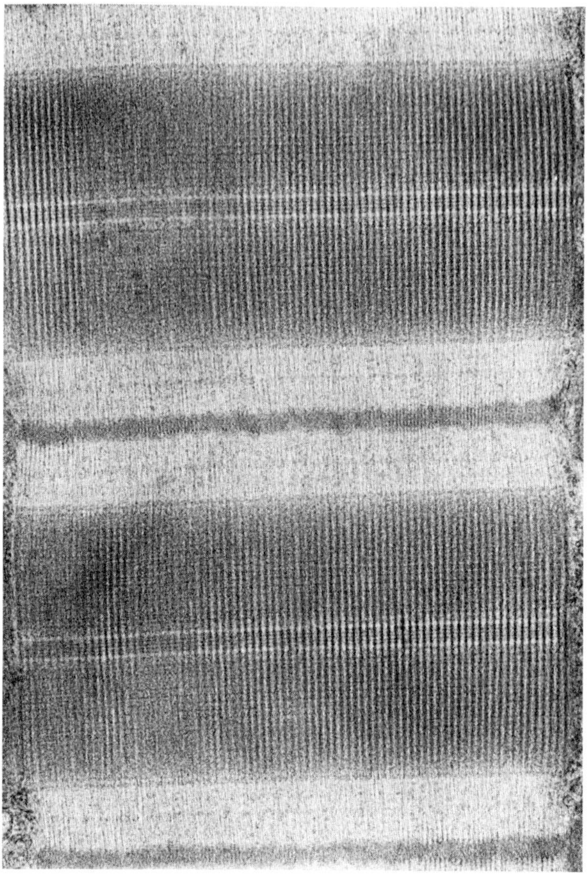

Figure 1–3. High-voltage electron micrograph of two cardiac sarcomeres in a 160 nm thick section photographed at 1000 KV. Sarcomere length is 2.12 u. Filaments are in precise register and reinforce through the thickness of the section. X46,000. (From Goldstein, M.A. 1984, in press. The cardiac sarcomere. In *Atherosclerosis reviews*. New York: Raven Press.)

values within the first week of life [30]. Myosin heavy chains (HC alpha and HC beta) from rabbit ventricle have been studied during development using monoclonal antibodies. HC alpha comprises approximately 50% of the ventricular myosin between birth and four weeks of age, whereas HC beta is the primary myosin variant at one year of age [31].

The time course for the ultrastructural organization of the myocyte coincides with the time course for functional maturity of the heart [32]. In the isolated perfused rat heart, active and passive tension, and peak isovolumetric pressure increase between 16 and 20 days. These physiological changes occur concurrently with increased structural organization of the myocyte. These changes are related to age and not cardiac mass, since increasing mass at a young age does not hasten physiological development.

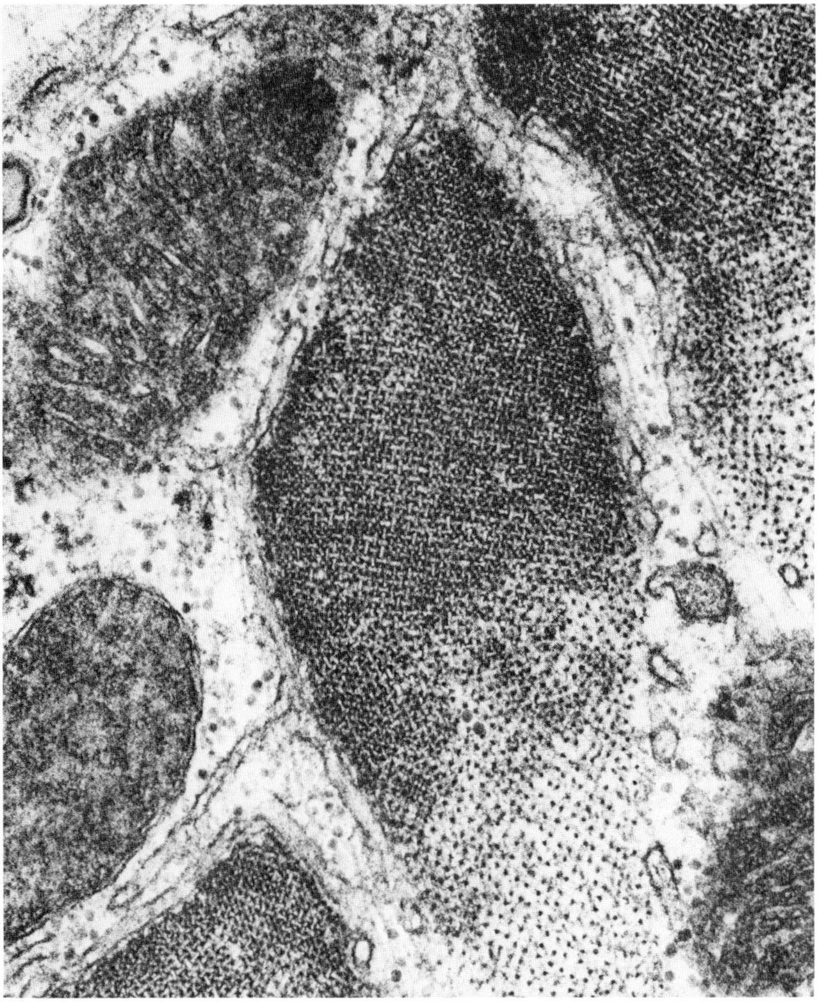

Figure 1–8. Cross section of a Z band from a normal canine papillary muscle which shows both the basket-weave and the small-square pattern of the Z lattice. X84,000. (From Goldstein, M.A.; Schroeter, J.P.; and Sass, R. L. 1979. The Z lattice in canine cardiac muscle. *J. Cell Biol.* 83:187–204.)

contain amorphous Z material [1, 4]. By two to three months in dog ventricle, the intercalated disk is oriented perpendicular to the cell axis with specialized regions seen in the adult [4].

Early in postnatal life the gap junction area increases three-fold relative to cell volume in rabbit [38] and relative to cell surface in cat [2, 3]. The area of individual gap junctions also increases two-fold within the first three days in rabbit left ventricle. Total gap junction area remains constant between three and 30 days, and then decreases to the adult value [38].

A detailed freeze fracture study of rabbit left ventricle gap junction growth shows that the packing of particles changes as gap junctions grow [38]. From one to three days the particles lose their dense packing, exhibiting an "aisle configuration." By five days, when rapid growth of gap junctions has slowed, the particles are again densely packed. The packing varies inversely with gap junction size, as density decreases in large junctions and increases in small junctions. Interparticle spacing is inversely related to density of packing. Shibata et al. [38] suggested that the large gap junctions represent growth of smaller junctions which occurs by an initial synthesis of lipid bilayer and later insertion of proteins (particles). It was also suggested that the increase in gap junction growth should result in a change in electrical coupling between cells.

INTERMEDIATE FILAMENTS

In the adult cardiac myocyte, 10 nm filaments course between Z bands of adjacent sarcomeres, from Z band to sarcolemma, and into the nuclear membrane and desmosomes [39] (figure 1–9). Cardiac intermediate filaments contain desmin (skeletin) [1]. The distributions of desmin-containing filaments changes during postnatal development as seen with indirect immunofluorescence. Newborn rat atrial and ventricular myocytes show a diffuse desmin distribution, with subsarcolemmal patches and some Z band staining. By four weeks desmin is localized to the Z band and intercalcated disk [1]. There are currently no reports of direct immunocytochemical or morphometric data on intermediate filaments in postnatal heart.

MICROTUBULES

Microtubules have been observed in adult heart [39, 40] and in rat heart during postnatal development [41, 42] (figure 1–10). In adult rat papillary muscle, microtubules are axially oriented between myofibrils and around the nucleus [40]. In postnatal heart, microtubule orientation at birth is more random, consistent with the irregular myofibril arrangement [41]. At five days the microtubules are oriented primarily in the direction of the long axis of the cell along the myofibrils [41]. The distribution of tubulin antibody as shown by indirect immunofluorescence is diffuse in rat cardiac myocytes at birth but by 5 to 9 days shows an intense pattern of fluorescence predominantly along the myofibril axis.

Microtubule density, the number of microtubules per um^2 of cell cross sectional area, increases between one and five days, peaking at day five [41]. Microtubule density then decreases between 5 and 21 days, and remains constant throughout adulthood. Microtubules can be subdivided into perinuclear and intermyofibrillar microtubules [42]. The perinuclear microtubule density remains constant during postnatal development, whereas the intermyofibrillar microtubule density increases between one and five days and decreases after five days. The changes in intermyofibrillar microtubule density closely parallel changes in total cellular microtubule density. The decrease in microtubule density does not necessarily mean a decrease in microtubule number. In fact, it may mean a stable number of micro-

Figure 1–9. Intermediate filaments running at the Z-line level are superimposed with longitudinally sectioned vesicles of sarcoplasmic reticulum between mitochondria in this EM of normal dog papillary. X56,000.

tubules with an increasing cell volume, particularly since Bishop et al. have shown that cell volume is increasing during this time period (personal communication). It was suggested that perinuclear microtubules help to maintain nuclear shape or function throughout the life of the cell. Intermyofibrillar microtubules may be involved in assembly and orientation of newly synthesized myofibrils during postnatal development when growth rate is high.

ATRIUM

The ultrastructure of atrium is qualitatively similar to that of ventricle, but there are some quantitative differences. Myofibril volume density is similar, but mitochondrial volume density is lower in atrium than in ventricle [16, 34]. Atrial myo-

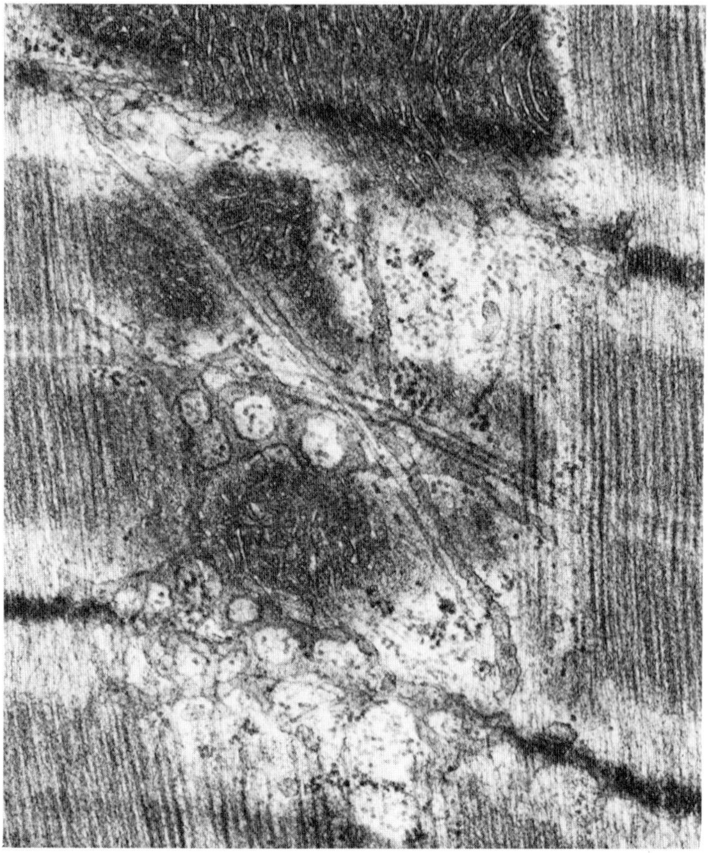

Figure 1–10. Longitudinal section of normal canine cardiac muscle showing a long straight SR tubule and a microtubule running parallel diagonally across the sarcomere. Three microtubule profiles sectioned tangentially run approximately along the M band. X56,000. (From Goldstein, M.A., and Entman, M.L. 1979. Microtubules in mammalian heart muscle. *J. Cell Biol.* 80:183–195.)

cyte diameter is smaller, and the cell grows more slowly. The differences between atrium and ventricle become greater with age [16, 34].

Electron dense granules are commonly seen in atrial myocardial cells. These granules contain glycoprotein [43] and may represent cytoplasmic transport between the Golgi and cell surface. Granules are seen early in fetal development in the human [44] and at birth in rat [1, 14] and cat [14]. There are no changes in atrial granules during postnatal development of rat or cat heart [14, 16].

EXTRACELLULAR AND NON-MUSCLE COMPONENTS

The collagen network typical of the adult [45, 46] develops in the first 20 days after birth in both rat and hamster. The weave of endomysium is not seen at birth but is visible at 4 days and appears similar to that of the adult by 15 days. The collagen struts from myocyte to myocyte and from myocyte to capillary form gradually as the heart grows. The myocyte-myocyte struts are more numerous in hamster and may provide a mechanical coupling consistent with the higher observed heart rate in hamster.

The perinatal change in circulation is a decisive factor in postnatal heart growth [47]. An increase in capillary density occurs together with the increase in myocardial mass [48]. Impaired nutrition can retard the postnatal growth rate [47]. Hollenberg et al. [49] have shown that rats raised in litters of 6 show more hyperplasia and hypertrophy than those raised in litters of 18. Hemodynamic function is also closely coupled to cardiac growth [22]. Exercise, both continual and intermittent, can also retard organ and cellular growth rate of the heart [48].

The density of intramural innervation increases gradually in the rat heart from day 1 postnatally until 22 days postnatally when the innervation of the adult animal is observed. In the rabbit adrenergic fibers are observed much earlier on the 22 day of gestation in the atrioventricular junction and on 24th day among the myocardial cells [50]. Postnatal biochemical and histochemical comparisons have been carried out in the rabbit [51]. Both the concentration of catecholamines and the anatomic distribution of the sympathetic nerves as detected by the monoamine fluorescence technique show that the levels increase progressively from birth to adult levels by three weeks. In the dog the conduction velocity in the Purkinje fiber at eight weeks postnatally is not as fast as in the adult [52].

The structure of the developing cardiac conduction system has been studied [53]. The sinus node consists of P cells, transitional cells, a central nodal artery, and a supporting collagen network. There are fewer P cells in the adult than in the newborn. During development both the artery and the collagen framework model the shapes of the sinus node. Collagen content progressively increases and plays a role in stabilizing the normal pacemaker. The internodal pathways and the bundle of His are defined during the postnatal period as well.

The exception to the general picture of growth in postnatal development is the closing of the ductus arteriosus. Degeneration and decrease in smooth muscle cells with a concurrent increase in collagen and breakdown of the distinct layers of the vessel wall characterize this process [54]. The following postnatal changes occur. Connective tissue occupies the lumen: smooth muscle cells increase and collagen fibers in the intima increase. Extracellular matrix decreases and collagen increases in the media. Smooth muscle cells degenerate and vacuoles increase. By three months the ductus is completely closed and appears macroscopically as a collagenous cord. Beyond three months the entire thickness consists of collagen fibers, degenerated and dead smooth muscle cells, and some tiny elastic fibers.

CONCLUDING REMARKS

As more new structural features of adult heart are discovered, we will be curious about their appearance in development. For example, very little attention was paid to cardiac microtubules and intermediate filaments ten years ago. Yet sufficient information has been obtained recently to warrant a review [55] and a separate heading in this review. A study of the cytoskeleton in development will help answer questions about myofibril formation. More than likely there will be other components of the cytoskeleton, the myofibril, and the regulatory components of the cardiac cells that will require further study. The heart muscle is capable of enormous plasticity and flexibility. Structural changes at all levels are occurring during development, and yet the heart keeps beating throughout the process. Different functional requirements are imposed at different ages and physiological states, and this necessitates a well-regulated contraction phenomenon. Complex structural entities such as myofibrils in a cytoskeletal matrix or reticular membranes at one level down to multiple proteins in a contractile filament at another allow for multiple interventions in the contractile process. The heart can thus continually adjust.

In the present review, the studies are still largely descriptive, but a number of major quantitative studies have been completed. Such quantitation represents an important first step in formulating questions that relate structure to function. For example, how do changes in excitation-contraction coupling relate to observed structural changes such as increase in surface area in intimate contact between adjacent myocytes, increase in gap junctional area, formation of T tubules, and changes in dyadic junctional complexes between sarcoplasmic reticulum and plasmalemma? Active tension and rate of contraction increase with age [10] and are thought to be related to maturation of these membrane systems. How do changes in metabolic capacity of the tissue affect contractility? Do the amounts of contractile proteins per unit volume increase in a corresponding time and manner with increases in force development? What is the significance of different isozyme patterns for myosin and perhaps other proteins such as alpha-actinin? Future studies will compare morphological parameters to physiological and biochemical changes occurring in postnatal development.

Continued ultrastructural study of the postnatal heart will help us gain knowledge of normal cardiac development. There are many questions still to be asked that relate to (1) congenital anomalies, (2) impaired growth, (3) regeneration after ischemia, and (4) hypertrophy in the postnatal animal versus hypertrophy in the adult. Such questions will provide insights into disease processes in developing and adult hearts.

REFERENCES

1. Carlsson, E.; Kjorell, U.; Thornell, L.E.; Lambertsson, A.; and Strehler, E. 1982. Differentiation of the myofibrils and the intermediate filament system during postnatal development of the rat heart. *Eur. J. Cell Biol.* 27:62–73.
2. Sheridan, D.; Cullen, M.J.; and Tynan, M.J. 1977. Postnatal ultrastructural changes in the cat myocardium: a morphometric study. *Cardiovasc. Res.* 11:536–540.
3. Sheridan, D.; Cullen, M.; and Tynan, M. 1979. Qualitative and quantitative observations on ultrastructural changes during postnatal development in the cat myocardium. *J. Mol. Cell. Cardiol.* 11:1173–1181.
4. Legato, M. 1975. Ultrastructural changes during normal growth in the dog and rat ventricular myofiber. In *Developmental and physiological correlates of cardiac muscle,* ed. M. Lieberman and T. Sano, pp. 249–273. New York: Raven Press.
5. Legato, M. 1979. Cellular mechanisms of normal growth in the mammalian heart. I. Qualitative and quantitative features of ventricular architecture in the dog from birth to five months of age. *Circ. Res.* 44:250–262.

6. Page, E.; Early, J.; and Power, B. 1974. Normal growth of ultrastructures in rat left ventricular myocardial cells. *Circ. Res.* 34–35 (Supp. II) II-12-16.
7. Colgan, J.; Lazarus, M.; and Sachs, H. 1978. Post-natal development of the normal and cardiomyopathic syrian hamster heart: a quantitative electron microscopic study. *J. Mol. Cell. Cardiol.* 10:43–54.
8. Hoerter, J.; Mazet, F.; and Vassort, G. 1981. Perinatal growth of the rabbit cardiac cell: possible implications for the mechanism of relaxation. *J. Mol. Cell. Cardiol.* 13:725–740.
9. Legato, M. 1979. Cellular mechanisms of normal growth in the mammalian heart. II. A quantitative and qualitative comparison between the right and left ventricular myocytes in the dog from birth to five months of age. *Circ. Res.* 44:263–279.
10. Park, I.; Michael, L.H.; and Driscoll, D.J. 1982. Comparative response to the developing canine myocardium to inotropic agents. *Am. J. Physiol.* 242:H13–H18.
11. Anversa, P.; Olivetti, G.; and Loud, A. 1980. Morphometric study of early postnatal development in the left and right ventricular myocardium of the rat. I. hypertrophy, hyperplasia, and binucleation of myocytes. *Circ. Res.* 46:495–502.
12. Korecky, B., and Rakusan, K. 1978. Normal and hypertrophic growth of the rat heart: changes in cell dimensions and number. *Am. J. Physiol.* 234:H123–128.
13. Stewart, J., and Page, E. 1978. Improved stereological techniques for studying myocardial cell growth: application to external sarcolemma, T system and intercalated disks of rabbit and rat hearts. *J. Ultrastruct. Res.* 65:119–134.
14. Hirakow, R.; Gotoh, T.; and Watanabe, T. 1980. Quantitative studies on the ultrastructural differentiation and growth of mammalian cardiac muscle cells. I. The atria and ventricles of the rat. *Acta Anat.* 108:144–152.
15. Hirakow, R.; and Gotoh, T. 1980. Quantitative studies on the ultrastructural differentiation and growth of mammalian cardiac muscle cells. II. The atria and ventricles of the guinea pig. *Acta Anat.* 108:230–237.
16. Gotoh, T. 1983. Quantitative studies on the ultrastructural differentiation and growth of mammalian cardiac muscle cells. The atria and ventricles of the cat. *Acta Anat.* 115:168–177.
17. David, H., and Marx, I. 1977. Quantitative and qualitative changes in the heart muscle of the left ventricle in male rats during development and maturation in the first 6 months of life. *Z. Mikrosk. Anat. Forsch.* 91:709–736.
18. Rakusan, K.; Raman, S.; Layberry, R.; and Korecky, B. 1977. The influence of aging and growth on the postnatal development of cardiac muscle in rats. *Circ. Res.* 11:212–218.
19. Olivetti, G.; Anversa, P.; and Loud, A. 1980. Morphometric study of early postnatal development in the left and right ventricular myocardium of the rat. II. Tissue composition, capillary growth, and sarcoplasmic alterations. *Circ. Res.* 46:503–512.
20. Smith, H., and Page, E. 1977. Ultrastructural changes in rabbit heart mitochondria during the perinatal period. Neonatal transition to aerobic metabolism. *Dev. Biol.* 57:109–117.
21. Overy, H.R., and Priest, R.E. 1966. Mitotic cell division in postnatal cardiac growth. *Lab. Invest.* 15:1100–1103.
22. Zak, R. 1973. Cell proliferation during cardiac growth. *Am. J. Cardiol.* 31:211–219.
23. Bishop, S.P., and Hine, P. 1975. Cardiac muscle cytoplasmic and nuclear development during canine neonatal growth. In *Recent advances in studies on cardiac structure and metabolism.* Vol. 8, *The cardiac sarcoplasm,* ed. P. Roy and P. Harris, pp. 77–98. Baltimore: University Park Press.
24. Goldstein, M.; Claycomb, W.; and Schwartz, A. 1974. DNA synthesis and mitosis in well-differentiated mammalian cardiocytes. *Science* 183:212–213.
25. Claycomb, W. 1977. Cardiac-muscle hypertrophy. Differentiation and growth of the heart cell during development. *Biochem. J.* 168:599–601.
26. Rumyantsev, P.P. 1978. DNA synthesis and mitotic division of myocytes of the ventricles, atria and conduction system of the heart during the myocardial development in mammals. *Tsitologiia* 20:132–141.
27. Anversa, P.; Olivetti, G.; Bracchi, P-G.; and Loud, A. 1981. Postnatal development of the M-band in rat cardiac myofibrils. *Circ. Res.* 48:561–568.
28. Traeger, L., and Goldstein, M. 1983. Thin filaments are not of uniform length in rat skeletal muscle. *J. Cell Biol.* 96:100–103.
29. Traeger, L.; Mackenzie, J.; Epstein, H.; and Goldstein, M. 1983. Transition in the thin-filament arrangement in rat skeletal muscle. *J. Muscle Res. Cell Motility* 4:353–366.
30. Uchino, J., and Tsuboi, K. 1970. Actin accumulation in developing rat muscle. *Am. J. Physiol.* 219:154–158.

31. Chizzonite, R.A.; Everett, A.W.; Clark, W.A.; Jakovcic, S.; Rabinowitz, M.; and Zak, R. 1982. Isolation and characterization of two molecular variants of myosin heavy chain from rabbit ventricle. Change in their content during normal growth and after treatment with thyroid hormone. *J. Biol. Chem.* 25:2056–2065.
32. Hopkins, S.; McCutcheon, E.; and Wekstein, D. 1973. Postnatal changes in rat ventricular function. *Circ. Res.* 32:685–691.
33. David, H.; Meyer, R.; Marx, I.; Guski, H.; and Wenzelides, K. 1979. Morphometric characterization of left ventricular myocardial cells of male rats during postnatal development. *J. Mol. Cell. Cardiol.* 11:631–638.
34. Hirakow, R., and Krause, W. 1980. Postnatal differentiation of ventricular myocardial cells of the opossum (Didelphis virginiana Kerr) and T-tubule formation. *Cell Tiss. Res.* 210:95–100.
35. Forbes, M., and Sperelakis, N. 1976. The presence of transverse and axial tubules in the ventricular myocardium of embryonic and neonatal guinea pigs. *Cell Tiss. Res.* 166:83–90.
36. McNutt, N.S. 1970. Ultrastructure of intercellular junctions in adult and developing cardiac muscle. *Am. J. Cardiol.* 25:169–183.
37. Hagopian, M.; Anversa, P.; and Loud, A. 1975. Quantitative radioautographic localization of newly synthesized protein in the postnatal rat heart. *J. Mol. Cell. Cardiol.* 7:357–367.
38. Shibata, Y.; Nakata, K.; and Page, E. 1980. Ultrastructural changes during development of gap junctions in rabbit left ventricular myocardial cells. *J. Ultrast. Res.* 71:258–271.
39. Ferrans, V.; and Roberts, W. 1973. Intermyofibrillar and nuclear-myofibrillar connections in human and canine myocardium. An ultrastructural study. *J. Mol. Cell. Cardiol.* 5:247–257.
40. Goldstein, M., and Entman, M. 1979. Microtubules in mammalian heart muscle. *J. Cell Biol.* 80:183–195.
41. Cartwright, J., and Goldstein, M. in press. Microtubules in the heart muscle of the postnatal and adult rat. *J. Mol. Cell Cardiol.*
42. Park, R.; Legier, M.; Cartwright, J.; and Goldstein, M. 1984. Perinuclear microtubules in postnatal rat heart. *J. Morph.*, in press.
43. Yunge, L.; Benchimol, S.; and Cantin, M. 1979. Ultrastructural cytochemistry of atrial muscle cells. VII. Radioautographic study of synthesis and migration of glycoproteins. *J. Mol. Cell. Cardiol.* 11:375–388.
44. Saetersdal, T.; Rotevatn, S.; Myklebust, R.; and Odergarden, S. 1980. Development of specific granules in embryonic cardiac myocytes of the human auricular wall. *Anat. Embryol.* 160:1–9.
45. Borg, T.K., and Caulfield, J.B. 1981. The collagen matrix of the heart. *Fed. Proc.* 40:2037–2041.
46. Borg, T.K. 1982. Development of the connective tissue network in the neonatal hamster heart. *Am. J. Anat.* 165:435–443.
47. Martin, H., and Meyer, R. 1980. Morphometric studies of rat myocardium during early postnatal period. *Z. Mikrosk. Anat. Forsch.* 94:460–466.
48. Bloor, C.M.; Pasyk, S.; and Leon, A.S. 1970. Interaction of age and exercise on organ and cellular development. *Am. J. Pathol.* 58:185–168.
49. Hollenberg, M.; Honbo, N.; and Samorodin, A.J. 1977. Cardiac cellular responses to altered nutrition in the neonatal rat. *Am. J. Physiol.* 233:H356–360.
50. Papka, R.E. 1981. Development of innervation to the ventricular myocardium of the rabbit. *J. Mol. Cell. Cardiol.* 13:217–228.
51. Friedman, W.F.; Pool, P.E.; Jacobowitz, D.; Seagren, S.C.; and Braunwald, E. 1968. Sympathetic innervation of the developing rabbit heart biochemical and histochemical comparisons of fetal neonatal and adult myocardium. *Circ. Res.* 23:25–32.
52. Rosen, M.R.; Legato, M.J.; and Weiss, R.M. 1981. Developmental changes in impulse conduction in the canine heart. *Am. J. Physiol.* 240:H546–H554.
53. James, T.N. 1970. Cardiac conduction system: fetal and postnatal development. *Am. J. Cardiol.* 25:213–226.
54. Toda, T.; Tsuda, N.; Takasi, Nishimori, I.; Leszczynski, D.; and Kummerow, F. 1980. Ultrastructure of developing human ductus arteriosus. *J. Anat.* 131:25–37.
55. Forbes, M.S., and Sperelakis, N. 1983. The membrane systems and cytoskeletal elements of mammalian myocardial cells. In *Cell and muscle motility*, Vol. 3, ed. R. Dowben and J. Shay, pp. 89–155. New York: Plenum Press.

2. ELECTROPHYSIOLOGY OF THE FETAL AND NEONATAL HEART

PETER DANILO, JR.

The developing heart has been a subject of investigation since at least 1651 when Harvey described the area vasculosa of the chicken egg [1]. Since that time, numerous studies have been done in which the anatomical development of the heart as well as its responses to a variety of pharmacologic agents were studied [2]. The electrophysiology of the developing heart was investigated first by visual means, i.e., by observing its rate of contraction and, later, by recording the body surface electrocardiogram (ECG). With the development of the glass capillary microelectrode, the study of transmembrane potentials (i.e., the electromotive force across the cell membrane generated by the unequal distribution of ions across the cell membrane) of embryonic and fetal hearts was made possible. In 1952, Fingl, Woodbury, and Hecht [3] reported the transmembrane action potential characteristics of embryonic chick atrial and ventricular myocardium. Since then, numerous studies have been published on the development of normal cardiac action potentials as well as on the development of responsiveness to a wide variety of cardioactive drugs. In this chapter I will discuss the changes in cardiac electrical activity and the response to cardioactive drugs that occur with embryonic, fetal, and neonatal development. Where pertinent, comparisons will be made to adult hearts.

Supported in part by USPHS HD-13063, HL-28958, HL-23358, N.Y.H.A. Grant-in-Aid. The author is a recipient of a Research Career Development Award, 5 KO 4 HL 00853. The author gratefully acknowledges the editorial assistance of Dr. Michael Rosen and the secretarial assistance of Ms. Edda Semiday.

M.J. Legato (ed.) THE DEVELOPING HEART, pp. 21–38.

ONTOGENESIS OF CARDIAC ELECTRICAL ACTIVITY

Experimental Animal Models

The most popular animal model for the study of the development of cardiac electrophysiology has been the chick. The advantages of this model are obvious: it is inexpensive; it has a short time course of development; it can be studied easily at nearly all stages of development; and the genetic make-up of the embryo is known.

The earliest descriptions of cardiac activity in the embryo appeared in the latter half of the 19th century (e.g., [4, 5, 6]) and were provided by embryologists who happened also to observe the beating heart. The observations of Pickering [2] were reported in 1893 and were quite sophisticated for their time. Using no instrumentation, he described the effects of temperature on spontaneous atrial and ventricular rate and on atrial-ventricular (A-V) conduction. He also studied the actions on heart rate of a variety of pharmacologic agents and described their structure-activity relationships and mechanisms of action. Many of his observations have been confirmed in subsequent studies.

In the early 20th century the embryologist, Patten, recorded the initiation of contraction in chick embryos of 9 to 10 somites (29 to 30 hrs.) post-fertilization, a stage at which the heart is tubular [7]. At this stage of development contractions are not rhythmic and appear confined to the ventricular region. Slightly later, at 40 hours, the atria begin to contract, and the heart begins to pump blood. Soon thereafter, the sinus venosus is formed, ultimately to become the primary cardiac pacemaker.

In 1952, the first report of the cellular electrophysiology of the embryonic heart was published [3].

ONTOGENY OF THE CARDIAC ACTION POTENTIAL

Because resting potential is dependent on the extracellular potassium concentration, $[K^+]_o$, and because different investigators have used $[K^+]_o$ of 2.7 to 5.6 mM in their studies, reports of the relationship between ontogeny and resting membrane potential are not in full agreement. Nevertheless, it appears that, for the chick, on the second day post-fertilization the resting membrane potential is in the range of -35 to -40 mV [8, 9]. It increases to approximately -60 to -70 mV by days 4 to 7 [10, 11, 12, 13, 14] and reaches a maximum of -70 to -80 mV by days 16 to 19 [10, 11, 12, 13, 14, 15, 16, 17, 18].

Changes in action potential amplitude, overshoot and $\dot{V}_{max}$ occur concurrently with those in resting membrane potential. For the embryonic chick ventricle, action potential amplitude has been reported to be between 46 and 58 mV at the second day post-fertilization, and between 90 and 115 mV at or near the time of hatching. Overshoot ranges from 11 to 19 mV at day 2, to 28 to 38 mV at hatching. For $\dot{V}_{max}$, values of 7 to 20 V/sec have been reported at day 2 [8, 9] to greater than 200 V/sec, near the time of hatching [12, 13]. There is a controversy concerning the action potential upstroke: it is uncertain whether there is an ontogenetic change in upstroke from one which is dependent on "slow channels" to one which is dependent on "fast" (sodium) channels. This point will be discussed below.

Relatively little information is available concerning action potential duration of the embryonic chick ventricle. Vleugels, et al. [11] reported little change in duration (measured to 90% repolarization) between days 7 and 19; values were 154 and 165 msec, respectively. Yeh and Hoffman [10] found that, at a stimulus cycle length of 600 msec, action potential duration was the same (130 msec) at days 6 through 19.

The ontogeny of cardiac action potentials also has been reported for chick atria. Between days 6 and 8, resting potential and action potential amplitude are unchanged (approximately -67 mV and 90 mV, respectively), and V_{max} increases from 64 V/sec at day 6 to 94 V/sec at the day 18 [16].

Thus, the ontogeny of the avian embryonic action potential has been well described. The mammalian embryo is more difficult to study and is not as completely characterized. Bernard [17] reported on the ventricular myocardial action potential of the developing rat heart. Between days 10 and 21 post-conception, resting membrane potential increases from -48 mV to -82 mV. V_{max} also increases during this time, from 5–8 V/sec to approximately 50 V/sec.

Thus, it appears that although ontogenetic changes in resting potential, over a similar period of "gestation," are roughly equivalent in the chick and the rat, the chick myocardial action potential may develop a more rapid upstroke. This difference may be explained, at least in part, by different experimental conditions or different species. For example, in the studies of chick embryos, temperatures were generally 34° to 37°C, whereas in those of the rat, the superfusate temperature was considerably lower, 24.5°C.

In recent years, developmental changes in the electrophysiologic properties of the canine ventricular specialized conducting system have been studied [18]. The dog has a gestation of 60 to 63 days, and the embryo does not implant until approximately the 18th day post-conception.

For Purkinje fibers removed from fetuses from just after implantation to just prior to natural birth, a direct relationship exists between crown-rump length (CRL) and maximum diastolic potential. For least-developed (CRL = 6 to 8 mm) hearts, maximum diastolic potential was approximately -68 mV; for near-term hearts (CRL = 150 to 160 mm) it was approximately -80 mV. Action potential amplitude increases from approximately 98 to 120 mV over the same time frame, and V_{max} from 200 to 450 V/sec.

Action potential duration of fetal canine Purkinje fibers, apparently unlike that of either the rat of the chick myocardium, increases as development progresses. Action potential duration measured to 50% repolarization increases 15 to 120 msec for least and most developed hearts, respectively. Duration measured to full repolarization increases from 65 to 240 msec.

A limited number of studies of cellular electrophysiology of the human fetal have been reported [19, 20, 21]. Because of the difficulty in obtaining human fetal hearts, a complete developmental study has been impossible. Available data suggest that the resting membrane potential of mid-trimester human fetal ventricular myocardium is relatively high (-81 mV) and temperature-sensitive [19]. Action potential amplitude ranges from 95 to 120 mV at this time, and the voltage time course of repolarization is similar to that of adult tissue [20].

Although a number of studies have been published on the cellular electrophysiology of fetal atrial tissues, they have not been related to ontogeny. However, they suggest that, as with ventricular tissues, those of fetal atria have a lower resting membrane potential and action potential amplitude than the adult.

ONTOGENETIC CHANGES IN IONIC CURRENTS

A number of studies have been performed to determine developmental changes in the ionic currents responsible for the resting and action potentials. These studies have utilized alterations in the ionic composition of extracellular fluid, blockers of various ionic currents, and the measurement of intracellular ionic activities. As with other studies of embryonic hearts, most have been done using the chick embryo.

Resting Membrane Potential

As expected, the membrane potential of embryonic hearts is dependent on the K^+ gradient across cell membranes. For older chick embryos, near the time of hatching, the relationship between $[K^+]_o$ and resting membrane potential is similar to that for the adult at $[K^+]_o = 20$ mM; approximately a 50 to 60 mV change occurs per decade change in $[K^+]_o$ [8]. For less developed hearts, the slope is lower, 30 to 40 mV/decade for two- to three-day hearts, and 40 to 50 mV/decade for four- to seven-day hearts. The deviations at lower $[K^+]_o$ from values predicted from the Nernst equation are greatest in the least developed hearts, in which resting potentials are relatively low (figure 2–1). Younger hearts apparently also tolerate a higher $[K^+]_o$ before loss of excitability ensues [8]. Membrane permeability to K^+ (P_K) derived from extrapolated values of $[K^+]_i$, intracellular potassium concentrations, suggest that the ratio $^PNa/P_K$ is higher in less developed hearts: 0.2, 0.1, and 0.01 for 3-, 5-, and 15-day ventricle, respectively. Similar changes in PNa/P_K have been reported by others [8, 22, 23, 24]. Although decreases have been reported in intracellular K^+ concentration, $[K^+]_i$, as a function of development [23, 24], data from other studies suggest that $[K^+]_i$ increases [22, 25, 27]. Because of the technical limitations involved in the estimation of intracellular ions, however, it seems preferable to directly measure ionic activity. Fozzard and Sheu [22] studied the chick embryo, using both K^+-selective glass microelectrodes and K^+-selective liquid ion exchange microelectrodes, and found that the intracellular activity of K^+, A^i_K+, increased from 71 mM at day 4 to approximately 90 mM at day 18. The activity of intracellular Na^+, $A^i_{Na}+$, in contrast, decreased from 12.5 to 7.0 mM. Using these values to calculate equilibrium potentials for K^+ and Na^+, the following were reported: V_K increased from -73.3 mV at day 4 to -79.3 mV at day 18, while V_{Na} increased from 57.0 to 71.8 mV. These investigators suggest that a changing activity of a Na-K exchange pump can explain the concomitant increase

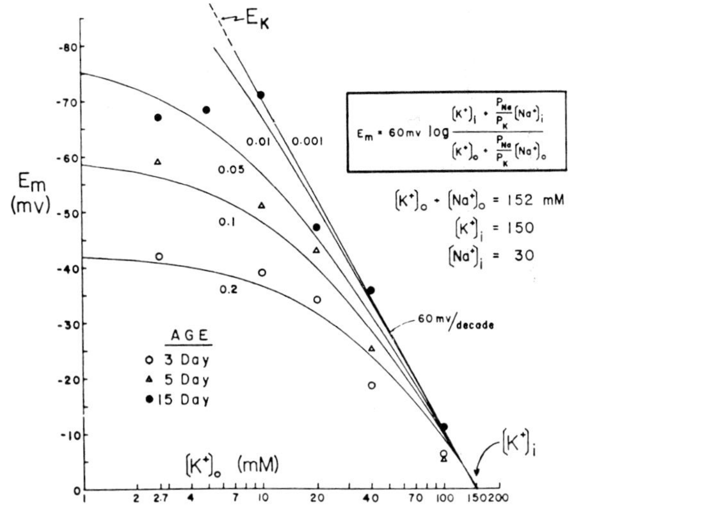

Figure 2–1. Resting potential (E_m) of chick embryonic ventricular myocardium plotted as a function of $[K^+]_o$ on a logarithmic scale. $[K^+]_o$ was elevated by substitution of K^+ for equimolar amounts of Na^+. Continuous lines give theoretical calculations from the constant-field equation (shown in inset) of resting E_m as a function $K^+{}_o$ for various assumed P_{Na}/P_K ratios of 0.001, 0.01, 0.05, 0.1, and 0.2. Calculations were made assuming a $[K^+]_i$ of 150 mM, $[Na^+]_i$ of 30 mM (estimated from $[Na^+]_o$ level at which excitability is lost); the sum of $[K^+]_o + [Na^+]_o$ was held constant at 152 mM, which was the method used to obtain the experimental data give here. As indicated by the equation, the shapes of these theoretical curves are dependent on the assumed values for $[Na^+]_i$. For a P_{Na}/P_k ratio of 0.001, the curve is linear over the entire range with a slope of 60 mv/decade, i.e., it closely follows E_k; at higher ratios, the slope continually diminishes as $[K^+]_i$ is lowered, and the curve flattens at low $[K^+]_o$ levels. Symbols give representative data obtained from embryonic chick hearts at days 3, 5, and 15. The data points for the 3-day heart follow the curve for a P_{Na}/P_k ratio of 0.2, those for the 5-day heart follow the theoretical curve for 0.1, and those for the 15-day heart follow the curve for a ratio of 0.01. The estimated intracellular K^+ activity ($[K^+]_i$) levels obtained by extrapolation to zero potential are nearly the same for hearts of all ages. (Taken from Sperelakis and Shigenobu, 1972. *J. Gen. Physiol.* 60:430–453.)

in A^i_{K+} and fall in A^i_{Na+}. In fact, such developmental changes in Na^+-K^+-ATPase have been reported for embryonic chick ventricle [26, 27].

Action Potential Upstroke

The study of the ionic basis for the embryonic and fetal action potential upstroke has involved changes in $[Na^+]_o$ and $[Ca^{++}]_o$, as well as agents which enhance (primarily catecholamines) or depress (Mn^{++}, tetrodotoxin, verapamil) fast and slow inward currents. In general, it is believed that early in cardiac development, the action potential upstroke is more dependent on an inward current carried by "slow channels" whereas later in development, inward current is carried through "fast channels" in the chick [8, 9, 10, 12, 13, 14, 15, 16], rat [17, 18], dog [18], and human [21]. Yeh and Hoffman [10] found that a reduction of $[Na^+]_o$ to 50% of normal reduced V_{max} to the same extent in chick ventricles at 6 and 19 days. Substituting choline, tetraethylammonium chloride (TEA), or tetramethylammonium chloride (TMA) for Na^+, results in the generation of "all or none" action potentials during the first ten days of incubation [29]. Choline or TMA substitution inhibited action potentials after day six. In the presence of tetrodotoxin (TTX), 1 mg/l hearts younger than six days had action potentials with a reduced amplitude and $\dot{V}_{max}$; complete abolition, however, was not observed. Hearts older than 10 days lost the ability to generate action potentials in the presence of TTX. Those hearts in which TTX did not abolish the action potential responded to Mn^{++} (a blocker of inward Ca^{++}) with a complete loss of excitability. Subsequent studies [8, 9] confirmed these results showing that TTX (as high as 20 mg/1) had no effect on the rate of spontaneously generated action potentials or on $\dot{V}_{max}$ of action potentials from hearts younger than < three days. The addition of La^{+++}, 1 mM, a putative blocker of inward Ca^{++} current, was without effect in these younger hearts, suggesting that despite TTX insensitivity, action potential upstroke was still dependent on Na^+ probably moving through a slow channel. Removal of extracellular Na^+ abolished excitability [9]. The relationships between $[Na^+]_o$ and action potential overshoot and $[Na^+]_o$ and $\dot{V}_{max}$ have been reported as markedly different in younger and older hearts [8]. With development, the curves relating $[Na^+]_o$ to each of these parameters shift upward to the left. For overshoot, the curves are linear between $[Na^+]_o$ = 40 to 50 mM for 4- and 17-day embryonic ventricles (slope ≅ 60 mV/decade). Above 50 mM, the curve for 4-day hearts attains a plateau and the slope approaches zero, whereas for the older hearts, the slope remains greater (figure 2–2). Similar qualitative changes occur for $\dot{V}_{max}$.

Pappano [16] studied the effects of TTX and $[Na^+]_o$ on embryonic chick atria. Action potentials from atria of ≥ 12 day embryos showed rapid reductions in amplitude and $\dot{V}_{max}$ following exposure to TTX, 5×10^{-8}M. Loss of excitability occurred within five minutes when extrinsically applied stimuli were used; spontaneously generated action potentials were similarly blocked. In contrast, the excitability of atria from 6-day hearts was not blocked by TTX, although amplitude and $\dot{V}_{max}$ were reduced. Reduction of $[Na^+]_o$ had different effects on atrial action potential overshoot in 6- and 18-day embryos; the slope of the line relating overshoot to Na^+_o was ≈ 60 mV/decade in 18-day embryos and only 10 to 20 mV in 6-day atria, suggesting that the Na^+-electrode characteristics of the membrane were augmented by increasing development. The effects of $[Na^+]_o$ on $\dot{V}_{max}$ (when corrected for age-related different control values) were equivalent in 6-, 12-, and 18-day embryonic atria.

In atria exposed to TTX, it was found that divalent cations (Ca^{++}, Ba^{++}, Sr^{++}) were less effective in restoring excitability in older than younger embryonic atria, an effect attributed to a decline in the ratio of inward current carried by divalent cations to outward K^+ [30]. A developmentally related increase in gK has been reported for embryonic hearts [8, 23].

The effect of TTX on fetal canine Purkinje fibers recently has been studied [18]. In fetuses obtained just after the time of implantation (≈ 18 days post-conception) through just prior to birth (normal gestation, 60 to 63 days) TTX has effects on

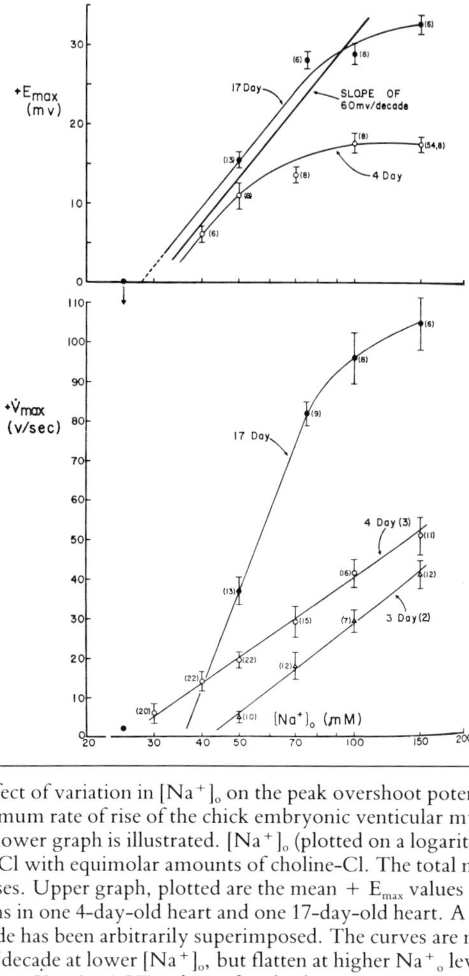

Figure 2–2. The effect of variation in $[Na^+]_o$ on the peak overshoot potential ($+E_{max}$) (upper graph) and on maximum rate of rise of the chick embryonic venticular myocardial action potential ($+V_{max}$) (lower graph is illustrated. $[Na^+]_o$ (plotted on a logarithmic scale) was lowered by replacing the NaCl with equimolar amounts of choline-Cl. The total number of impalements is given in parentheses. Upper graph, plotted are the mean $+E_{max}$ values ($\pm$ 1 SE bars) of multiple penetrations in one 4-day-old heart and one 17-day-old heart. A heavy line having a slope of 60mv/decade has been arbitrarily superimposed. The curves are nearly linear with a slope approaching 60 mv/decade at lower $[Na^+]_o$, but flatten at higher Na^+_o levels. Lower graph, plotted are the mean $+V_{max}$ ($\pm$ 1 SE) values of multiple penetrations into two 3-day-old hearts, three 4-day-old hearts, and one 17-day-old heart. The curve for the old heart is linear at lower $[Na^+]_o$ levels, and flattens at the higher levels; for the young hearts, the curves are linear over the entire range. (Taken from Sperelakis and Shigenobu, 1972. *J. Gen. Physiol.* 60:430–453.)

the Purkinje fiber action potential which are equivalent throughout this period of development (figure 2–3). TTX, 7×10^{-7} to 1.6×10^{-5}M, induces concentration-dependent decreases in action potential amplitude and V_{max}. The lack of a developmentally related effect of TTX on fetal canine fibers could be due to species differences, tissue differences, or to the inability to study TTX effects earlier than 18 days post-conception. Iijima and Pappano [12] found that the sensitivity to TTX of V_{max} of embryonic chick ventricle was equivalent between the 4th and 18th days of development. Similar results reported by Marcus and Fozzard [13] who also found that the heart of the adult chicken is of equivalent sensitivity to

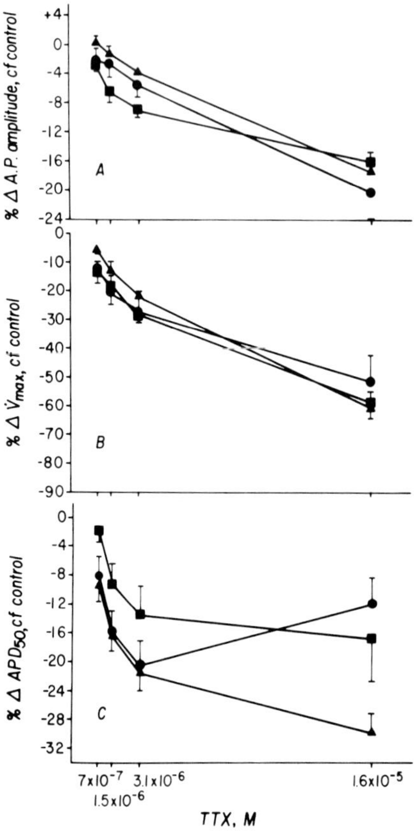

Figure 2–3. The effects of tetrodotoxin (TTX) on action potential amplitude (A), V_{max} (B), and APD_{50} (C). Data ($\bar{X} \pm SE$) are normalized with respect to control values. Abscissa, molar concentration of TTX; ordinate, percent change, compared to control. For all three panels, CRL is 28.8 ± 5.1mm (●———●) for group I 77.5 ± 4.5mm (■———■) for group II, and 142.0 ± 7.1mm (▲———▲) for group III.

Group	Control Values Action Potential Amplitude, mV	$\dot{V}_{max}$, V/sec	APD_{50}, msec
I n = 8	114.4 ± 2.5	315 ± 38	96.6 ± 13.6
II n = 6	121.3 ± 2.1	409 ± 33	107 ± 8.6
III n = 6	125.2 ± 3.8	509 ± 62	133.0 ± 6.4

Taken from [18], with permission.

TTX as that of the embryo. In these studies of chick embryos [12, 13], maximal depression of V_{max} was observed at TTX concentration of approximately 1×10^{-7}M. In contrast for fetal Purkinje fibers a 100- to 200-fold increase in TTX was required — 1.6×10^{-5}M. The extent to which these differences may be dependent on age is unknown.

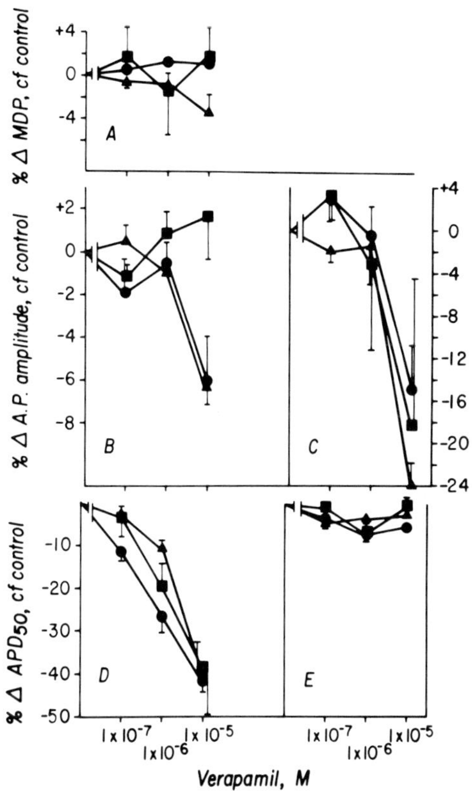

Figure 2–4. Effects of verapamil on maximum diastolic potential (A), action potential amplitude (B), $\dot{V}_{max}$ (C), and APD_{50} (D), and APD_{100} (E). Abscissa, molar concentration of verapamil; ordinate, percent change is given variable ($\bar{X} \pm SEM$), compared to control change. For all five panels, CRL is 38.0 ± 20mm (●———●) (group I), 83.3 ± 10.3mm (■———■ group II) and 132.5 ± 5.5mm (▲———▲, group III). Control values:

Group	Amplitude mV	MDP, mV	$\dot{V}_{max}$, V/sec	APD_{50} msec	APD_{100} msec
I n = 5	115.7 ± 3.5	−78.5 ± 0.9	323 ± 21	108.5 ± 0.9	181.7 ± 22.4
II n = 3	119.6 ± 1.8	−85.8 ± 2.1	455 ± 37	119.7 ± 7.9	191.7 ± 7.9
III n = 6	121.7 ± 2.9	−86.5 ± 1.4	481 ± 33	132.5 ± 7.0	210.5 ± 8.2

Taken from [18], with permission.

Inward Ca^{++} currents have been studied in fetal Purkinje fibers [18] using verapamil, a blocker of the slow inward current. Equivalent effects of verapamil were seen on fetal Purkinje fibers from early, middle, and late gestation (figure 2–4). The depression of $\dot{V}_{max}$ was maximal at verapamil 1×10^{-5}M. Action potential amplitude also was reduced by verapamil. Fibers from the least and most developed hearts were equivalent in their sensitivity to verapamil, while there was no effect on verapamil on fibers from mid-gestation. In addition, verapamil also decreased the duration of the plateau to an equivalent degree in all age groups. The effect of verapamil on $\dot{V}_{max}$ and action potential amplitude suggests that either these parameters depend to some extent on a slow inward current (Ca^{++} or Na^{+}) or that verapamil reduces the rapid inward Na^{+} current. Although there is evidence that verapamil blocks the movement of Na^{+} through a slow channel in embtyonic chick ventricle [31], the existence of such a current in fetal canine Purkinje fibers has not yet been studied. Effects of verapamil on adult Purkinje fibers similar to those on fetal fibers have been reported [32]. Verapamil in low concentrations ($< 1 \times 10^{-5}$M) has no effect on action potential amplitude, $\dot{V}_{max}$, or plateau duration. Higher verapamil concentrations do depress action potential amplitude, $\dot{V}_{max}$ and decrease plateau duration. Rosen et al. [32] found that verapamil, 1×10^{-5}M, reduced $\dot{V}_{max}$ from 545 to 334 V/sec, a reduction of nearly 40%. That such a decrease in inward current can be attributed to an effect of Ca^{++} current is unlikely since it has been suggested that in cardiac Purkinje fibers the peak inward Ca^{++} current is 1% to 2% of the peak inward Na^{+} current [33].

NEONATAL HEARTS

Studies of the cellular electrophysiology of neonatal hearts are fewer than those of fetal or embryonic hearts. The most detailed studies have been reported for the neonatal canine ventricular specialized conducting system. As a group, neonatal (0 to 10 days of age) Purkinje fibers continue the developmental changes in the transmembrane action potential begun during fetal life.

Mean values of action potential characteristics recorded for neonatal and adult Purkinje fibers may vary slightly from study to study, but there are statistically significant differences between these age groups (table 2–1).

Action potential amplitude increases in the neonatal Purkinje fiber, and this change appears to be part of the continuum occurring during intrauterine development. Action potential amplitude increases from approximately 120 mV in the term fetus to 123 mV within the first 10 days of life. This change, although small, is statistically significant. $\dot{V}_{max}$ is statistically equivalent over this time, being approximately 474 V/sec sec in the late-gestation fetus and 456 V/sec in the neonate. Changes in action potential duration that were evident in the fetal heart continue in the neonate. Action potential duration measured to 50% repolarization increases from 120 msec in late fetal life to 160 msec in neonates. Action potential duration (to full repolarization) increases from approximately 220 to 263 msec. When considering action potential duration, it should be kept in mind that for neonatal Purkinje fibers duration is equivalent in proximal subendocardial and free-running portions and decreases in distal locations [34]. In contrast, action potential duration in the adult Purkinje fiber is greatest in distal segments, at the ventricular muscle-Purkinje fiber junction [34, 35].

Other animal models have been used to study the effects of age on the transmembrane action potential. Cavoto et al. [36] found that the resting potential of rat atria increased from -74 mV at one month of age to -82 mV at three months (electrically stimulated), while action potential amplitude increased from 83 to 90

Table 2–1. Action Potential Characteristics of Canine Purkinje Fibers

	Mid-gestation Fetus	Late-gestation Fetus	Neonate	Adult
Amplitude mV	110.0 ± 1.3 (52)	119.8 ± 0.8 (40)	122.9 ± 0.7 (64)	127.6 ± 0.5 (151)
MDP mV	−78.6 ± 1.2 (52)	−83.7 ± 1.2 (38)	−88.6 ± 0.5 (63)	−90.5 ± 0.4 (147)
V_{max} V/sec	347 ± 15 (50)	474 ± 17 (44)	456 ± 12 (60)	544 ± 10 (130)
APD_{50} msec	104.8 ± 3.4 (53)	130.6 ± 4.5 (45)	160.0 ± 2.8 (62)	174.7 ± 2.4 (141)
APD_{100} msec	163.3 ± 7.8 (53)	214.8 ± 4.0 (45)	263.2 ± 3.1 (62)	294.9 ± 2.4 (137)

All values are mean ± S.E. Numbers in parentheses are number of data points. For neonates and adults, data were pooled from previously published studies. All action potential characteristics (except for V_{max}) for neonates and adults are statistically different ($P < .01$). Mid-gestation is defined as 30 to 40 days, post-conception; late gestation as 50 to 60 days. Mid-gestation and late gestation action potential characteristics are significantly ($P < .01$) different, as are late gestation and neonatal characteristics.
MDP, maximum diastolic potential.
V_{max}, maximum upstroke velocity.
APD_{50}, action potential duration measured to 50% repolarization.
APD_{100}, action potential duration measured to full repolarization.

mV. Increases of V_{max} in the young rat apparently do not parallel the increase in membrane potential. V_{max} in one-month-old rats was 445 V/sec and in three-month-old rats was 308 V/sec.

Developmental changes in the effects of pharmacologic modifiers of ionic currents underlying the Purkinje fiber action potential have been reported [25]. TTX decreases the amplitude and duration of the action potential as well as V_{max} of neonatal and adult fibers. However, the neonate requires somewhat less TTX than the adult to depress the action potential. The duration of neonatal action potentials is significantly decreased by TTX 7×10^{-7}M, whereas that of the adult requires a higher threshold concentration, 1.5×10^{-6}M. Action potential amplitude of neonatal fibers required about 1/5 as much TTX as adults for a threshold reduction. V_{max} of neonatal fibers was significantly decreased by TTX, $\geqslant 1.5 \times 10^{-6}$M, whereas V_{max} of adult fibers required a slightly greater threshold concentration of 3.1×10^{-6}M. For all three variables the maximum effect of TTX was greater in the neonate than in the adult. The reasons for these age-dependent effects of TTX are not known. A developmental increase in the number and/or density of Na^+-channels is suggested by the increase in V_{max} occurring between neonatal and adult life, but there are no data to support this hypothesis. Because the relationship between membrane potential and V_{max} (i.e., membrane responsiveness) is equivalent in neonatal and adult fibers, it is likely that fast channel kinetics do not change significantly. The effect of TTX on action potential duration suggests that there is a slow inward Na^+ current operating during repolarization [37]. This current is more sensitive to TTX than that of the action potential upstroke, both in adults [26, 37] and neonates [25]. Voltage clamp studies of adult sheep Purkinje fibers have shown that there is a significant sodium current operative at plateau potentials (i.e., a "window current") and that this current is probably more sensitive to the effects of tetrodotoxin than that underlying the upstroke of the action potential [38]. The fact that the duration of the fetal Purkinje fiber action potential is more sensitive to tetrodotoxin than V_{max} suggests that this window current exists relatively early in fetal life.

Repolarizing currents have been studied in neonatal and adult Purkinje fibers [25] using a variety of blocking agents including verapamil and tetraethylammonium (TEA) chloride.

The effect of verapamil on the amplitude of slow response action potentials (produced in Na^+ – free, Ca^{++} – enriched solutions) is greater in the adult than the neonatal Purkinje fiber. These results suggest that the adult Purkinje fiber has a larger inward Ca^{++} current than does that of the neonate which is attenuated by blocking agents. In contrast, sensitivity to tetrodotoxin is greater in the neonate than in the adult.

Thus, despite the lack of quantification of age-related changes in ionic conductances, it is possible to demonstrate changes in the sensitivities to blocking agents of cardiac action potentials and to hypothesize that age and development may be responsible for changes in cellular electrophysiology.

EFFECTS OF AUTONOMIC NERVOUS SYSTEM AGONISTS AND ANTAGONISTS

Ontogenetic changes in cardiac autonomic innervation and responsiveness have been studied extensively, primarily in the embryonic chick heart (see review by Pappano [39]). The following work was done on Purkinje fibers obtained from fetal, neonatal, and adult canine hearts.

Purkinje fibers obtained from fetal hearts during early (18 to 30 days post-conception), middle (31 to 45 days), and late (46 to 63) gestation (superfused with a physiologic salt solution) undergo an increase in spontaneous rate from 0 during early gestation to approximately 10 beats/min during late gestation. The spontaneous rate of fibers from near-term fetuses is equivalent to that of the neonate. For adult fibers, spontaneous rate is approximately 12 beats/min and is statistically equivalent to that of the neonate as well as the near-term fetus.

Responsiveness of the ventricular specialized conducting system to adrenergic agonists is also a function of ontogenesis [40]. Purkinje fibers from fetal hearts initially show little response to epinephrine (figure 2–5). During early gestation relatively high ($\geq 1 \times 10^{-6}$M) concentrations of epinephrine are required to increase spontaneous rate. With development, responsiveness increases and becomes similar to that seen for neonatal fibers. The threshold concentration of epinephrine as well as that required to elicit a maximal increase in rate decrease. Of further interest is the observation of a negative chronotropic response to epinephrine in fibers from near-term fetal hearts. Thus, not only are there quantitative changes in the dose-response relationship for epinephrine and spontaneous rate but a qualitative change also occurs — a negative chronotropic response to epinephrine.

The response to adrenergic agonists and antagonists of neonatal and adult Purkinje fibers has been studied [41]. Significant differences exist between the two age

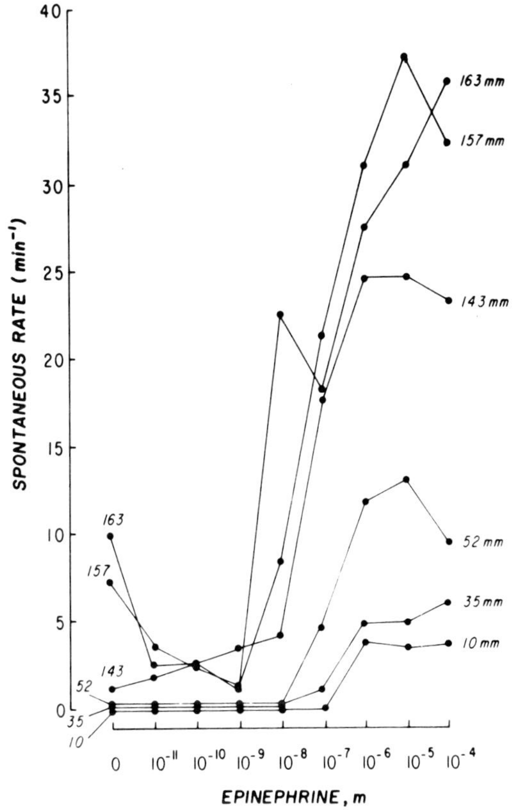

Figure 2–5. The relationship between fetal development and the effect of epinephrine on Purkinje fiber spontaneous rate. Abscissa: molar concentration of epinephrine; ordinate: spontaneous rate, beats/min. For each of the six curves is shown the crown-rump length (mm) of each fetus from which Purkinje fibers were studied. It can be seen that the intrinsic spontaneous rate was 0 for fetuses from early to mid-gestation (i.e., CRL = 10, 35 and 52mm). Rate thereafter increased such that for near-term fibers spntaneous rate was eight to ten beats/min. Also note the increase in the maximum positive chronotropic effect of epinephrine. For fibers from near-term hearts, low epinephrine concentrations (1×10^{-11} to 1×10^{-9}M) induced negative chronotropic effects, whereas higher concentration caused positive chronotropic effect.

groups with respect to both the negative and positive chronotropic response to adrenergic stimulation. The agonists studied were isoproterenol, a relatively pure β-agonist, and epinephrine, an agonist with both α- and β-agonistic properties. Initial studies with adult Purkinje fibers yielded two distinct response patterns (figure 2–6). Approximately two-thirds of adult fibers responded to low concentrations of epinephrine (1×10^{-11} to 1×10^{-8}M) with a decrease in spontaneous rate, while at higher concentrations ($\geqslant 1 \times 10^{-7}$M) an increase in rate occurred. This type of response was termed a biphasic response. A second group of adult Purkinje fibers responded to all concentrations of epinephrine with increase in rate only; this response was called monophasic. These two groups of adult fibers also

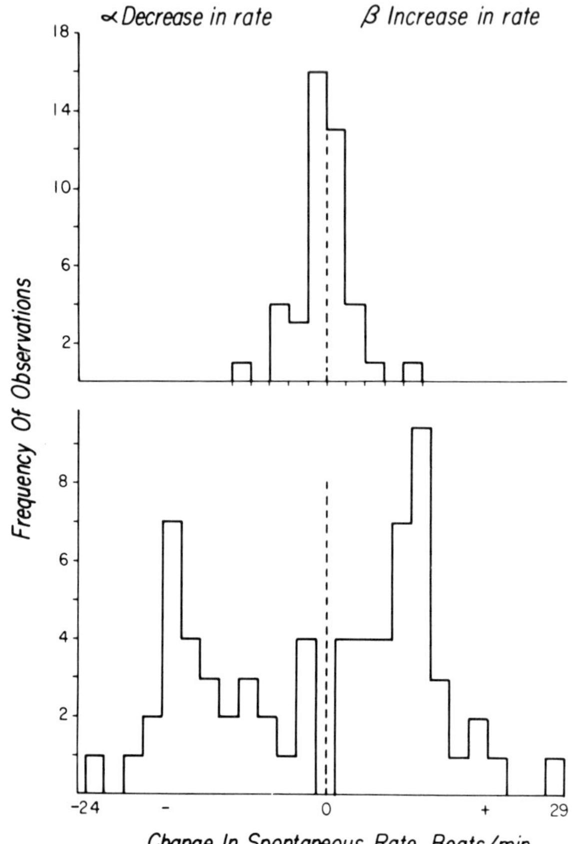

Figure 2–6. Histograms of epinephrine-induced changes in the spontaneous rate of adult Purkinje fibers. In the upper panel are shown the responses of fibers to epinephrine, 1×10^{-11}M, a concentration which was associated with no statistically significant effect on rate. The abscissa shows the interval of changes in spontaneous rate; the ordinate shows the frequency of observations for each interval. Note that in the upper panel a unimodal distribution of responses was recorded with the mean centered at 0 and the distribution approximately normal. In the bottom panel are shown the responses of adult Purkinje fibers to 1×10^{-9}M epinephrine. In this case, the population of response is bimodal, each mode being a good approximation of the normal distribution.

differed qualitatively from each other. The maximum positive chronotropic response of the biphasic group was approximately 75% above control, whereas that of the monophasic group was 25% above control. Beta-blockade, using propranolol, and α-blockade, using phentolamine, suggested that the negative chronotropic effect of epinephrine was α-adrenergic and the positive chronotropic effect, β-adrenergic. In contrast to biphasic adult fibers, where the proportion was two-thirds, only one-half of the fibers from neonatal hearts showed a biphasic response. The maximum positive chronotropic effect for biphasic neonatal fibers

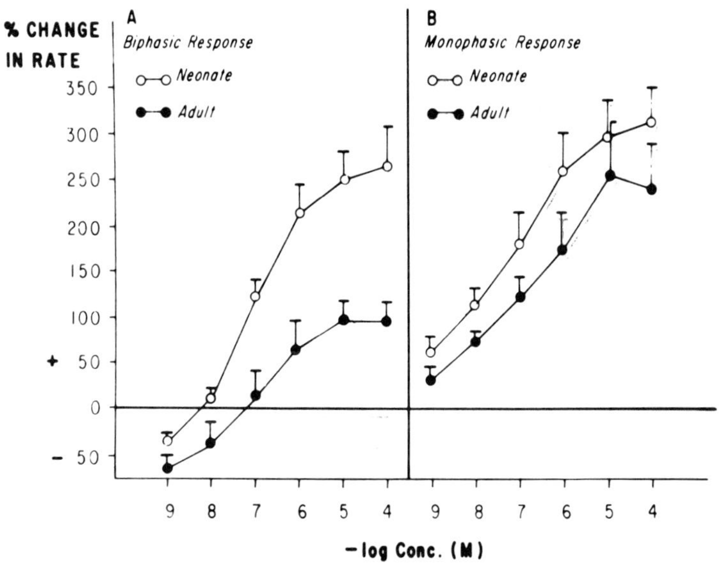

Figure 2–7. Effects of epinephrine on the biphasic and monophasic dose-response curves for neonatal and adult fibers. The biphasic neonatal and adult curves differ significantly at epinephrine, 10^{-7} through 10^{-4}M. There is no significant difference between the monophasic curves. Taken from [41], with permission.

was significantly greater than that for adults — 250% versus 75%, respectively, above control. However, for those fibers responding with only an increase in spontaneous rate (i.e., monophasic) the responses were statistically equivalent. Figure 2–7 illustrates these differences between neonatal and adult Purkinje fibers.

These age-dependent effects of adrenergic amines on Purkinje fiber spontaneous rate require further study to determine the underlying mechanism. There may be age-related differences in adrenergic receptor characteristics such as receptor number, density, affinity, or specificity. There also may be differences in the rates of development of α- and β-receptors during fetal and neonatal life. The presence of α-receptors in cardiac tissues has been demonstrated using radioligand binding technique [42, 43, 44, 45]. These receptors are apparent during fetal life and have been reported to decrease in number with ontogenesis in the fetal lamb [44]. For the dog, it has been reported that specific binding can be demonstrated in the neonate (one to five weeks of age) but not in the adult [45]. For canine Purkinje fibers, where the magnitude of the epinephrine-induced negative chronotropic effect is greater in the adult than the neonate, one might expect to demonstrate α-receptors more easily in the adult than the neonate.

Recent studies [46] have shown that in the first two days of life α-adrenergic stimulation may induce an increase in spontaneous rate in some fibers. The α-blocking agent, phentolamine, prevented this effect. This suggests that the α-adrenergic component of catecholamine responsiveness undergoes marked qualitative changes during the first two days of neonatal life. Although the underlying reasons for such a change are unknown, an increase in cardiac innervation is occurring during this time, indicated by the increase in norepinephrine content of neonatal hearts that occurs then. Norepinephrine content increases from approximately 100 mg/g tissue (wet weight) at day 2 to 1600 mg/g at day 10.

The effects of parasympathetic agonists on fetal and embryonic hearts have been extensively reviewed [39]. The effects of this class of drugs on the developing ventricular conducting system have been less well characterized. For adult canine Purkinje fibers, the major effect of acetylcholine is to decrease spontaneous rate. The effects of acetylcholine on action potential duration are variable and depend, at least in part, on the species of animal studied, on $[K^+]_o$, and on the rate of stimulation.

Neonatal Purkinje fibers respond differently than adult fibers to acetylcholine [47]. At low concentrations (1×10^{-11} to 1×10^{-9}M) spontaneous rate increases, whereas higher concentrations ($> 1 \times 10^{-9}$M) have a negative chronotropic effect. Both the negative and positive effects are blocked by atropine, suggesting there is a common muscarinic receptor. The possibility of an acetylcholine-induced release of endogenous catecholamine was eliminated by the use of propranolol. Neither aspect of the chronotropic response of neonatal Purkinje fibers was altered by beta-blockade. Similarly, the alpha-blocking agent, phentolamine, was without effect. The possibility that histamine may have been released secondary to exposure of neonatal fibers to acetylcholine was eliminated when it was found that histamine, 1×10^{-9}M to 1×10^{-4}M, was without effect on spontaneous rate. Thus, the likelihood is very low that the positive chronotropic effect of acetylcholine in the neonate is mediated by other than a muscarinic receptor. Although the ionic mechanisms for the positive chronotropic effect of acetylcholine on neonatal Purkinje fibers are not known, such an effect has been studied and demonstrated in adult sheep Purkinje fibers which also develop a positive chronotropic response to acetylcholine [48]. Here, the positive chronotropic effect results from a decrease in potassium conductance. Whether such a phenomenon occurs in neonatal canine fibers remains to be determined.

CONCLUSIONS

A great deal of work on fetal hearts has demonstrated developmental changes in the characteristics of transmembrane action potentials, both in working myocardial cells and in the ventricular specialized conducting system. The relative contributions of fast (TTX-sensitive) and slow (TTX-insensitive) inward currents to maximum upstroke velocity require further study. Information on this subject will allow a better understanding of fetal electrophysiology which, in turn, will result in a better understanding of that of the adult.

Because of increased use during pregnancy of cardioactive drugs such as β-adrenergic blocking agents, the need has increased to know not only how electrophysiologic properties of the fetal and neonatal heart change during development but also how these agents affect the immature heart. Because many of the drugs used exert their effects either directly or indirectly through the autonomic nervous system, it is important to understand better the effects of autonomic agents on fetal hearts. Further, in light of recent developments in the intrauterine treatment of fetal cardiac disorders, this need is underscored. Although there have been many recent gains made in our understanding of how age can modify drug action in developing hearts, additional information is required to guide the development of pharmacological agents which may be used more safely during pregnancy. Future developments are anticipated from studies of adrenergic receptor binding characteristics (e.g., specificity, density, etc.). This information will ultimately result in more rational drug development and safer application of those cardioactive agents already in use.

REFERENCES

1. Harvey, Wm. 1653. Anatomical exercitationes concerning the generation of living creatures. *Exercitation* XVII. London.
2. Pickering, J.W. 1893. Observations on the physiology of the embryonic heart. *J. Physiol. (Lond.)* 14:383–466.
3. Fingl, E.; Woodbury, L.A.; and Hecht, H.H. 1952. Effects of innervation and drugs upon direct membrane potentials of embryonic chick myocardium. *J. Pharm. Exp. Ther.* 104:103–114.
4. His, W. 1868. *Untersuchungen uber die erste anlage des Wirbelthierleibes.* Leipzig: Vogel.
5. Wernicke, R. 1878. *Zur physiologie des embryonalen Herzens.* Jena: Neuenhahn.
6. Preyer, W. 1885. *Specielle physiologie des embryo.* Leipzig: Th Grieben.
7. Patten, B.M., and Kramer, T.C. 1933. The initiation of contraction in the embryonic chick heart. *Amer. J. Anat.* 53:349–374.
8. Sperelakis, N., and Shigenobu, K. 1972. Changes in membrane properties of chick embryonic hearts during development. *J. Gen. Physiol.* 60:430–453.
9. Shigenobu, K., and Sperelakis, N. 1971. Development of sensitivity to TTX of chick embryonic hearts with age. *J. Mol. Cell. Cardiol.* 3:271–286.
10. Yeh, B.K., and Hoffman, B.F. 1968. The ionic basis of electrical activity in embryonic cardiac muscle. *J. Gen. Physiol.* 52:666–681.
11. Vleugels, A.; Carmeliet, E.; Bosteels, S.; and Zaman, M. 1976. Differential effects of hypoxia with age on the chick embryonic heart. *Pflüg Arch.* 365:159–166.
12. Iijima, T., and Pappano, A.J. 1979. Ontogenetic increase in the maximal rate of rise of the chick embryonic heart action potential. Relationship to voltage, time and tetrodotoxin. *Circ. Res.* 44:359–367.
13. Marcus, N.C., and Fozzard, H. 1981. Tetrodotoxin sensitivity in the developing and adult chick heart. *J. Mol. Cell. Cardiol.* 13:335–340.
14. DeHaan, R.L.; McDonald, T.F.; and Sachs, H.G. 1975. Development of embryonic chick heart cells in vitro. In *Developmental and physiological correlates of cardiac muscle,* ed. M. Lieberman and T. Sano, p. 161. New York: Raven Press.
15. Shimizu, Y., and Tasaki, K. 1966. Electrical excitability of developing cardiac muscle in chick embryos. *Tohoku J. Exp. Med.* 88:49–56.
16. Pappano, A.J. 1972. Action potentials in chick atria. Increased susceptibility to blockade by tetrodotoxin during embryonic development. *Circ. Res.* 31:379–338.
17. Bernard, C. 1975. Establishment of ionic permeabilities of the myocardial membrane during embryonic development of the rat. In *Developmental and physiological correlates of cardiac muscle,* ed. M. Lieberman and T. Sano, pp. 169–184. New York: Raven Press.
18. Danilo, P.; Reder, R.F.; Binah, O.; and Legato, M.J. 1984. Fetal canine cardiac Purkinje fibers: electrophysiology and ultrastructure. *Am. J. Physiol.,* 246:H250–H260.
19. Gennser, G., and Nilson, E. 1970. Excitation and impulse conduction in human fetal heart. *Acta Physiol. Scand.* 79:305–320.
20. Tuganowski, W., and Cekanski, A. 1971. Electrical activity of a single fiber of the human embryonic heart. *Pflüg Arch.* 323:21–26.
21. Janse, M.J.; Anderson, R.H.; van Capelle, F.J.L.; and Durrer, D. 1976. A combined electrophysiological and anatomical study of the human heart. *Am. Heart J.* 91:556–562.

22. Fozzard, H.A., and Sheu, S.S. 1980. Intracellular potassium and sodium activities of chick ventricular muscle during embryonic development. *J. Physiol.* 306:579–586.
23. Carmeliet, E.E.; Horres, C.R.; Lieberman, M.; and Vereecke, J.S. 1976. Developmental aspects of potassium flux and permeability of the embryonic chick heart. *J. Physiol.* 254:673–692.
24. McDonald, T.F., and DeHaan, R.L. 1973. Ion levels and membrane potential in chick heart tissue and cultured cells. *J. Gen. Physiol.* 61:89–109.
25. Reder, R.F.; Miura, D.S.; Danilo, P.; and Rosen, M.R. 1981. The electrophysiological properties of neonatal and adult canine cardiac Purkinje fibers. *Circ. Res.* 48:658–668.
26. Sperelakis, N. 1972. (Na^+, K^+)-ATPase activity of embryonic chick heart and skeletal muscle as a function of age. *Biochem. Biophys. Acta* 266:230–237.
27. Klein, R.L. 1963. The induction of a transfer adenosine triphosphate phosphohydrolase in embryonic chick heart. *Biochem. Biophys. Acta* 73:488–498.
28. Couch, J.R.; West, T.C.; and Hoff, H.E. 1969. Development of the action potential of the prenatal rat heart. *Circ. Res.* 24:19–31.
29. Ishima, Y. 1968. The effect of tetrodotoxin and sodium substitution on the action potential in the course of development of the embryonic chicken heart. *Proc. Japan. Acad.* 44:170–175.
30. Pappano, A.J. 1976. Action potentials in chick atria. Ontogenetic changes in the dependence of tetrodotoxin-resistant action potentials on calcium, strontium and barium. *Circ. Res.* 39:99–105.
31. Shigenobu, K.; Schneider, J.A.; and Sperelakis, N. 1974. Verapamil blockade of slow Na^+ and Ca^{++} response in myocardial cells. *J. Pharmacol. Exp. Ther.* 190:280–288.
32. Rosen, M.R.; Ilvento, J.P.; Gelband, H.; and Merker, C. 1974. Effects of verapamil on electrophysiologic properties of canine cardiac Purkinje fibers. *J. Pharmacol. Exp. Ther.* 189:414–422.
33. Fozzard, H.A. 1983. Calcium channels in the heart. 1983. In *Calcium antagonists: the state of the art and role in cardiovascular disease. Proceedings of the Calcium Antagonists Symposium of the Amer. Coll. Clinical Pharmacol. Symposium on Frontiers of Pharmacology,* vol. III, pp. 7–12.
34. Untereker, W.J.; Danilo, P.; and Rosen, M.R. 1984. Developmental changes in action potential duration, refractoriness and conduction in the canine ventricular conducting system. *J. Ped.*, in press.
35. Myerburg, R.J.; Stewart, J.W.; and Hoffman, B.F. 1970. Electrophysiological properties of the canine peripheral AV conducting system. *Circ. Res.* 26:361–378.
36. Cavoto, F.V.; Kelliher, G.J.; and Roberts, J. 1974. Electrophysiological changes in the rat atrium with age. *Am. J. Physiol.* 226:1293–1297.
37. Coraboeuf, E.; Deroubaix, E.; and Couloumbe, A. 1979. Effect of tetrodotoxin on action potentials of the conducting system in the dog. *Am. J. Physiol.* 236:H561–H567.
38. Atwell, D.; Cohen, L.; Eisner, D.; Ohba, M.; and Ojeda, C. 1979. The steady state TTX-sensitive ("window") sodium current in cardiac Purkinje fibers. *Pflüg Arch.* 379:137–142.
39. Pappano, A.J. 1977. Ontogenetic development of autonomic neuro effect or transmission and transmitter reactivity in embryonic and fetal hearts. *Pharmacol. Rev.* 29:3–33.
40. Danilo, P.; Reder, R.; Mill, J.: and Petrie, R. 1979. Developmental changes in cellular electrophysiologic charcteristics and catecholamine content of fetal hearts. *Circulation* 59,60:II–50.
41. Rosen, M.R.; Hordof, J.P.; Ilvento, J.P.; and Danilo, P. 1977. Effects of adrenergic amines on electrophysiological properties and automaticity of neonatal and adult canine Purkinje fibers. *Circ. Res.* 40:390–400.
42. Wei, J.W., and Sulakhe, P.V. 1979. Regional and subcellular distribution of beta and alpha adrenergic receptors in the myocardium of different types. *Gen. Pharmacol.* 10:263–267.
43. Yamada, S.; Yamamura, H.I.; and Roeske, W.R. 1980. Ontogeny of mammalian cardiac α,-adrenergic receptors. *Eur. J. Pharmacol.* 68:217–221.
44. Cheng, J.B.; Cornett, L.E.; Goldfein, A.; and Roberts, J.M. 1980. Decreased concentration of myocardial α-adrenergic with increasing age in fetal lambs. *Br. J. Pharmacol.* 70:515–517.
45. Felder, R.A.; Calcagno, P.L.; Eisner, G.M.; and Jose, P.A. 1982. Ontogeny of myocardial adrenoceptors. II. Alpha adrenoceptors. *Ped. Res.* 16:340–342.
46. Reder, R.; Danilo, P.; and Rosen, M.R. 1984. Developmental changes in adrenergic effects on canine Purkinje fiber automaticity. *Dev. Pharmacol. Therap.*, 7:94–108.
47. Danilo, P., Rosen, M.R., and Hordorf, A.J. 1978. Effects of acetyl choline on the ventricular conducting system of neonatal and adult dogs. *Circ. Res.* 43:777–784.
48. Carmeliet, E., and Ramon, J. 1980. Effect of acetylcholine on time-independent currents in sheep cardiac Purkinje fibers. *Pflüg Arch.* 387:207–216.

3. THE DEVELOPMENT OF THE ELECTROCARDIOGRAM

M. IRENÉ FERRER

The electrophysiologic events responsible for the generation of the clinical electrocardiogram represent a conglomerate of behaviors stemming from diverse and intrinsically quite different tissues. In reviewing, therefore, the development of the final product — namely, the adult human electrocardiogram — one must in turn outline the development of the contributing parts. The inscription of P, QRS, and T complexes has its origin in the sequential contributions of the following tissues: the sino-atrial node and its surrounding approaches; the atrial myocardium and its specialized preferential intranodal pathways; the AV node and junctional tissue; the penetrating portion or common bundle of His and its branches; the Purkinje network; and the ventricular myocardium.

The information relative to the development of these tissues is almost entirely anatomic. There is a paucity of electrophysiologic information. Both will be reviewed.

THE SA NODE: THE PRIMARY PACEMAKER

The normal pacemaker of the heart does not control the cardiac rhythm from the onset of cardiac development. Current anatomic information as recently reviewed [1, 2] reveals that at the very early stages of development, when the heart is in the "straight-tube" position, there already is a differentiation of the specialized conducting tissues from working myocardial cells. This occurs by a separate migration of conducting cells from the myoepicardial mantle. Thus in the early stages of development there are probably two distinct populations of cells within the

M.J. Legato (ed.) THE DEVELOPING HEART, pp. 39–46.

heart. Evidence is presented by these investigators [2] that in the straight-tube stage "both phylogenetic and ontogenetic studies indicate that specialized zones of cells exist at the junctions between the primordial chambers of the heart tube." These are actually junctional rings of specialized tissues existing at the sino-atrial (between the sinus horn and atrium), atrio-ventricular, ventriculobulbar, and bulbo-truncal junctions. These rings of tissue represent the primordia of conducting tissues.

During the first six to seven weeks of life there is as yet no evidence of a discrete structure comparable to the definitive SA node [2]. At this early stage only the primordia of the AV bundle and its ventricular branches are present astride the ventricular septum, but the AV node as such is not yet formed. Hence it can be said that the specialized tissues are not yet ready at six to seven weeks for sinus rhythm, if it existed, to reach the ventricles.

Information from fetal electrocardiograms indicates that definable complexes, usually only QRS complexes, first appear on the fetal electrocardiogram between 10 and 12 weeks of embryonic age [3]. This suggests that the cardiac rhythm in the early stages of development of the human embryo originates in junctional tissue.

When the embryo is about eight weeks old, a structure approximating the mature SA node is first visualized [2]. By 10 weeks of age, when the right superior cardinal vein evolves to become the superior vena cava, the specialized tissue which remains at its base gradually becomes the normal pacemaker. The cells of the SA ring at the junction of the superior vena cava and the right auricle aggregate around a developing artery; by 10 weeks of age a discrete SA node is found surrounding this vessel. The tripartite cell population typical of the adult node — P cells, transitional cells, and working myocardial cells — is seen at birth; but earlier, at 10 weeks of age in the embryo, the cellular arrangement is more dense, being characterized by a tightly packed mass of small cells with more P cells and transitional cells than in the newborn [2]. Clearly then, there are ongoing developmental changes in this SA node, and these continue, especially with regard to increase in collagen in the node, until adult life [4].

Given this changing anatomy, one must also consider the development of the autonomic nervous system in relation to cardiac tissue. The fetal heart has varying rates before it develops autonomic innervation; the cardiac nerve supply develops rather late in cardiac development [1]. Once any neural control develops it is predominantly cholinergic [1, 5]. Thus the sinus rate in the fetus and newborn is rapid and relatively unstable, probably, in part at least, due to the unbalanced autonomic innervation. Adrenergic innervation appears later and is complete only some months after birth. It is now known that heart rate fluctuates widely in newborn infants [5] and indeed is varying all the time. Immediately after birth it is 135 to 140/minute; during the first postnatal hour it lies between 75 and 200/minute; the third hour averages 115/minute and slowly increases to 120 to 140/minute at the end of the first 24 hours of extrauterine life [5]. In healthy premature infants during their first week of life the heart rate is labile and rapid, reaching 200/minute with slight stimuli and falling to 100/minute during sleep. Sinus arrhythmia occurs at any age, and phasic variations in rate have even been seen in fetuses; however, fixed rates in sinus rhythm may occur suddenly, replacing variation, and the cause is unknown. The reason for sudden periods of sinus bradycardia in healthy babies is also unclear, but these are usually short-lived. Sinus tachycardia between 175 and 190/minute are seen between birth and one month of age, and rates of 200 to 220/minute can occur. There are reports of rates as high as 260/minute in neonates [5]. These tachycardias are probably due to increase in adrenergic activity. Sinus rates gradually decrease as the child ages, and values range between 60 and 80/minute in adults at rest. The mechanism of this gradual decrease in intrinsic rhythmicity of the SA node has not been studied but is presumed to be related to autonomic factors.

The Appearance of Sinus Rhythm

The question should be asked, in view of our present knowledge, when does sinus rhythm control the human heart? The information from fetal electrocardiograms unfortunately is very incomplete. The reason, of course, is the technical difficulties inherent in obtaining this fetal electrocardiogram. Recordings from the infant's scalp cannot be obtained until labor begins, and recordings via the mother's abdominal wall are unsatisfactory because deflections are blunted and usually only QRS complexes can be found. There are now newer technologies which can augment the fetal action potentials so that recordings via the maternal abdomen are accurate; when these are clinically available longitudinal serial observations of fetal electrocardiograms in utero will answer the question. These augmenting techniques have recently successfully recorded the action potential intrinsic to and inside the SA node — the true "firing" of this pacemaker. This event, which precedes the P wave, is, of course, not visible on the clinical electrocardiogram.

At present, one can only opine that sinus rhythm begins early in the middle trimester of gestation. This opinion is reinforced by the fact that a definitive or completed AV node can be recognized only after 10 weeks in humans, and only at midterm are definitive bundle branches recognizable even though these specialized tissues are distinguishable at six weeks [2]. Both these tissues are needed, of course, to convey sinus rhythm to the ventricles.

ATRIAL MYOCARDIUM, INTERNODAL TRACTS: THE P WAVE

The spread of the SA node impulse as it enters and then passes over the atrial myocardium creates the P wave. The internodal tracts connecting the SA to the AV node offer direct access to the lower node from the pacemaker [6]; however, some investigators [2] have questioned the special character of these tracts, believing that conduction to the AV node is over homogeneous working atrial myocardium. The electrophysiologic studies, however, tend to support specific tracts between the nodes. There are no developmental data about them.

The form of the atrial excitation wave (P) changes during early neonatal life. At birth the P wave is tall (3 mm. or more) and narrow with a duration of 0.03 to

0.08 sec. [3]. Between one month and one year it measures 0.04 to 0.08 sec.; between one and three years, 0.06 to 0.09 sec.; and after that, it slowly widens until it reaches adult values (0.08 to 0.11 sec.) somewhere between 5 and 10 years of age. The height gradually lessens with age, and by 10 years of age 3 mm. values are no longer seen. The development of a wider, shorter P wave occurs during the normal increase in atrial size. There are no serial measurements of atrial conductivity during these developmental changes, and these are needed to explain almost a doubling of atrial excitation time and a decline in P voltage with maturation.

AV NODE AND HIS BUNDLE (PURKINJE SYSTEM): THE PR INTERVAL

Developmental anatomic studies have recently emphasized the individuality of this node, the penetrating bundle or common bundle of His, and the bifurcating bundle and its branches [2]. Each of these has a separate origin embryologically, hence different physiologic behaviors are no surprise. Specialized structures which can be considered analogous to both node and bundle are present in the earliest embryologic human specimens (five to six weeks) which suggests development of both structures in situ. The formation of the AV node in its adult form is a late event embryologically. The node is composed developmentally of two parts: the deep and the superficial. The deep component appears to be of AV canal origin while the superficial, more atrial, component is derived from the atrial venous valves. When the two parts are joined in the developmental shifts of atrial growth, the definitive AV node is established. No electrophysiologic data yet exist at these early stages, but it is likely that only the deep component of the node has pacemaker capability. This idea is reinforced by the fact that developmentally the penetrating bundle (also a potential area, of self-excitation) and the deep nodal component are one and the same tissue structure [2]. By 10 weeks the AV node is recognizable as a separate entity in the developing heart. The penetrating bundle is formed from invaginating AV ring tissue, but the bundle branches probably arise from the bulbo-ventricular ring. These branches, though present at six weeks, are still difficult to distinguish from myocardial cells until midterm when they are recognizable in their right and left positions.

To summarize, anatomic and electrophysiologic studies on human fetuses have recently shown that by 10 to 16 weeks of embryonic life the conducting system of the heart functions as it does in the adult heart [7].

Very recent investigations on the development of normal impulse initiation and conduction in the Purkinje fiber [8] have shown in fetal canine tissue that the action potential "undergoes a continuum of change from early fetal life through term." Increase in maximum diastolic potential, action potential amplitude and duration, and V_{max} all occur as the fetus ages. The changes in Purkinje fiber conduction are a consequence of membrane maturational changes, changes in cell size, shape, and the variety and extent of intercellular connections. Since the Purkinje system is the first clearly differentiated conduction tissue to function in fetal development (the SA and AV nodes develop later), such data are crucial to our understanding of arrhythmias.

The PR Interval

This interval includes atrial activity (the P wave duration) and the AV node-His bundle conduction time to the ventricular myocardium (PR segment). The P wave changes have already been discussed above. Information concerning the total PR interval in infants and children reveals [3] that it may be quite short at birth and during the first week (0.08 to 0.10 sec.) and later lengthens to 0.12 to 0.14 sec. by one to three years and 0.12 to 0.17 by five years or later. There are no data to state where this lengthening of conduction occurs — in the node, in the main bundle, or in its branches.

VENTRICULAR MYOCARDIUM: THE QRS-T COMPLEX

QRS Duration

It has long been known that the duration of the QRS complex — the ventricular myocardial excitation complex — may measure 0.08 or 0.09 sec. at birth; yet at the end of the first, second, or third day of life the duration of this QRS may decrease by as much as 0.03 to 0.04 sec. Hence after age two, three, or four days the QRS complex is usually 0.05 to 0.06 sec. in duration. In view of recent work on the Purkinje fibers [8] and the developmental changes in their electrical properties, one looks for a solution of this unusual feature in the future as well as clarification of the cause for the larger QRS voltage in children as compared to adults. Perhaps the explanations will also follow concerning the electrocardiogram of premature infants which differ from normal ones in that the QRS voltage may be quite low and does not reach normal size until the first postnatal month. Noteworthy in this regard is the fact that studies of the developing working ventricular myocardial cells have shown that there are significant although transient differences in the quantitative internal composition of the myocyte itself as growth proceeds [9]. If intracellular components in the myocyte are changing in early growth, perhaps the membrane characteristics responsible for QRS generation also alter with development.

Electrical Axis

Changes in the frontal plane electrical axis of the QRS have long been known to be associated with the growth of the normal postnatal heart [3, 10]. These shifts from a right-axis deviation at birth (+90° to +110°) leftward to the adult-axis orientation (+30° to +90°) take place in the first half year of infancy and are completed usually by age six to eight months. The classical explanation for the changing axis is a rotational one: there is a shift from a vertical to a more horizontal position of the whole heart in the chest. Developmental changes within the myocytes of the ventricles as growth proceeds [9] have recently become known and therefore the classical rotational concept may have to be reviewed.

Precordial Leads

Right-sided predominance of the ventricles at birth is demonstrated by the developmental changes that occur in the precordial leads of the electrocardiogram. The large R wave in right V leads gradually decreases in size, and this change may begin around six months or one year of age and continue gradually over the first four to six years of life. Cellular and electrophysiologic data are not yet available to clarify all these changes, but it is likely that simple hemodynamic changes will not fully explain them.

T Wave Changes

There are constant changes in T waves during early cardiac development. Electrophysiologically the T wave is the resultant of the recovery phase of the action potential as it returns to resting potential. The well-known transmembrane ionic shifts (sodium, potassium, calcium) have elucidated much of the formation of this wave in cellular preparations. However, there are yet no data to explain the marked changes in T direction or polarity in the clinical electrocardiogram with growth. In the first 24 hours of life the T waves may be almost flat in contour in all leads and then become larger in a few days. The T wave in the right V leads may be upright in the first 24 hours but after the first few days, or occasionally even as long as three weeks after birth, the T waves in the right V leads normally become quite negative and remain so up to approximately one year of age [3]. Thus after age three weeks and up to one year, the normal T wave is negative in the right and positive in the left precordial leads. It is usual for the negative T waves in V_1-V_3 to become upright in the early years of life, but the change may not take place until the late teens or early twenties — the persistence of the "juvenile" pattern. The details of this developmental shift are unknown.

NOTE ON ABNORMAL RHYTHMS

In view of our present knowledge of the continuum of change occurring in the developing heart's specialized tissues, the evaluation of the effects of disease and drugs becomes difficult. The same pathologic process may conceivably produce quite different effects depending on the stage of development existing in the tissues. The effect of drugs has recently been shown to be diametrically opposite depending on the age of the tissues being treated. A study (in dogs) of amrinone, a substance in the adult shown to have a positive inotropic action, demonstrated that the effects of the drug on ventricular muscle contractility were age-related. [11]. There was a decrease in contractility in the newborn (0 to 3 days old), and by day 4 to 10 the medication increased contractility. There was no difference in the action potential plateau in these two ages. The effects of amrinone in the young dog are most likely due to a difference in membrane composition and function compared to the adult animal, and not to differences in intracellular composition.

With regard to the generation of arrhythmias, very little is known of fetal and infant electrophysiology. Instability of the sinus rate in early life has been noted [5]. Furthermore, remnants or "rests" of primordial or Purkinje-like tissues noted in such places as the central fibrous body and anulus fibrosus [2] and in both atrial and ventricular tissues [6] conceivably could become self-excitatory pacemakers under certain circumstances, especially in response to catecholamines. Re-entrant tachycardias may travel over accessory AV bundles which remain instead of being absorbed during development [6].

THE ELECTROCARDIOGRAM OF TWINS

An interesting aspect of the development of the electrocardiogram is that concerned with twins. Although investigation into this subject is still scanty, some rather unexpected facts have come to light [12]. The incidence of twins is 1 in 86 pregnancies, and of the total number of twins about one-third are identical twins (monozygotic). The electrocardiograms in monozygotic twins are identical in only 52%, with dissimilar tracings in 48%. In dizygotic or non-identical twins the electrocardiograms are identical in 42%, and in 58% the tracings are different. Thus identical electrocardiograms are seen in both identical and non-identical twins. This has suggested the need for an in-depth genetic evaluation of these subsets using all the modern genetic approaches. In this connection it is interesting to note that the fingerprints of identical twins are not the same [13]. Each one has his or her own individual fingerprints. The ridges and furrows of fingerprints are now thought to be due to external positional pressures on the skin of the fingers in utero. The positions of twin fetuses are constantly shifting and are not identical for each fetus [13].

In summary, we are entering a new phase of understanding of the development of the electrocardiogram by virtue of the ability to examine electrophysiology at the heart cell level. It will be exciting to watch the unfolding of this knowledge as observations continue on the minute details of myocardial and specialized tissue growth and development.

REFERENCES

1. Brooks, C.McC., and Lu, H.H. 1972. *The sino-atrial pacemaker of the heart.* Springfield, IL: Charles C Thomas.
2. Anderson, R.H.; Becker, A.E.; and Wenink, A.C.G. 1977. The development of the conducting tissues. In *Cardiac arrhythmias in the neonate, infant and child,* ed. N.K. Roberts and H. Gelband pp. 1–28. New York: Appleton-Century-Crofts.
3. Ferrer, M.I. 1973. The electrocardiogram in normal children. In *Electrocardiographic notebook,* pp. 31–33. Mt. Kisco, NY: Futura Publishing Co.
4. James, T.N. 1970. Cardiac conduction system, fetal and postnatal development. *Amer. J. of Cardiol.* 25:213.
5. Ferrer, P.C. 1977. Arrhythmias in the neonate. In *Cardiac arrhythmias in the neonate, infant and child,* ed. N.K. Roberts and H. Gelband, pp. 265–268. New York: Appleton-Century-Crofts.
6. Lev, M., and Bharati, S. 1977. Anatomy of the conduction system in normal and congenitally abnormal hearts. In *Cardiac arrhythmias in the neonate, infant and child,* ed. N.K. Roberts and H. Gelband, pp. 29–42. New York: Appleton-Century-Crofts.
7. Janse, M.J.; Anderson, R.H.; Van Capelle, F.J.L.; and Durrer, D. 1976. A combined electrophysiological and anatomical study of the human fetal heart. *Am. Heart J.* 91:556.
8. Danilo, P.; Reder, R.F.; Binah, O.; and Legato, M.J. 1984. Fetal canine cardiac Purkinje fibers: electrophysiologic and ultrastructural characteristics. *Am. J. Physiol.* 246:H250–H260.
9. Legato, M.J.; Mulieri, L.A.; and Alpert, N.R. 1983. Parallels between normal growth and compensatory hypertrophy in the rabbit. In *Perspectives in Cardiovascular Research, Vol. 7. Myocardial Hypertrophy and Failure.* ed. N.R. Alpert pp. 111—126. New York. Raven Press.
10. Davignon, A.; Rautaharju, P.; Boisselle, E.; Soumis, F.; Mégalas, M.; and Choquette, A. 1979/80. Normal ECG standards for infants and children. *Ped. Cardiol.* 1:123–131.
11. Binah, O.; Legato, M.J.; Danilo, P.; and Rosen, M.R. 1983. Developmental changes in the cardiac effects of amrinone in the dog. *Circ. Research,* 52:747–752.

12. Ferrer, M.I. 1983. Identity and the electrocardiogram, with a note on twins and altered identities. *Amer. J. of Cardiol.* 51:1784–1785.
13. Baden, M. Associate Chief Medical Examiner, Suffolk County, New York. Personal communication.

4. THE DIFFERENTIATION OF THE PURKINJE FIBERS IN THE MAMMALIAN HEART — COMPARISONS WITH THE ORDINARY MYOCYTES

STURE FORSGREN

INTRODUCTION

Despite extensive studies, the origin and development of the different parts of the conduction system is still a controversial topic (see [1, 2, 3, 4, 5]). Another problem deals with the identification of conduction cells and the existence of interspecies variations: e.g., in man it is difficult to distinguish Purkinje fibers (ventricular conduction cells) from ordinary myocytes with conventional histological techniques [6, 7, 8]. Yet another controversial question concerns the degree of differentiation of the conduction cells: are they highly differentiated cells or embryonic remnants? One reason for describing conduction cells as "embryonic" and "retarded in development" is the fact that many of the biochemical and metabolic characteristics distinguishing conduction tissue from ordinary myocardium in the adult heart also differentiate embryonic from adult myocardium (cf. [9]). In addition, conduction cells are anatomically similar to undifferentiated ordinary myocytes [10, 11, 12, 13]. Moreover, some investigators believe they represent remnants of embryonic tissue on the basis of their physiological properties [14].

What has rendered the interpretation of the structure and function of adult conduction cells difficult is the fact that information on their developmental pattern has been very limited, especially in the mammalian heart. Recently, we have studied the development of the Purkinje fibers in cow and man using immuno- and enzyme histochemistry and electron microscopy [15, 16, 17, 18, 19, 20]. The data show that the Purkinje fibers differentiate along a line separate from ordinary

M.J. Legato (ed.) THE DEVELOPING HEART, pp. 47–67.

working ventricular myocytes and that they do not develop by a failure to differentiate form primitive cardiac cells. The use of enzyme and immunohistochemistry enabled us to identify fetal Purkinje fibers and thus a basis for ultrastructural identification of human fetal Purkinje fibers at mid-gestation was established. Further, it was observed that the Purkinje fibers in several respects are well differentiated early in fetal development. In this chapter I will summarize these observations in relation to those of others on developing and adult Purkinje fibers and compare them with ordinary working ventricular myocytes.

ORIGIN AND IDENTIFICATION OF THE PURKINJE FIBERS

The Embryonic Myocardium

The mature myocardium has a heterogeneous cell population. In addition to ordinary working myocytes it contains conduction cells, fibroblasts, nerves, vascular endothelium, vascular smooth muscle, and endocardial and epicardial cells. In the earliest stages of development, however, the developing myocytes are anatomically homogeneous [21, 22].

There is some controversy about when in development conduction cells appear. The reason for this is that it is difficult to make a distinction between conduction tissue and working myocardium in embryonic hearts because of the ultrastructural similarity of these two cell types early in development. It is clear that a conduction system cannot be identified morphologically at the stage when cardiac contractions start. Furthermore, most authors agree that the Purkinje fibers appear at a later stage than the AV node and the AV bundle.

Origin of the Purkinje Fibers

Shaner and Walls have proposed that the His bundle and bundle branches develop from new growth at the lower pole of the AV node [23, 24]. Others believe that the peripheral portions of the conduction system originate separately from the AV node in situ in the ventricular trabeculae [2, 25, 27, 28, 29, 30]. We have identified bundles of Purkinje fibers superficially in the trabeculae at early fetal stages of bovine hearts [15, 16]. The presence of a distinct subendocardial cell population in the ventricles at ten weeks of gestation in man (cf below) lends support to the concept that Purkinje fibers develop *in situ*: in sections incubated with antibodies against the intermediate filament protein skeletin (desmin) (cf. below) two cell types were observed, the shift occurring proximally in left-bundle branch and distally in the intramural right-bundle branch [18]. The cells in the peripheral parts showed more intense fluorescence (cf. figure 4–2c) than those in the proximal parts. These data suggest that the peripheral parts of the conduction system originate separately from the proximal parts.

Another explanation of how the conduction tissues develop is based on the assumption that four myocardial rings — the "sinuatrial," "atrioventricular," "bulboventricular," and "truncobulbar" — occur between the four different parts of the heart tube. The final distribution of these rings and the structures which are derived from them, i.e., the SA node, the AV node, the His bundle, and the pe-

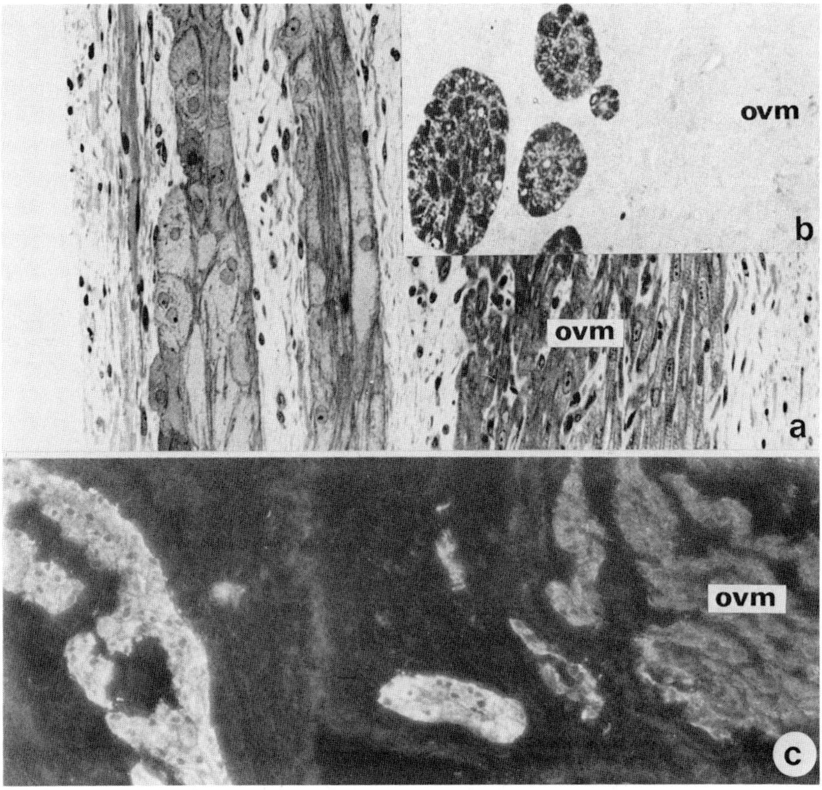

Figure 4–1. This figure demonstrates typical features by which Purkinje fibers (to the left, *a–c*) are distinguished in semithin plastic sections stained with toluidine blue (*a*) and PAS (*b*) and in a cryostat section incubated with antibodies against the intermediate filament protein skeletin (desmin) (*c*) of specimens of hearts of large bovine fetuses. The Purkinje fibers form bundles that are surrounded by connective tissue. The Purkinje fibers are larger than the ordinary myocytes and have few myofibrils in the central parts (*a*). The Purkinje fibers show a strong PAS-reaction (*b*) and a higher labeling of antiskeletin (*c*) as compared with the ordinary myocytes. OVM = ordinary ventricular myocardium. (*a* ×320, *b* ×180, *c* ×160)

ripheral portions of the conducting system, has been related to the development of the heart [31, 32, 33]. The Purkinje fibers are thought to develop from the bulboventricular ring. Further studies are needed in order to verify these theories.

Identification of the Fetal Purkinje Fibers

Considerable inter-species variations exist in Purkinje fiber morphology. In the cow, the Purkinje fibers are anatomically distinctive: they form bundles of cells surrounded by a connective tissue sheath [6, 8] and have a high content of the intermediate filament protein skeletin (desmin), as demonstrated by immunohistochemical techniques [34]. Even at an early fetal stage, subendocardial bundles of cells can be distinguished in semithin plastic sections [15]. Somewhat later in development, these bundles are distinguished from ordinary myocardium by their skeletin-positivity and strong PAS-reaction (figure 4–1) [15].

In man, bundles of cells surrounded by connective tissue can be seen subendocardially in fetal hearts in semithin plastic sections (figure 4–2a, -b) [19]; as in adult hearts [6, 7, 8] these cells are, however, not very easily distinguished from ordinary myocardium [cf. 24, 25, 35]. On the other hand, as mid-gestation and even at ten weeks of gestation, a subendocardial cell population can be clearly distinguished from the main myocardial mass in the ventricles by its high content of skeletin, as demonstrated immunohistochemically (figure 4–2c) [18]. However, it must be stressed that the skeletin reactivity in this cell population is not of the same magnitude as in the bovine fetal Purkinje fibers.

There are some distinctive ultrastructural differences between Purkinje fibers and ordinary ventricular myocytes [see, e.g., 36, 37, 38, 39]. In particular the absence of T-tubules in the Purkinje fiber in contrast to the ventricular myocyte, which has a well-developed transverse tubular system, has been stressed [36, 40]. In studies on bovine fetal hearts, we also observed that M-bands appeared earlier in the Purkinje fibers than in the ordinary ventricular myocytes [16]. This was corroborated by the presence of myofibrillar MM-creatine kinase, as detected immunohistochemically, at an earlier stage in the Purkinje fibers that in the ordinary ventricular myocytes [17]. MM-creatine kinase is known to make an essential contribution to the electron density of the M-band [41, 42]. On the basis of this, the appearance of the M-line region in the human fetal heart at mid-gestation was studied. In the ventricles, only a subendocardial cell population was labelled by antibodies against MM-creatine kinase and showed dense M-bands (figure 4–2d–g) [19]. As the subendocardial cell population was located in regions where conduction cells were to be expected, we concluded that it corresponded to the Purkinje fiber cell population. This was further supported by the fact that these cells generally showed a more intense fluorescence than the other myocytes in sections incubated with anti-skeletin antibodies. The appearance of the cells in the subendocardial regions of the human fetal heart with respect to electron density at M-line regions will be described in more detail later in this chapter.

CELL PROLIFERATION AND GROWTH

In early studies [cf. 43, 44] cell division was hardly ever observed in conduction cells. The "specialized structure" of conduction cells was postulated to explain the absence of mitosis [44].

On the whole, surprisingly little is known about the proliferation and regenerative capacities of conduction cells. We have, however, observed mitotic figures in the Purkinje fiber populations of both the bovine (15) and human fetal hearts, as otheres have in the conducting tissue of the rat and mouse [45, 46]. In the rat and mouse it was found that the conducting cells replicate much less intensely than the ordinary ventricular and atrial myocytes at embryonal stages [46]. On the other hand, the withdrawal of myocytes from the mitotic cycle proceeds more rapidly in the ordinary ventricular myocytes than in the conducting cells and in the atrial myocytes [46]. Thus, in the embryo the rate of mitotic division is high in the working myocardium but declines rapidly after birth [47, 48, 49]. Further-

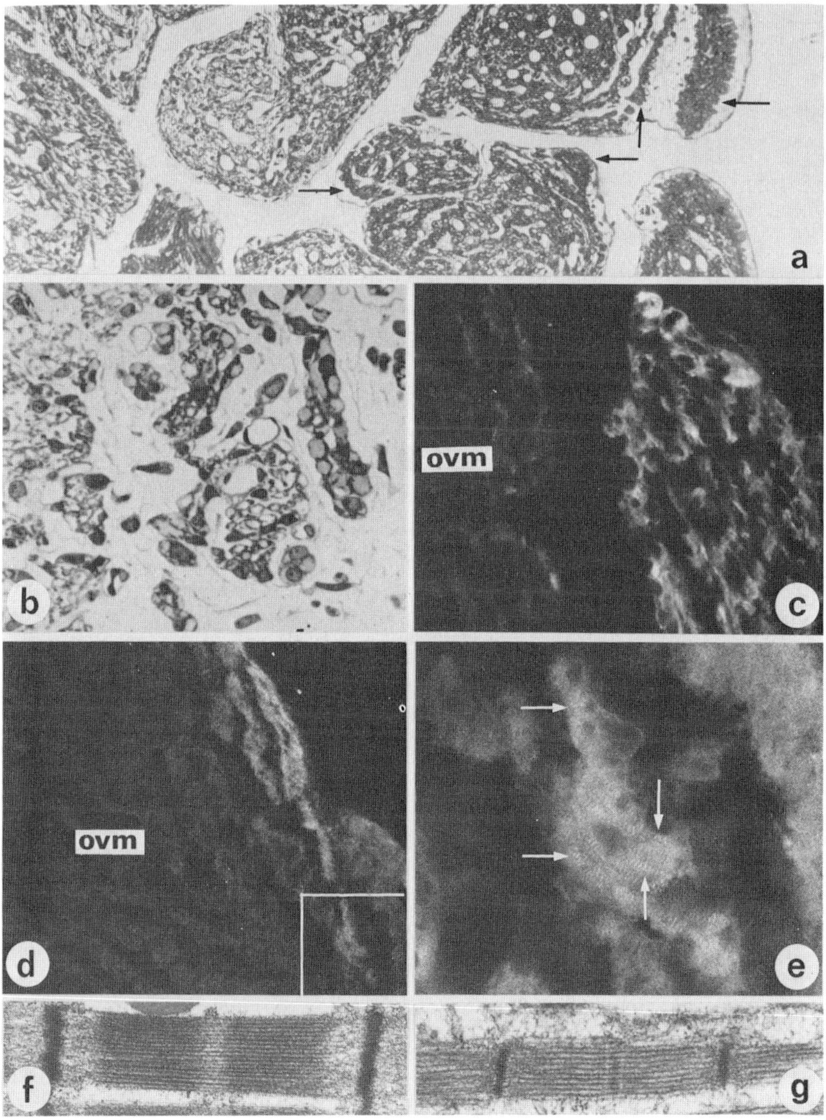

Figure 4–2. Features related to the identification of a subendocardial cell population in the ventricles in human fetal hearts at mid-gestation are illustrated. Cells in the superficial parts of the trabeculae are partly arranged into bundles (arrows in *a*, to the right in *b*), around which capillaries appear fewer than in the main myocardial mass, in semithin plastic sections stained with toluidine blue; nevertheless a clear distinction between two different cell types (conduction cells/working myocytes) is difficult to make. In crystat sections incubated with antibodies against skeletin (desmin) (*c*) a subendocardial cell population shows a more intense fluorescence than the myocytes in the main myocardial mass. This subendocardial cell population conforms largely to that one which is labelled by antibodies against MM-creatine kinase (*d*) and which shows a high activity of cholinesterase (not illustrated, cf. figure 4–3a). Cross-striations (arrows) in sections incubated with antibodies against MM-creatine kinase appear only in this subendocardial cell population (*e*, cf. *d*). In accordance with this the great majority of ventricular myocytes do not show electron dense M-bands (*f*), but only those located in subendocardial regions (*g*). OVM = ordinary ventricular myocardium. (*a* ×32, *b* ×320, *c* ×500, *d* ×160, *e* ×640, *f*, *g* ×28800)

more, cardiac muscle shows a pattern different from skeletal muscle. Synthesis of DNA and of cell specific proteins are not mutually exclusive in cardiac muscle, which is the case for skeletal muscle. Furthermore, skeletal muscle is capable of restoring innumerable muscle fibers, while adult cardiac muscle is unable to regenerate necrotic parts by proliferation of the remaining myocytes [see, e.g., 50, 51]. Differences exist, however, between different types of myocytes: atrial myocytes in mammals and both ventricular and atrial myocytes in lower vertebrates display a much greater capacity for reactivated hyperplasia than ventricular myocytes in mammals [51]. To the best of my knowledge the only study on the regenerative capacity of the conduction system is that by Rumyantsev and Kassem [52], in which they describe a scarcity of DNA synthesizing myocytes in the AV node and AV bundle of rats with left ventricular myocardial infarction. Certainly there is a great need for further studies on the proliferative capacity of conduction tissues in the adult heart. This becomes particularly evident as the "ease" of DNA synthesis and mitosis reactivation is a property of cells which are different from mammalian ventricular myocytes; i.e., they lack T tubules and have fewer sarcomeres than ventricular working cells [52].

The diameter of the ordinary myocytes increases from about 5 to 6 μ at birth to 10 to 15 μ in the adult heart [38, 50]. Purkinje fibers and ordinary ventricular myocytes are of approximately the same size in young ungulate fetuses, while the Purkinje fibers are larger in old fetuses (figure 4–1a) [15, 23, 44]. In the adult ungulate heart, the diameters of the Purkinje fibers are three to four times larger than those of the ordinary myocytes [6, 8, 53].

Postnatally the heart increases in size due not only to hypertrophy of the myocytes but also to an increase in the non-muscle cell population. In fact, in the adult heart, non-muscle cells constitute half or even more of the total cell population [50, 54]. An increase of the interstitial connective tissue also occurs in the AV junction during aging [26, 55, 56, 57]. This is proposed to be a critical process in normal development; an overproduction of connective tissue components, giving rise to an abnormal dispersion of AV node and AV bundle cells, can lead to heart block or other arrhythmias [57].

ANATOMIC-ELECTROPHYSIOLOGICAL CORRELATIONS

The early embryonic heart exhibits the properties of spontaneity, conductivity, and contractility. According to Lieberman [14], two developmental patterns may then follow: in some cells a process is triggered which causes the cardiac cell membrane to loose the "embryonic" ability to pacemake, while others maintain the properties of automaticity and conductivity.

Sarcolemmal Appearance: Bundle Formation

Conduction through the Purkinje fiber system of the adult heart occurs at a rate of 1.5 to 5.0 m/sec. [8, 58, 59]. The rapid conduction velocity of the Purkinje fibers may be partly the result of their large diameters and their aggregation into tightly packed bundles [40, 60]. Other factors contribute to the speed of conduction in

Purkinje fibers. Human Purkinje fibers at mid-gestation are folded and have numerous thin projections (cf. figure 4–6a, -b), as well as narrow infoldings [61]. These features are not seen in ordinary myocytes at mid-gestation and may significantly increase the surface area of the Purkinje cells. These phenomena have been described for the Purkinje fibers of other species. Stretching of Purkinje fiber bundles to a certain degree leads to an increase in conduction velocity, which has been related to "unfolding" and "unbuckling" of the fibers [65, 66].

Nexuses

Nexuses, which are frequent between adult Purkinje fibers [67], may be the site of functional electrical coupling between cells [for review see 68]. The sparsity of nexuses in the AV mode may partly explain the very low conduction velocity of this region [69]. Even at mid-gestation, nexuses are frequent between human Purkinje fibers [19]. However, it should be stressed that the cells at this stage also have a large portion of their surface area facing the extracellular space, even within bundles [61].

Nerves

The sympathetic and parasympathetic nervous system are important modulators of cardiac activity. In Purkinje fibers, there is documented cholinergic antagonism of the electrophysiological effects of catecholamines [70]. Further, the electrophysiological effects of acetylcholine on Purkinje fibers are somewhat different from or opposite to those on the SA node and atrium [71].

In our studies on fetal hearts we have used the histochemical method for demonstration of cholinesterase activity. This method has been very frequently used for demonstration of cholinergic nerves in the heart. Cholinesterase activity can, however, also be demonstrated in some adrenergic nerves and in non-nervous tissue. Our most noteworthy observation is the clearly higher cholinesterase activity in the peripheral Purkinje fibers than in the ordinary ventricular myocytes in the human fetal heart (figure 4–3a) [18]. That was the case both at ten weeks gestational age and at midgestation. In both periods of development, the cholinesterase-positive cell population largely corresponded to the skeletin-rich one (cf. above). Some cholinesterase-positive nerves were seen in the ventricles, particularly at mid-gestation, but few if any were seen close to the peripheral Purkinje fibers. Elias described cholinesterase-positive nerves in frequent association with Purkinje cells in human heart [72]. In contrast, the bovine fetal Purkinje fibers show only a slightly higher activity or, more often, an equal activity to the ordinary ventricular myocytes (figure 4–3c).

With respect to the cholinesterase reaction of fetal Purkinje fibers, a few comments can be made. First, it is evident that there are interspecies differences with respect to the degree of cholinesterase activity of conduction cells in the ventricles [cf. 55, 74]. Second, the high cholinesterase activity of human fetal Purkinje fibers may reflect an early step in neurogenesis. A high cholinesterase activity is detected in tissue prior to ennervation [75]. Third, a high cholinesterase activity may indi-

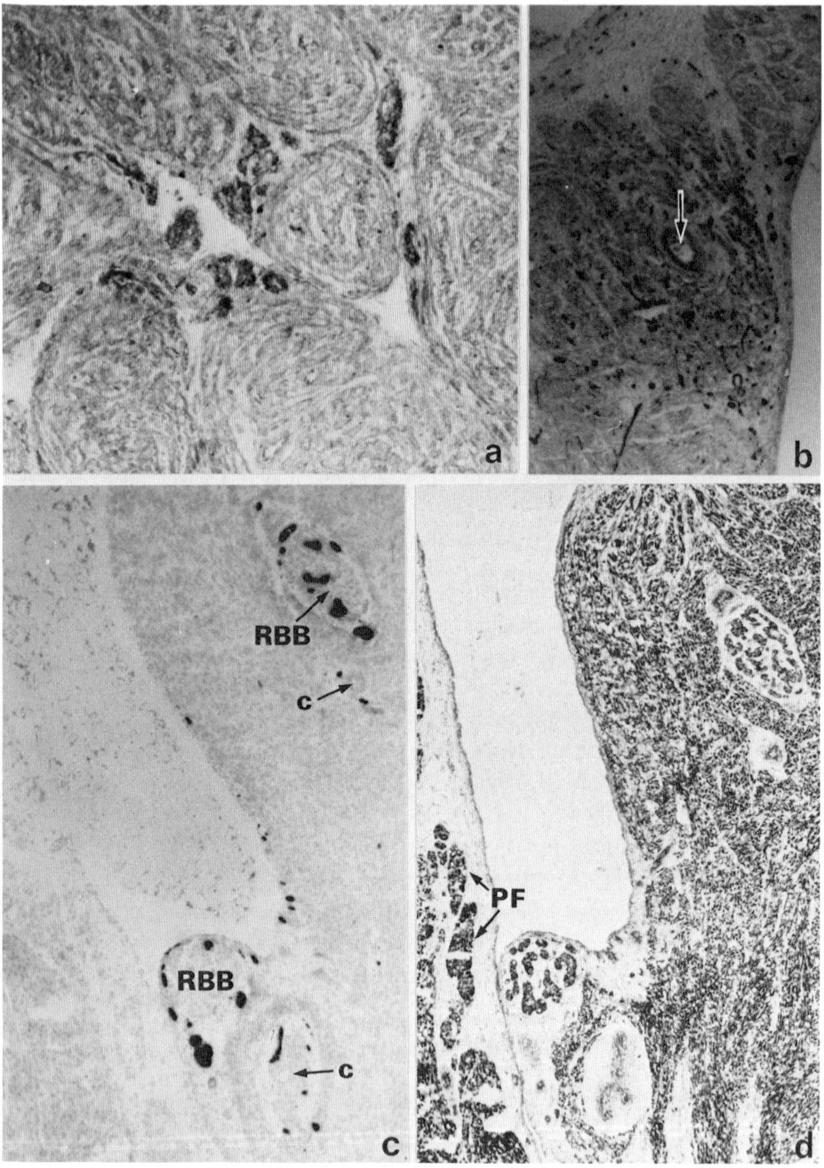

Figure 4–3. The demonstration of cholinesterase activity in specimens of human fetal hearts at mid-gestation (*a, b*) and a specimen of a bovine fetal heart (CR-length 15 cm) (*c*). Bundles of cells in the superficial parts of ventricular trabeculae (cf. figure 4–2) show a clearly higher activity than the main myocardial mass (*a*). Few if any cholinesterase-positive nerves are seen in relation to these bundles: in contrast such nerves are abundant in the SA node (*b*, SA node artery at arrow). In the right ventricle of a bovine fetal heart several cholinesterase-positive nerves are seen, particularly in relation to both levels of the right bundle branch (RBB) here shown and capillaries (*c*) (*c*, cf. *d*). (*D*) is a serial section to (*c*) stained for the demonstration of NADH-TR activity. Septum is to the right, right ventricular free wall to the left. Note that the cells of the right bundle branch and also the Purkinje fibers (PF) located subendocardially on the ventricular wall do not show a higher activity of cholinesterase than the ordinary myocytes. (*a* ×160, *b* ×40, *c* ×80)

cate a differentiated state of the cells: cholinesterase is thought to be involved in the differentiation of cardiac myocytes and also in the appearance of automaticity [76]. Patches of cholinesterase and also acetylcholine receptors develop on the surface of embryonic Xenopus muscle cells in the absence of nerves. These patches are proposed to reflect intrinsic muscle mechanisms whose operation is linked with the differentiation of the cells and subject to external regulating influences such as innervation [77].

M-Band Maturation

The developmental patterns of immunohistochemical detection of myofibrillar MM-creatine kinase in the different cardiac regions of the bovine fetal heart are shown in table 4–1. It is evident that the M-band matures early in regions which in the adult show high conduction velocities and vice versa. The reason for this is unclear. One possibility is that early M-band maturation is related to an early general differentiation process of fast conducting cells. It has been proposed that the function of M-band bound MM-creatine kinase is involved in the regulation of the ATP metabolism related to muscular contraction [42, 81, 81].

Using the presence of distinct M-bands in the majority or myofibrils as the criterion for identification of the Purkinje fibers in the human fetal heart at midgestation, it was observed that some myocyte bundles in the subendocardial regions do not show M bands at this stage [61]. These may represent ordinary myocytes that will become part of the ordinary traebecular tissue during the compactization of the subendocardium. Using this same criterion, it was also observed that the majority of the Purkinje fibers on the endocardial surface of the right ventricular free wall formed large areas of a continuous layer of cells [61].

Table 4–1.

	Immunohistochemical Detection of MM-CK During Fetal Development	Conduction Velocity in Adult Hearts (meters/sec)
Right atrium	Early	1,0–1,8
AV node	Late	0,02–0,05
AV bundle	Early	1,2–2,0
Right and left bundle branches	Early	2,0–4,0
Peripheral Purkinje fiber system	Early	1,5–5,0
Ordinary ventricular myocardium	Late	0,3–1,0

Table showing the differences between different cardiac regions with respect to (1) the development of visible M-band regions containing MM-creatine kinase (MM-CK, a protein that is known to make an essential contribution to the electron density of the M-band; see [41, 42, 78]). in bovine fetal hearts (cf. [17,79]) and (2) the mean conduction velocity of adult hearts (cf. [8, 58, 59, 80]). "Early" and "late" do not refer here to specific gestational ages but to the fact that there were distinct *differences between* different cardiac regions [17, 79]. It should further be stressed that AV node and AV bundle refer here to the regions above and below, respectively, the level where a morphological transition between "AV node cells" and AV bundle cells" occurred, as seen in the serial sections stained with enzyme histochemical stains (cf. [79]). For details in the development pattern in the "transitional region" between the AV node and the right atrium, further studies are needed.

The Purkinje fibers in this layer as well as the rest of the Purkinje fiber population were largely separated from the ordinary myocardial tissue, although occasionally small nests of M band containing cells were found among the ordinary working ventricular myocytes [61].

Even the subendocardial bundles of myocytes that occurred at an early fetal stage of bovine hearts (cf. above) showed at least signs of dense M bands [16]. By using the M-band criterion, these bundles can be defined as Purkinje fiber bundles. The myofibrils in these bundles were very thin. In comparison, in myocytes which doubtless constitute working myocardium, dense M bands appear late; e.g., in the ordinary ventricular myocardium in the rat M bands do not appear until after birth [83, 84].

AEROBIC AND ANAEROBIC METABOLISM

It is frequently proposed that conduction tissues have a higher rate of anaerobic metabolism but a lower rate of aerobic metabolism than ordinary myocardium in adult hearts. This concept is based upon findings of a higher amount of glycogen, a higher activity of glycolytic and gluconeogenetic enzymes, fewer mitochondria, and a lower activity of respiratory enzymes in the conduction cells than in the ordinary myocytes [85, 86, 87, 88]. This difference between conduction cells and ordinary myocytes is also an important difference between embryonic and adult myocardium. Thus, the fetal heart has a greater glycolytic capacity than the adult heart: during development of the heart the oxidative capacity increases due to an increase in both mitochondrial content and mitochondrial activity [89, 90, 91, 92]. The specific development pattern of the conducting cells in this respect has, however, not been extensively studied.

We have studied the development of the bovine Purkinje fibers and ordinary ventricular myocytes with respect to activities of succinate dehydrogenase and mitochondrial glycerol-3-phosphate dehydrogenase by means of enzyme histochemistry [17]. Succinate dehydrogenase is an oxidative enzyme, participating in the Krebs cycle, while mitochondrial glycerol-3-phosphate dehydrogenase is most active in tissues with high anaerobic capacities. With respect to succinate dehydrogenase, the Purkinje fibers showed a higher activity in small fetuses, an equal activity in large ones, and a lower activity in adult hearts, as compared with the ordinary myocytes (figure 4–4). In both fetal and adult hearts, the Purkinje fibers showed a higher activity of mitochondrial glycerol-3-phosphate dehydrogenase than the ordinary myocytes, the exception being the hearts of rather small and intermediate-sized fetuses in which the ordinary myocytes eventually showed activity equal to that of Purkinje fibers. The enzyme activities in the Purkinje fibers seemed to remain unchanged during the fetal stages studied, as shown by simultaneous sectioning of blocks of samples from different fetal stages. Ultrastructurally both Purkinje fibers and ordinary ventricular myocytes are rich in glycogen and have few and scattered mitochondria at the early fetal stage of bovine hearts,

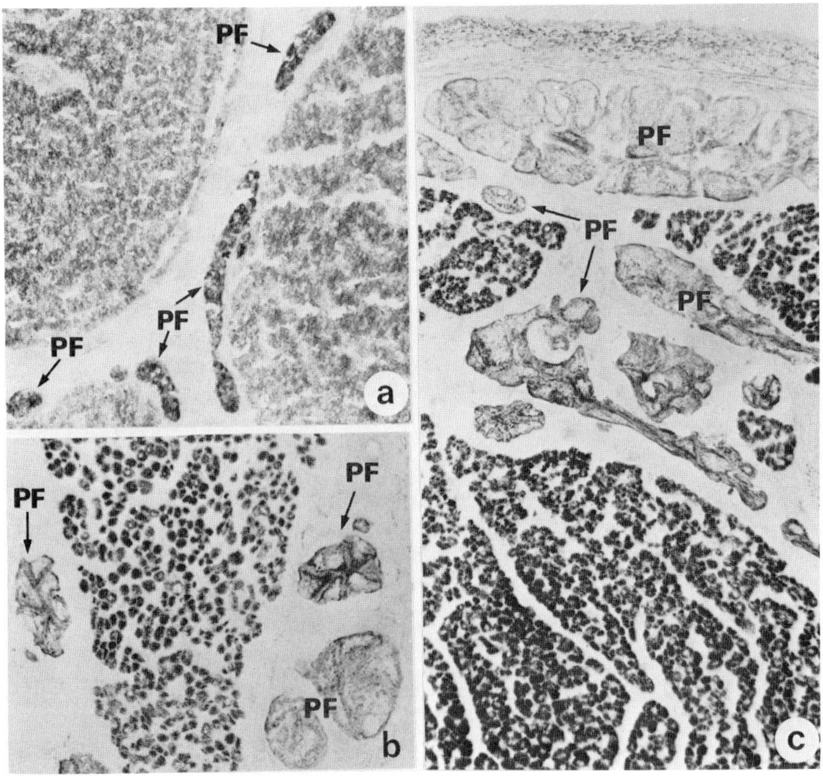

Figure 4–4. The changing pattern of activity of succinate dehydrogenase between Purkinje fibers (PF) and ordinary ventricular myocytes is illustrated. Specimens of hearts of a bovine fetus of CR-length 17 cm (*a*) and an adult cow (*b, c*). Note that adult Purkinje fibers may show a rather strong staining reaction in cell peripheries. (*a* ×130, *b* ×145, *c* ×130)

while at the very late fetal stage a difference exists which in principle is comparable to that of the adult organism [16].

From the above, it is evident that a change in the metabolic pattern occurs between Purkinje fibers and ordinary ventricular myocytes during development. Recently we have also observed that the typical differences in metabolic properties between the AV node, the AV bundle, and the proximal parts of the bundle branches and the ordinary myocardium in adult hearts, as revealed by enzyme histochemistry, are not established at early stages [79]. The observation of a high activity of succinate dehydrogenase in the Purkinje fibers in small bovine fetuses indicate that the Purkinje fibers are more differentiated than the ordinary myocytes in these fetuses; taken together with other observations this suggests that the energy demand is high in Purkinje fibers at early fetal stages. For appropriate quantitative comparisons between Purkinje fibers of fetal and adult stages, biochemical studies are needed (cf. figure 4–4). In contrast to the mammalian Purkinje fibers, avain cells appear to be depleted of glycogen during development [93, 94].

FILAMENT SYSTEMS

Filament Systems in Adult Hearts

The filamentous material of the cardiac myocytes consists of myofibrils, the intermediate filaments, leptofibrils, and the so-called myofilament-polyribosome complexes [cf. 95]. In adult hearts the Purkinje fibers have myofibrils which appear ultrastructurally identical to ordinary myocytes with respect to banding pattern. However, myofibrillar "alterations" frequently occur in the Purkinje fibers, including a highly variable Z-disk morphology and insertion of "extra sarcomeres." Typical features of adult Purkinje cells are also an abundance of intermediate filaments, the occurrence of leptofibrils, and complexes of myofilaments and polyribosomes. However, large inter-species variations occur. All these aspects of Purkinje fiber morphology have recently been reviewed by Thornell and Eriksson [96], and it was concluded that the various filament systems of the Purkinje cell together with desmosomes which link them together and connective tissue wrapping cells into bundles are all important factors in the distribution of mechanical stress in the conduction cells during the heart beat.

In adult ordinary myocytes, filaments other than the thick and thin filaments also exist: intermediate filaments encircle and interconnect Z-disks, insert into desmosomes and into portions of the intercalated disks [97, 98]. In some species, leptofibrils are frequent [99]. On the other hand, myofilament-polyribosome complexes typical of Purkinje fibers have not been observed in ordinary myocytes [95].

Developmental Patterns in Purkinje Fibers

The appearance of the myofibrils in conducting cells has often been described as similar to that of very young cells [10, 12, 13] and some investigators regard them as developed [11]. In fact, one interpretation of the appearance of the Purkinje fiber myofibrils has been that they are vestigial organelles [100].

We have studied bovine hearts and found that myofibrils in the Purkinje fibers were randomly oriented throughout the fetal stages, while those in the ordinary ventricular myocytes progressively became aligned in parallel [16]. As described earlier in this chapter, electron dense M-bands appeared sooner in the Purkinje fibers than in the ordinary ventricular myocytes. Furthermore, immunohistochemical detection of MM-creatine kinase, which is related to the appearance of electron dense M-bands (cf. above), occurred earlier in atrial myocytes than in ordinary ventricular myocytes [17]. MM-creatine kinase was detectable equally early in the cells of the AV bundle and the proximal parts of the bundle branches as in those of the Purkinje fiber system [79]. On the other hand, the AV node cells were labelled at a later stage [79].

Leptofibrils occurred occasionally in the Purkinje fibers at late fetal stages [16, 94, avian heart]. On the other hand, myofilament-polyribosome complexes typical of adult cow Purkinje fibers did not occur in the fetal hearts. Intermediate filaments were abundant in the fetal Purkinje fibers [16]: confirming this, the Purkinje fibers showed an intense skeletin reactivity [15]. However, at the very early fetal stage there was no difference in the amount of skeletin between the bundles of cells that occurred subendocardially and the myocytes in the rest of the ventricular myocardium [101].

All Purkinje fibers were stained in sections incubated with antibodies against either ventricular or atrial myosin heavy chains in early fetal stages [91, 92]. With increasing fetal age, however, the pattern changed; several of the Purkinje fibers were not labelled by the antibodies to atrial myosin heavy chains.

Our observations show that the Purkinje fibers go through a distinct differentiation process. The early appearance of electron dense M-bands in Purkinje fibers indicates early differentiation of the cells. Further, it is clear that the notion that leptofibrils are embryonic remnants [102] and that the organization of myofilaments in the Purkinje fibers is consistent with an embryonal organization [12] do not hold true. Rather, it is more likely that the appearances of the filament systems, particularly the intermediate filaments and the leptofibrils, are related to changes in the need for cell support.

In collaboration with S. Schiaffino's group in Padua, Italy, we have recently begun to study Purkinje fiber differentiation with immunohistochemical techniques and antibodies against ventricular and atrial myosin heavy chains. An important observation has been that the pattern of staining of the cells with respect to labeling by antibodies against atrial myosin heavy chains changes during fetal development: in early fetal stages all Purkinje fibers are stained, but with increasing fetal age several of the cells are not labelled [101]. Studies on embryonic hearts are currently being undertaken.

Myofibrillar imperfections in the Ventricular Myocardium in the Human Fetal Heart at mid-gestation

Myofibrillar "alterations," particularly those affecting the Z-disks, have been reported frequently in the diseased heart. For example, widening, streaming [e.g., 103, 104] and splitting [103, 105] of the Z-disk is seen in the hypertrophic heart [see also 106]. Furthermore, myofibrillar alterations including Z-disk changes and disorientation of the myofilaments occur as a result of unloading; this has been induced experimentally in animal hearts [107, 108]. To what degree such phenomena normally occur in the human heart is largely unknown.

Imperfections in the myofibrils are frequently observed in the *Purkinje fibers* in the human fetal heart at mid-gestation. These include occurrence of widened and irregular Z-disks (figure 4–5), streaming of Z-disk material, lack of electron dense material at Z-disk levels (figure 4–6a), and the occurrence of isolated sarcomeres which lie adjacent to formed myofibrils and which are devoid of electron dense material at Z-disk levels (figure 4–6b). Sarcomeres inserting into intercalated disks may be less well organized than other sarcomeres (figure 4–6a). An increased amount of electron dense material can sometimes be seen where the myofibrils insert into fasciae adherentes (figure 4–5). Also some ordinary ventricular myo-

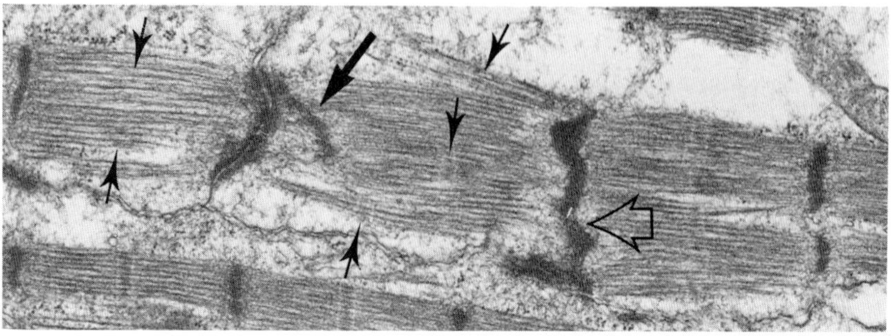

Figure 4–5. Purkinje fiber myofibrils of a human fetal heart at mid-gestation are shown. There is an increase of electron dense material ("extra Z-disk" material) (large arrow) at the fascia adherens to the left. Signs of dense M-bands (small arrows) occur also in the sarcomeres which insert into this fascia adherens. One of the Z-disks (open arrow) is broad and irregular and is in contact with the fascia adherens below. (×28800)

cytes show myofibrillar imperfections; i.e., streaming and scattering of Z-disk substance and insertion of "extra sarcomeres" in parallel with Z-disk variations (figure 4–7) [109]. The great majority of ordinary ventricular myocytes do not have these features, however. Those ordinary myocytes that occupy the inner aspect of the free walls, septum, and papillary muscles and the subepicardial regions have "extensive" Z-disc variations. They show parallelly aligned myofibrils, an abundance of electron dense accumulations in which ribosomes and filaments can be seen (figure 4–8a), and intermediate filaments usually cross-linking the myofibrils at Z-disc levels. Intermediate filaments also connect myofibrils and link myofibrils to the sarcolemma (figure 4–8b) [109].

The observations on the mid-gestation hearts show that myofibrillar alterations normally occur not only in Purkinje fibers but also in interkinetic (mitotic myocytes show Z-disc disintegration) ordinary ventricular myocytes. These are scarce, however, at this stage of development in man. Such changes may be related to altered loading conditions, which in turn probably are secondary to genetically programmed events [109]. Furthermore, it cannot be excluded that Z-disc streaming and extra sarcomeres in elongated myocytes containing broad myofibrils and large masses of ribosome-filamentous accumulations are related to enhanced myofibrillar growth. Myofibrillar imperfections may also be related to the occurrence of a high rate of myofibrillar protein degradation in the cells. This is supported by the fact that [109] as the myocytes showing Z-disc and other myofibrillar alterations also showed variations in other filament systems (intermediate filaments, leptofibrils); these structures are believed to have a mechanical role in the cell [96, 110]. These myocytes could lie adjacent to dividing and dying myocytes, as some of them showed appearances suggesting an enchanced activation of the lysosome system. (According to a two-step model of myofibrillar degradation, the myofibrillar material is in the second step degraded in lysosomes) [111]. The dependence

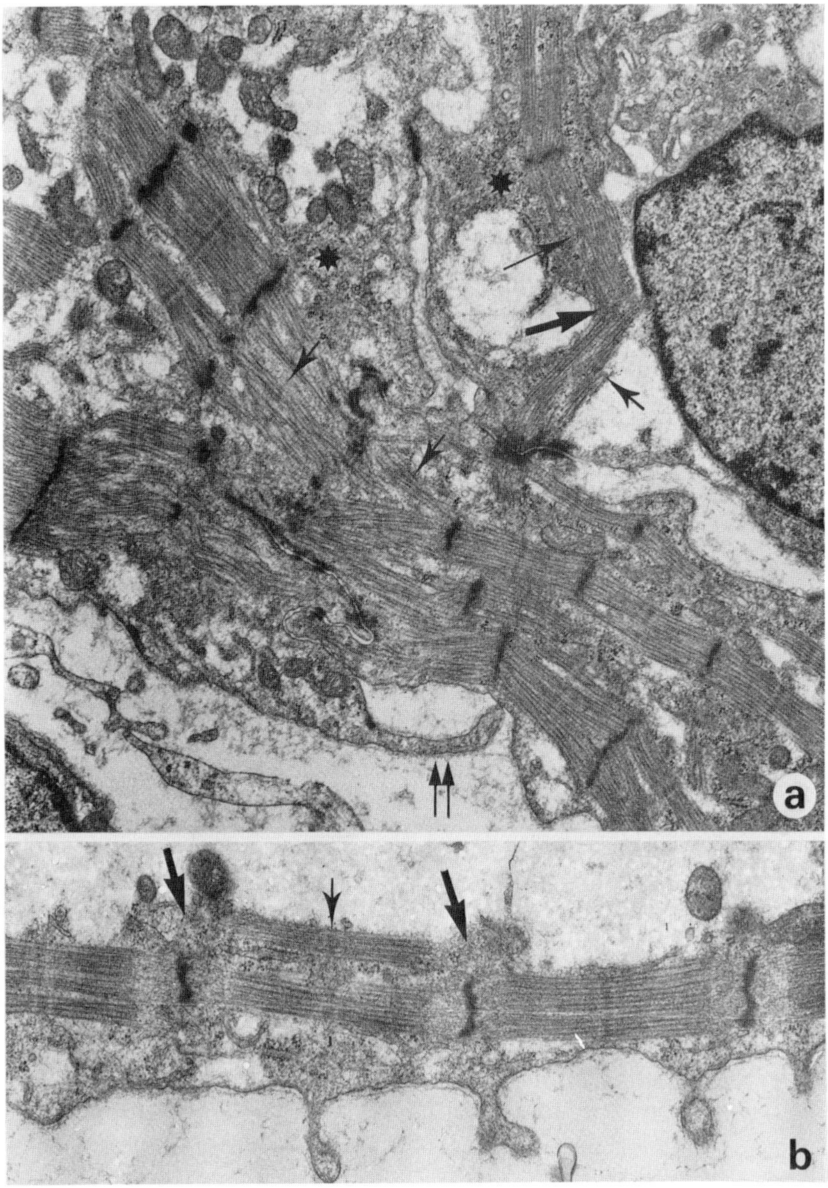

Figure 4–6. Myofibrillar appearances in Purkinje fibers of human fetal hearts at mid-gestation are shown. There is a large and irregular area for cell contact in the middle in (*a*), containing several intercellular junctions. Note the differences in sarcomere organization, that the direction of the myofibril may change from sarcomere to sarcomere, and that electron dense material may be missing at Z disk levels (large arrows). Despite the occurrence of myofibrillar imperfections there are at least some signs of dense M-bands (cf. small arrows). Electron-dense accumulations, in which ribosomes can be distinguished, are observed (stars). Note also the presence of thin (stalked) projections (double arrow in *a*, below in *b*). (*a* ×20000, *b* ×28800)

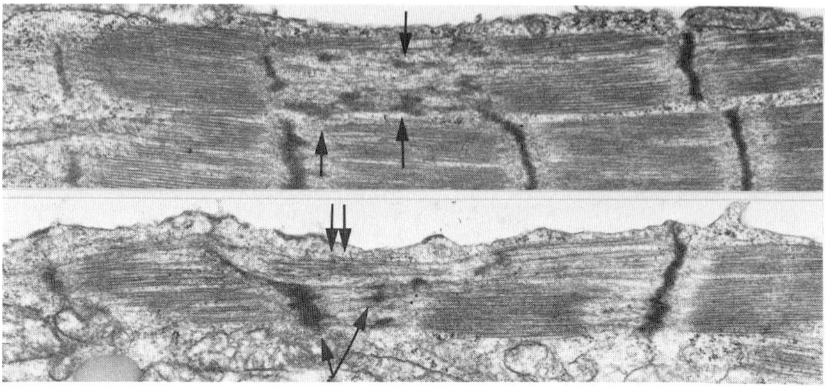

Figure 4–7. Myofibrils of myocytes of a papillary muscle of a human fetal heart at mid-gestation. Streaming and scattering of Z-substance (single arrows) is observed, and further, an "extra sarcomere" (double arrow) appears to be inserted. (×20000)

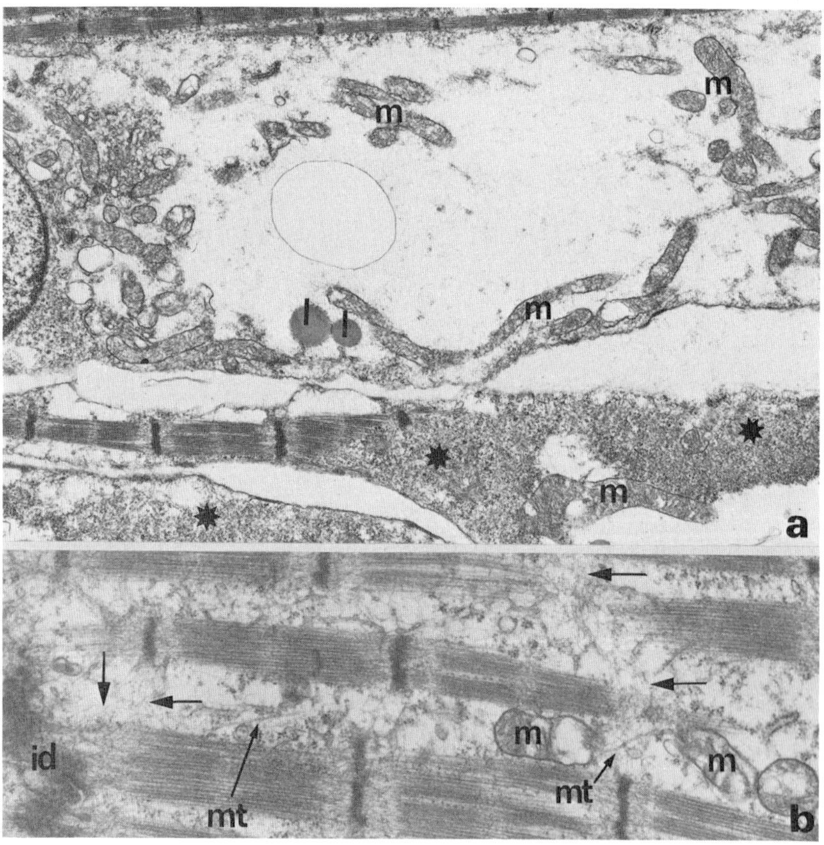

Figure 4–8. Ordinary ventricular myocytes of a human fetal heart at mid-gestation. The central parts of the myocytes appear empty, but are filled with densely stained particles in sections stained with periodic acid-thiosemicarbazide-silverproteinate for demonstration of glycogen (not illustrated). Intermediate filaments (arrows in *b*) are abundant at Z-disk levels and in relation to an intercalated disk (id). Accumulations, in which ribosomes are abundant, occur (stars in *a*). M = mitochondria. L = lipid droplets. Mt = microtubules. (*a* ×13300, *b* ×23600)

of "normal" ultrastructure on a normal loading condition is frequently documented in the literature [108, 112].

ACKNOWLEDGMENTS

I am grateful to my collaborators in the investigations which are the basis for this chapter. That includes particularly Drs. A. Eriksson and E. Strehler and most particularly Dr. L.-E. Thornell. I am also indebted to B. Berggren for photographic work and to L. Carlsson, M. Enerstedt, and B. Holmbom for technical assistance. Financial support has been given by the Swedish Medical Research Council (12X-3934), the Swedish Society of Medical Sciences, and the Faculty of Medicine, University of Umeå.

REFERENCES

1. Robb, J.S. 1965. Embryology of atrioventricular conducting system. In *Comparative basic cardiology,* pp. 335–348. New York: Grune & Stratton.
2. Virágh, S., and Challice, C.E. 1977. The development of the conduction system in the mouse embryo heart. *Develop. Biol.* 56:397–411.
3. Marino, T.A.; Truex, R.C.; and Marino, D.R. 1979. The development of the atrioventricular node and bundle in the ferret heart. *Am. J. Anat.* 154:135–150.
4. van Mierop, L.H.S. 1979. Morphological development of the heart. In *Handbook of physiology,* section 2, The cardiovascular system, vol. I, ed. R.M. Berne, N. Sperelakis, and S.R. Geiger, pp. 1–28.
5. Virágh, S., and Challice, C.E. 1982. The development of the conduction system in the mouse embryo heart. *Develop. Biol.* 89:25–40.
6. Nandy, K., and Bourne, G.H.A. 1963. A study of the morphology of the conducting tissue in mammalian hearts. *Acta. Anat.* 53:217–226.
7. Truex, R.C., and Smythe, M.Q. 1965. Comparative morphology of the cardiac conduction tissue in animals. *Ann. N.Y. Acad. Sci.* 127:19–33.
8. Davies, M.J. 1971. *Pathology of conducting tissue of the heart.* London: Butterworth & Co.
9. DeHaan, R.L. 1961. Differentiation of the atrioventricular conducting system of the heart. *Circulation* 24:458–470.
10. Page, E.; Power, B.; Fozzard, H.A.; and Meddoff, D.A. 1969. Sarcolemmal evaginations with knob-like or stalked projections in Purkinje fibers of the sheep's heart. *J. Ultrastruct. Res.* 28:288–300.
11. Virágh, S., and Challice, C.E. 1969. Variations in filamentous and fibrillar organization, and associated sarcolemmal structures, in cells of the normal mammalian heart. *J. Ultrastruct. Res.* 28:321–334.
12. Arluk, D.J., and Rhodin, J.A.G. 1974. The ultrastructure of calf heart conducting fibers with special reference to nexuses and their distribution. *J. Ultrastruct. Res.* 49:11–23.
13. Mochet, M.; Moravec, J.; Guillemot, H.; and Hatt, P.Y. 1975. The ultrastructure of rat conductive tissue: an electron microscopic study of the atrioventricular node and the bundle of His. *J. Mol. Cell Cardiol.* 7:879–889.
14. Lieberman, M. 1970. Physiologic development of impulse conduction in embryonic cardiac tissue. *Am. J. Cardiol.* 25:279–284.
15. Forsgren, S.; Thornell, L.-E.; and Eriksson, A. 1980. The development of the Purkinje fiber system in the bovine fetal heart. *Anat. Embryol.* 159:125–135.
16. Forsgren, S., and Thornell, L.-E. 1981. The development of Purkinje fibers and ordinary myocytes in the bovine fetal heart: an ultrastructural study. *Anat. Embryol.* 162:127–136.
17. Forsgren, S.; Strehler, E.; and Thornell, L.-E. 1982. Differentiation of Purkinje fibers and ordinary ventricular and atrial myocytes in the bovine heart: an immuno- and enzyme histochemical study. *Histochem. J.* 14:929–942.
18. Forsgren, S.; Eriksson, A.; Kjörell, U.; and Thornell, L.-E. 1982. The conduction system in the human heart at midgestation — immunohistochemical demonstration of the intermediate filament protein skeletin. *Histochem.* 75:43–52.
19. Forsgren, S.; Carlsson, E.; Strehler, E.; and Thornell, L.-E. 1982. Ultrastructural identification of human fetal Purkinje fibers — a comparative immunocytochemical and electron mi-

croscopic study of composition and structure of myofibrillar M-regions. *J. Mol. Cell Cardiol.* 14:437–449.

20. Forsgren, S. 1982. Differentiation of heart Purkinje fibers: an immuno- and enzyme histochemical and ultrastructural study. Umeå University Medical Dissertations, New Series No. 83.
21. Manasek, F.J. 1968. Embryonic development of the heart. I. A light and electron microscopic study of myocardial development in the early chick embryo. *J. Morph.* 125:329–365.
22. Manasek, F.J. 1970. Histogenesis of the embryonic myocardium. *Am. J. Cardiol.* 25:149–168.
23. Shaner, R.F. 1929. Development of the atrioventricular node, bundle of His and sino-atrial node in the calf; with description of a third embryonic node-like structure. *Anat. Rec.* 44:85–99.
24. Walls, E.W. 1947. The development of the specialized conducting tissue of the human heart. *J. Anat.* 81:93–100.
25. Navaratnam, V. 1965. The development of the specialized musculature of the human heart. *Ceylon J. Med. Sci.* 14:68–85.
26. James, T.N. 1970. Cardiac conduction system: fetal and postnatal development. *Am. J. Cardiol.* 25:213–226.
27. Anderson, R.H., and Taylor, I.M. 1972. Development of atrioventricular specialized tissue in human heart. *Brit. Heart J.* 34:1205–1214.
28. Truex, R.C.; Marino, T.A.; and Marino, D.R. 1978. Observations on the development of the human atrioventricular node and bundle. *Anat. Rec.* 192:337–350.
29. Mall, F.P. 1912. On the development of the human heart. *Am. J. Anat.* 13:249–298.
30. Sanabria, T. 1936. Recherches sur la différenciation du tissu nodal et connecteur du coeur des mammifères. *Arch. Biol. Paris* 47:1–70.
31. Anderson, R.H.; Becker, A.E.; Wenink, A.C.G.; and Janse, M.J. 1976. The development of the cardiac specialized tissue. In *The conduction system of the heart: structure, function and clinical implications,* ed. J.H.H. Wellens, K.I. Lie, and M.J. Janse, pp. 1–28. Leiden: Stenfert HE, Kroese BV.
32. Anderson, R.H.; Becker, A.E.; and Wenink, A.C.G. 1977. The development of the conducting tissues. In *Cardiac arrhythmias in the neonate, infant and child,* ed. N.K. Roberts and H. Gilband, pp. 1–28. New York: Appleton-Century-Crofts.
33. Wenink, A.C.G. 1976. Development of the human cardiac conducting system. *J. Anat.* 121:617–631.
34. Eriksson, A.; Thornell, L.-E.; and Stigbrand, T. 1979. Skeletin immunoreactivity in heart Purkinje fibers of several species. *J. Histochem. Cytochem.* 27:1604–1609.
35. Robb, J.S.; Kaylor, C.T.; and Turman, W.G. 1948. A study of specialized heart tissue at various stages of development of the human fetal heart. *Am. J. Med.* 5:324–336.
36. Sommer, J.R., and Johnson, E.A. 1968. Cardiac muscle: a comparative study of Purkinje fibers and ventricular fibers. *J. Cell. Biol.* 36:497–526.
37. Martinez-Palomo, A.; Alanis, J.; and Benitez, D. 1970. Transitional cardiac cells of the conductive system of the dog heart. *J. Cell. Biol.* 47:1–17.
38. Legato, M.J. 1973. Ultrastructure of the atrial, ventricular and Purkinje cell with special reference to the genesis of arrythmias. *Circulation* 47:178–189.
39. Thornell, L.-E., and Eriksson, A. 1976. The ventricular conduction system — ultrastructure and function. *Forensic Sci.* 8:97–102.
40. Sommer, J.R., and Waugh, R.A. 1978. Ultrastructure of heart muscle. *Envir. Hlth. Perspect.* 26:159–167.
41. Strehler, E.E.; Pelloni, G.; Heizmann, C.W.; Eppenberger, H.M. 1979. M-protein in chicken cardiac muscle. *Exp. Cell. Res.* 124:39–45.
42. Strehler, E.E.; Pelloni, G.; Heizmann, C.W.; and Eppenberger, H.M. 1980. Biochemical and ultrastructural aspects of M_r 165,000 M-protein in cross-striated chicken muscle. *J. Cell. Biol.* 86:775–783.
43. Field, E.J. 1951. The development of the conducting system of the heart of the sheep. *Brit. Heart J.* 13:129–147.
44. Muir, A.R. 1954. The development of the ventricular part of the conducting tissue in the heart of the sheep. *J. Anat.* 88:381–391.
45. Erokhina, I.L., and Rumyantsev, P.P. 1974. The autoradiographic study of the synthesis of DNA and RNA in cells of the conductive system of the developing rodent hearts. In *Functional morphology, genetics and biochemistry of the cell,* ed. A.S. Troshin, pp. 45–48. Leningrad: Institute of Cytology.

46. Rumyantsev, P.P. 1978. DNA synthesis and mitotic division of myocytes from ventricles, atria, and conductive system of the heart during myocardial development in mammals. *Tsitologia* 20:132–141.
47. Grohmann, D. 1961. Mitotische Wachstumintensität des embryonalen und fetalen Hunchenherzens und ihre Bedeutung für die Entstehung von Herzmissbildungen. *Z Zellforsch* 55:104–122.
48. Klinge, O., and Stocker, E. 1968. DNS-synthese in Rattenherzen als Funktion des Lebensalters: Autoradiographische Untersuchungen mit H^3-Thymidin. *Experientia* 24:167–168.
49. Sasaki, R.; Morischita, T.; and Yamagata, S. 1968. Mitosis of heart muscle cells in normal rats. *Tohoku J. Exp. Med.* 96:405–423.
50. Zak, R. 1973. Cell proliferation during cardiac growth. *Am. J. Cardiol.* 31:211–219.
51. Rumyantsev, P.P. 1977. Interrelations of the proliferation and differentiation processes during cardiac myogenesis and regeneration. *Int. Rev. Cytol.* 51:187–273.
52. Rumyantsev, P.P., and Kassem, A.M. 1976. Cumulative indices of DNA synthesizing myocytes in different compartments of the working myocardium and conductive system of the rat's heart muscle following extensive left ventricle infarction. *Virchows Arch. B. Cell. Path.* 20:329–342.
53. Truex, R.C. 1961. Comparative anatomy and functional considerations of the cardiac conduction system. In *The specialized tissues of the heart,* ed. A. deCarvalho, W.C. DeMello, and B.M. Hoffman, pp. 22–43. Amsterdam: Elsevier Publishing Co.
54. Zak, R. 1974. Development and proliferative capacity of cardiac muscle cells. *Circ. Res.* Suppl. II, 34 and 35:11–26.
55. Becker, A.E., and Anderson, R.H. 1978. Morphology of the human atrioventricular junctional area. In *The conduction system of the heart,* ed. H.J.J. Wellens, K.I. Lie, and M.J. Janse, pp. 263–286. The Hague: Martinus Nijhoff Medical Division.
56. Hecht, F.M. 1980. Studie über quantitative Altersveränderungen am Hisschen Bündel des Menschen. *Virchows Arch. A Path. Anat. and Histol.* 386:343–356.
57. Marino, T.A.; Biberstein, D.; and J.B. Severdia. 1981. The ultrastructure of the atrioventricular junctional tissues in the newborn ferret heart. *Am. J. Anat.* 161:383–392.
58. Draper, M.H., and Weidmann, S. 1951. Cardiac resting and action potentials recorded with an intracellular electrode. *J. Physiol.* 115:74–94.
59. Hoffman, B.F., and Cranefield, P.F. 1960. The Purkinje fibers. In *Electrophysiology of the heart,* p. 175. New York: McGraw-Hill.
60. Sommer, J.R., and Johnson, E.A. 1979. Ultrastructure of cardiac muscle. In *Handbook of physiology,* section 2, The cardiovascular system, ed. R.M. Berne, N. Sperelakis, and S.R. Geiger, pp. 113–186. Bethesda, MD: American Physiological Society.
61. Forsgren, S. The Purkinje fibers in the human right ventricle. An electron and light microscopic study on the fetal heart at mid-gestation. Submitted for publication.
62. Canale, E.; Campbell, G.R.; Uehara, Y.; Fujiwara, T.; and Smolich, J.J. 1983. Sheep cardiac Purkinje fibers: Configurational changes during the cardiac cycle. *Cell. Tiss. Res.* 232:97–110.
63. Mobley, B.A., and Page, E. 1972. The surface area of sheep cardiac Purkinje fibers. *J. Physiol.* 220:547–563.
64. Rybicka, K. 1978. Ultrastructural study on the extrusion of multivesicular bodies from cardiac cells. *Virchows Arch. B. Cell. Path.* 28:119–133.
65. Dominguez, G., and Fozzard, H.A. 1979. Effect of stretch on conduction velocity and cable properties of cardiac Purkinje fibers. *Am. J. Physiol.* 237:C119–C124.
66. Rosen, M.R.; Legato, M.J.; and Weiss, R.M. 1981. Developmental changes in impulse conduction in the canine heart. *Am. J. Physiol.* 240:H546–H554.
67. Barr, L.; Dewey, M.M.; and Berger, W. 1965. Propagation of action potentials and the structure of the nexus in cardiac muscle. *J. Gen. Physiol.* 48:797–828.
68. De Mello, W.C. 1982. Intercellular communication in cardiac muscle. *Circ. Res.* 51:1–9.
69. Tranum-Jensen, J. 1978. The fine structure of the atrial and atrioventricular (AV) junctional specialized tissues of the rabbit heart. In *The conduction system of the heart,* ed. H.J.J. Wellens, K.I. Lie, and M.J. Janse, pp. 55–81. The Hague: Martinus Nijhoff Medical Division.
70. Bailey, J.C.; Watanabe, A.M.; Besch, Jr., H.R.; and Lathrop, D.A. 1979. Acetylcholine antagonism of the electrophysiological effects of isoproterenol on canine cardiac Purkinje fibers. *Circ. Res.* 44:378–383.
71. Carmeliet, E., and Ramon, J. 1980. Electrophysiological effects of acetylcholine in sheep cardiac Purkinje fibers. *Pflügers Arch.* 387:197–205.
72. Elias, E.A.; de Vries, G.P.; Elias, R.A.; Tigges, A.J.; Meijer, A.E.F.H. 1980. Enzyme histo-

chemical studies on the conducting system of the human heart. *Histochem. J.* 12:577–589.
73. Kent, K.M.; Epstein, S.E.; Cooper, T.; and Jacobowitz, D.M. 1974. Cholinergic innervation of the canine and human ventricular conducting system. *Circulation* 50:948–955.
74. Finlay, M., and Anderson, R.H. 1974. The development of cholinesterase activity in the rat heart. *J. Anat.* 117:239–248.
75. Karczmar, A.G.; Srinivasan, R.; and Bernsohn, J. 1973. Cholinergic function in the developing fetus. In *Fetal pharmacology,* ed. L.O. Boréus, pp. 127–177. New York: Raven Press.
76. Pappano, A.J. 1977. Ontogenetic development of autonomic neuroeffector transmission and transmitter reactivity in embryonic and fetal hearts. *Pharmacol. Rev.* 29:3–33.
77. Moody-Corbett, F., and Cohen, M.W. 1981. Localization of cholinesterase at sites of high acetylcholine receptor density on embryonic amphibian muscle cells cultured without nerve. *J. Neurosci.* 1:596–605.
78. Strehler, E., Carlsson, E.; Eppenberger, H.M.; and Thornell, L.-E. 1983. Ultrastructural localization of M-band proteins in chicken breast muscle as revealed by combined immunocytochemistry and ultramicrotomy. *J. Mol. Biol.* 166:141–158.
79. Forsgren, S.; Strehler, E.; and Thornell, L.-E. 1983. Differentiation of the atrioventricular node, the atrioventricular bundle and the bundle branches in the bovine heart: an immunohistochemical and enzyme histochemical study. *Histochem. J.* 15:1099–1111.
80. Katz, A.M. 1977. *Physiology of the heart.* New York: Raven Press.
81. Eppenberger, H.M.; Wallimann, T.; Kuhn, H.J.; and Turner, D.C. 1975. Localization of creatine kinase isoenzymes in muscle cells: physiological significance. In *Isozymes II,* ed. C. Markert, pp. 409–424. New York: Academic Press.
82. Wallimann, T.; Turner, D.C.; and Eppenberger, H.M. 1977. Localization of creatine kinase isoenzymes in myofibrils. I. Chicken skeletal muscle. *J. Cell. Biol.* 75:297–317.
83. Anversa, P.; Olivetti, G.; Bracchi, P.-G.; and Loud, A. 1981. Postnatal development of the M-band in rat cardiac myofibrils. *Circ. Res.* 48:561–568.
84. Carlsson, E.; Kjörell, U.; Thornell, L.-E.; Lambertsson, A.; and Strehler, E. 1982. Differentiation of the myofibrils and the intermediate filament system during postnatal development of the rat heart. *Eur. J. Cell. Biol.* 27:62–73.
85. Schiebler, T.H. 1961. Histochemische Untersuchungen am Reizleitungssystem tierischer Herzen. *Naturwissenschaften* 14:502–503.
86. Pape, C.; Kübler, W.; and Smekal, P.V. 1969. Morphometrie am Reizleitungssystem und Arbeitsmyocard des Kalbherzens. *Beitr. Path. Anat.* 140:23–37.
87. Snijder, J., and Meijer, A.E.F.H. 1970. Enzyme histochemical studies on the Purkinje fibers of canine, bovine and porcine hearts. *Histochem. J.* 2:395–409.
88. Meijer, A.E.F.H., and de Vries, G.P. 1978. Enzyme histochemical studies on the Purkinje fibers of the atrioventricular system of the bovine and porcine hearts. *Histochem. J.* 10:399–408.
89. Warshaw, J.B. 1969. Cellular energy metabolism during fetal development. I. Oxidative phosphorylation in the fetal heart. *J. Cell. Biol.* 41:651–657.
90. Smith, H.E., and Page, E. 1977. Ultrastructural changes in rabbit heart mitochondria during the perinatal period. *Dev. Biol.* 57:109–117.
91. Hoerter, J., and Opie, L.H. 1978. Perinatal changes in glycolytic function in response to hypoxia in the incubated or perfused rat heart. *Biol. Neonate* 33:144–161.
92. Glatz, J.F.C., and Veerkamp, J.H. 1982. Postnatal development of palmitate oxidation and mitochondrial enzyme activities in rat cardiac and skeletal muscle. *Biochem. Biophys. Acta.* 711:327–335.
93. Hirakow, R. 1966. Fine structure of Purkinje fibers in the chick heart. *Arch. Histol. Jap.* 27:485–499.
94. Bogusch, G. 1979. Electron microscopic investigations on the differentiation of Purkinje cells in the ontogenetic development of the chicken heart. *Anat. Embryol.* 155:259–271.
95. Thornell, L.-E. 1972. Myofilament-polyribosome complexes in the conducting system of hearts from cow, rabbit and cat. *J. Ultrastruct. Res.* 41:579–596.
96. Thornell, L.-E., and Eriksson, A. 1981. Filament systems in the Purkinje fibers of the heart. *Am. J. Physiol.* 241:H291–H305.
97. Ferrans, V.J., and Roberts, W.C. 1973. Intermyofibrillar and nuclear-myofibrillar connections in human and canine myocardium. An ultrastructural study. *J. Mol. Cell. Cardiol.* 5:247–257.

98. Thornell, L.-E.; Eriksson, A.; Stigbrand, T.; and Sjöström, M. 1978. Structural proteins of cow Purkinje and ordinary ventricular fibers — a marked difference. *J. Mol. Cell. Cardiol.* 10:605–616.
99. Myklebust, R., and Jensen, H. 1978. Leptomeric fibrils and T-tubule desmosomes in the Z-band region of the mouse heart papillary muscle. *Cell. Tiss. Res.* 188:205–215.
100. Oliphant, L.W., and Loewen, R.D. 1976. Filament systems in Purkinje cells of the sheep heart: possible alterations of myofibrillogenesis. *J. Mol. Cell. Cardiol.* 8:679–688.
101. Thornell, L.-E.; Forsgren, S.; Gorza, L.; Sartore, S.; and Schiaffino, S. In press. Differentiation of fiber types in cardiac muscle. Proceedings of the 2nd symposium on etiology and morphogenesis of congenital heart disease, ed. S. Takao and A.H. Nora. New York: Futura.
102. Caesar, R.; Edwards, G.A.; and Ruska, H. 1958. Electron microscopy of the impulse conducting system of the sheep heart. *Z. Zellforsch.* 48:698–719.
103. Hatt, P.-Y. 1975. Cellular changes and damage in mechanically overloaded hearts. In *Recent advances in studies on cardiac structure and metabolism, vol. 6, Pathophysiology and morphology of myocardial cell alteration,* ed. A. Fleckenstein and G. Rona, pp. 325–333. Baltimore: University Park Press.
104. Goldstein, M.A.; Sordahl, L.A.; and Schwartz, A. 1974. Ultrastructural analysis of left ventricular hypertrophy in rabbits. *J. Mol. Cell. Cardiol.* 6:265–273.
105. Maron, B.J.; Ferrans, V.J.; and Roberts, W.C. 1975. Ultrastructural features of degenerating cardiac muscle cells in patients with cardiac hypertrophy. *Am. J. Pathol.* 79:387–414.
106. Maron, B.J., and Ferrans, V.J. 1978. Ultrastructural features of hypertrophied human ventricular myocardium. In *Progress in cardiovascular diseases,* vol. XXI, pp. 207–238. New York: Grune & Stratton.
107. Tomanek, R.J., and Cooper, N.G. 1981. Morphological changes in the mechanically unloaded myocardial cell. *Anat. Rec.* 200:271–280.
108. Cooper, N.G., and Tomanek, R.J. 1982. Load regulation of the structure, composition, and function of mammalian myocardium. *Circ. Res.* 50:788–798.
109. Forsgren, S., and Thornell, L.-E. Z disk variations in human ventricular myocytes: an ultrastructural study on the fetal heart at midgestation. Submitted for publication.
110. Mary-Rabine, L.; Albert, A.; Pham, T.D.; Hordof, A.; Fenoglio, Jr., J.J.; Malm, J.R.; and Rosen, M.R. 1983. The relationship of human atrial cellular electrophysiology to clinical function and ultrastructure. *Circ. Res.* 52:188–199.
111. Zak, R., and Galhotra, S.S. 1983. Contractile and regulatory proteins. In *Cardiac metabolism,* ed. A.J. Drake-Holland and M.M. Noble, pp. 339–364. New York: John Wiley & Sons.

5. MODELS OF CARDIAC DEVELOPMENT: TRANSPLANTS, ORGAN CULTURE, CELL DISPERSION, AND CELL CULTURE

RICHARD B. ROBINSON

The ability of the heart to survive outside the body has been recognized at least since the time of Galen in the second century [1]. In modern times, a variety of preparations have been developed that permit heart tissue or cells to survive outside their natural environment. These include homotopic and heterotopic transplants, organ and tissue culture, cell dispersion, and cell culture. In homotopic transplantation, the heart is surgically removed and reimplanted in situ for the purpose of achieving total cardiac denervation (or the heart is replaced with a donor heart, such as in clinical cardiac transplantation surgery). Heterotopic transplantation of the intact heart involves placing the donor heart in the abdominal cavity or another location of the recipient and allowing perfusion of the donor heart by the recipient's blood supply. Another form of heterotopic transplantation involves the implantation of a small piece of (generally) fetal or neonatal heart tissue in one of a number of anatomical locations of the recipient animal, without maintaining coronary circulation.

Organ culture involves removing an intact organ or tissue fragment and maintaining it in vitro in a nutrient-rich medium. If differs from tissue culture or explantation in that measures are taken to preserve tissue organization. Unlike homotopic and some heterotopic transplants, the blood supply is not maintained, so the tissue section must be thin enough to allow diffusion of oxygen from the surface to the deepest cells.

Cell dispersion protocols use enzyme solutions to weaken the intercellular matrix and separate the tissue into single cells. The single cells may be studied acutely,

M.J. Legato (ed.) THE DEVELOPING HEART, pp. 69–94.

as is generally the case when adult cardiac tissue is dispersed, or grown in culture for many days or weeks. To grow the cells in culture one usually starts with fetal or newborn tissue and carries out the dispersion under sterile conditions. The single cells are plated in plastic dishes or other chambers containing nutrient medium, where they attach and constitute a primary culture. If, after several cell divisions the culture is passaged by resuspending and replating the cells, then it is defined as a primary cell line. After many (>70) such passages, if a culture continues to proliferate it is designated an established cell line. (For a more detailed discussion of the various culture preparations see [2].) Most cardiac cultures are primary cultures.

The purpose of this chapter is to review these preparations and evaluate their applicability to studies of cardiac cell development. The rationale for using such preparations in developmental studies is that they may provide a defined and adjustable environment in which cardiac cell development can be studied and modified. In addition, the unrestricted access to the tissue made possible by some of these techniques often simplifies the process of collecting experimental data. However, given the unphysiological nature of the preparations, great care must be taken in evaluating the extent to which these models actually mirror developmental processes. Unavoidably, one's conclusion in this regard will depend on what aspect of cardiac development is of greatest interest, e.g., electrophysiology, ultrastructure, metabolism, etc.

INTACT TISSUE

Homo- and Heterotopic Transplants

Clinical reports on cases of homotopic transplantation have concentrated on the role of cardiac innervation, e.g., in coronary spasm [3] and cardiac arrhythmias [4, 5]. The technique has not been applied directly to the question of the role of innervation in controlling cardiac development. This may in part be due to the technical difficulty of performing cardiac transplantation on an animal prior to the development of cardiac innervation. In addition, even in the case of an adult animal, interpretation would be complicated because not all cholinergic nerves are removed [6] and the heart is reinnervated, beginning as early as 26 days for the cholinergic and 74 days for the adrenergic nerves [7]. Furthermore, cardiac denervation would not eliminate the influence of circulating catecholamines on cardiac tissue. On the other hand, for those studies relating explicitly to the reasonably short term (< one month) effects of removal of sympathetic innervation (but not neurohormones), the preparation much more closely approximates the normal physiological state than any in vitro model. Although beyond the scope of this review, there has been some effort made to achieve similar but more widespread effects without surgery by chemical sympathectomy, using reserpine [8], guanethidine [9], 6-hydroxydopamine [10], or antibodies to nerve growth factor [11].

Heterotopic cardiac transplantation has been employed by a number of laboratories in immunological studies [12, 13, 14, 15]. One study which compared 50-

day-old rat transplants to the recipient hearts found statistically insignificant differences for all parameters measured other than the degree of fibrosis. Nuclear, mitochondrial, and sarcomeric appearance were normal, and both baseline maximum tension and isoproterenol stimulated tension were unchanged [16]. Another study reported that levels of choline acetyl transferase eight days after heterotopic transplantation were reduced to uniformly low levels throughout the heart, although control hearts had much greater levels in the atria and SA node than in the ventricles [6]. This may reflect the non-uniform distribution of pre-ganglionic as compared to post-ganglionic parasympathetic neurons.

All the intact heart transplantation studies, taken together, suggest that removal of innervation from a mature heart, in which the working state is approximately maintained, does not significantly alter the differentiated state of the tissue. That is, any direct effect of innervation on cardiac cell development is irreversible over the time studied in these reports. This is not in conflict with reports that chemical sympathectomy alters cardiac adrenergic function [8], since in that case both peripheral and circulating catecholamine levels are reduced.

To clarify the role of innervation during cardiac development, transplantation of fetal or newborn cardiac tissue might prove more informative. Fetal or newborn cardiac tissue has been transplanted into the brain ventricle [17], cheek pouch [18], and anterior eye chamber [19, 20]. Injection of myocardial cells into the brain ventricle demonstrated the ability of these cells to survive and raised the possibility of culturing hormone-secreting cells in the brain near potential sites of action. Gilmour and Zipes reported on the electrophysiologic characteristics of pieces of neonatal hamster atrium and ventricle transplanted to the cheek pouch and allowed to revascularize. They found normal electrophysiological characteristics four to seven days later and could document effects of ischemia on the transplants. Transplantation of cardiac tissue to the anterior eye chamber resulted in reinnervation by the host animal, as demonstrated by the effect of ambient light on spontaneous contractile rate. Although none of these methods has yet been pursued with respect to cardiac development, the latter has been used to study the role of innervation in the development of the membrane potential of vascular muscle transplants from normal (KNR) versus hypertensive (SHR) rats (figure 5–1). It was found that the reduced membrane potential characteristic of the caudal artery of spontaneously hypertensive rats could be prevented if the tissue were transplanted into the anterior eye chamber of normal control rats in the first two weeks after birth, before the difference in membrane potential had developed. Likewise, caudal arteries from young normal rats failed to develop the increased membrane potential if transplanted into spontaneously hypertensive rats, but not if transplanted into normotensive rats [21]. When the eye of the host animal was first sympathetically denervated, these changes in membrane potential of the transplanted tissue were prevented [22]. Thus, this transplantation model has proven useful in confirming a direct role of peripheral sympathetic innervation in the development of (vascular) muscle function and its phenotypic modification. A similar approach would appear promising in exploring the role of innervation in cardiac cell development.

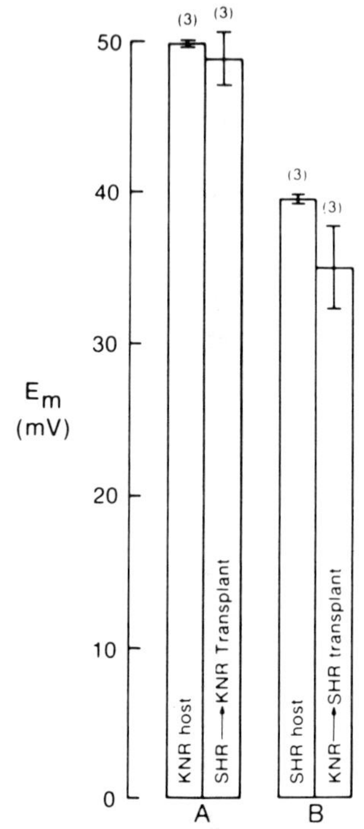

Figure 5–1. Membrane potentials of caudal arteries from two-week-old rats transplanted into host animals and from the host animals. SHR stands for spontaneously hypertensive rats and KNR for genetically matched Kyoto-Wistar normotensive rats. Notice that the young animal artery transplants took on the membrane electrical characteristic of the hosts into which they were transplanted. For each experiment, seven impalements were obtained from each artery, all at 16°C. The number of tissues is shown above each bar, as are standard errors for the number of impalements [$n = 21$]. All membrane potential [E_m] values are negative with reference to the superfusion solution. Both groups in *B* are significantly less negative than both groups in *A* [$P<0.05$]. Within *A* and within *B*, there are no significant differences. Reprinted from [21].

Organ Culture

Cardiac organ culture can be said to date from the 1912 tissue culture experiments of Burrows [23] in which fragments of embryonic heart tissue were explanted in a plasma clot. In the modern technique, whole fetal hearts are floated on mesh at the interface between a nutrient medium and an oxygen enriched gas phase and maintained in vitro for several days [24]. In organ culture the medium surrounding the tissue is under total control of the investigator. Although the force-generating function of the heart is not preserved to the extent that it is in homo- or heterotopic transplantation, the three-dimensional geometry of the intact heart is better preserved than is the case with either tissue or cell culture. In addition, one need not worry about possible effects of dispersing enzymes on cardiac cell characteristics, which is a potential problem with single cell dispersion or cell culture. However, individual cell surfaces are less accessible in organ culture than in single cell dispersion or cell culture.

The characteristics of the fetal mouse heart in organ culture have been extensively studied by Wildenthal, who has recently reviewed progress in this field [25]. His laboratory has found that the in vitro hearts are in a state of negative protein balance, with degradation exceeding new synthesis [26, 27]. This has limited the maximum time for maintaining the preparation to about one week. If fetal hearts of 18 days or younger are used, at least some specific enzymes (e.g., creatine kinase) maintain their in vivo levels of activity, but they do not continue to increase in activity as they would in vivo. If older hearts are employed, the appropriate in vivo level of activity is not maintained [28]. Electrophysiological studies from one laboratory, using embryonic chick hearts in organ culture, indicate that any age tissue will maintain the action-potential characteristics of the in vivo source tissue, but will not continue to develop in vitro [29, 30]. In contrast, DeHaan et al. reported that organ cultured embryonic chick heart does continue to develop electrophysiologically in vitro, provided that protein synthesis is not inhibited [31].

Thus far this organ culture preparation has proven useful in several types of developmental studies. By adjusting the nutrient medium in an attempt to prolong in vitro survival, much has been learned about the metabolic requirements of the developing heart [32]. Also, short-term organ culture of hearts of different fetal ages has permitted direct measurement of several characteristics of the developing heart that would have otherwise been difficult to obtain. This includes data on the onset of pharmacological responsiveness [33, 34, 35] and its dependence on exogenous hormone levels [36].

ADULT SINGLE CELLS

Acute

Dispersed adult ventricle cells have been prepared from rat [37], rabbit [38], guinea pig [39, 40], cow [41], cat [42], and dog [43, 44]. The majority of recent papers have favored a protocol that uses collagenase as the dispersing enzyme, introduced into the coronary circulation. Using this technique, it is possible to obtain single cells with action-potential characteristics remarkably similar to those recorded from intact ventricular tissue (figure 5–2). Work in this field has been recently reviewed [45, 46]. There are two possible reasons for using such a preparation in developmental studies. If the cells can be maintained in vitro for several days or weeks, then one would have access to a controlled environment for such investigations. Work in this area is discussed in subsequent sections. Alternatively, single cells from animals of different ages or animals whose myocardial development has been altered by raising them under special experimental conditions could be used acutely for those experimental protocols where whole hearts or homogenates are unsuitable. For example, there are well-known electrophysiological differences between the newborn and adult heart [47, 48, 49, 50]. To characterize fully the individual ionic currents which contribute to the action potential at each age it

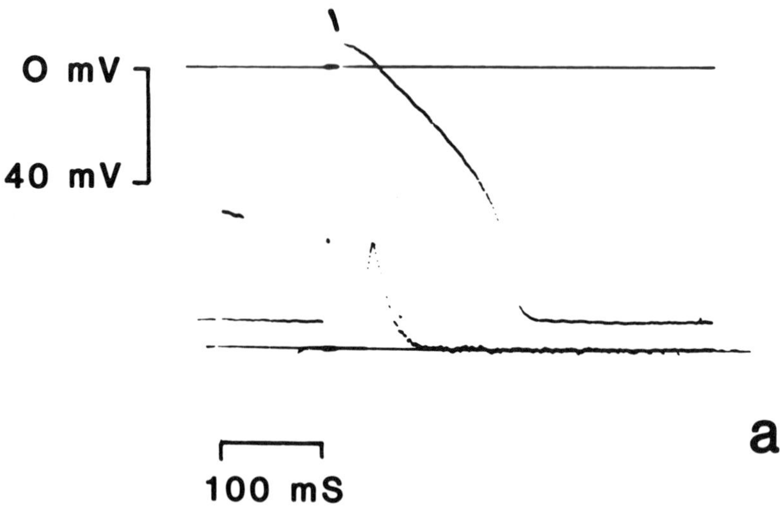

Figure 5–2. Transmembrane action potential recording from a single adult canine ventricle cell, electrically driven at a cycle length of 1500 mS. The upper trace represents the zero potential level, the middle trace is the action potential recording, and the lower trace is the first-time derivative of the action potential displayed at a faster sweep speed. In the time derivative trace, the square pulse at the start of the trace represents the derivative of a 200 V/S calibrating sawtooth.

would be desirable to use the voltage clamp or recently developed patch clamp techniques [51, 52], in which the membrane potential is controlled in order to determine the voltage and time dependence of the ionic currents. The former technique requires a reasonably small membrane surface area without restricted extracellular space, while the latter requires direct access to a clean cell surface. In the case of the heart these requirements mean using either cell cultures [53] or acutely dispersed single cells [54]. As another example, radioligand binding studies to characterize the density and affinity of specific hormone receptors generally are conducted on tissue homogenates containing many cell types. This can complicate interpretation of experimental results. For instance, both muscle and non-muscle cells of the heart possess specific beta adrenergic binding sites [55] (figure 5–3). If homogenates of newborn and adult ventricle are compared and found to differ in terms of number of sites per mg protein it may be difficult to determine if this reflects a difference in the number of sites per muscle cell as a function of age.

The available dispersion techniques result in preparations that are composed of up to 90% viable rod-shaped muscle cells exhibiting good preservation of ultrastructural detail (figure 5–4). Although this largely eliminates contaminant cells, so far it has not proven possible to entirely remove them from an acute preparation or to determine to what extent they are dead or damaged muscle cells (which may or may not still possess specific receptor binding sites) and to what extent they are non-muscle cells. Also, the enzyme exposure may alter receptor characteristics, with no opportunity for repair in an acute preparation. In developmental studies there is the additional problem that the enzyme protocol may be more damaging

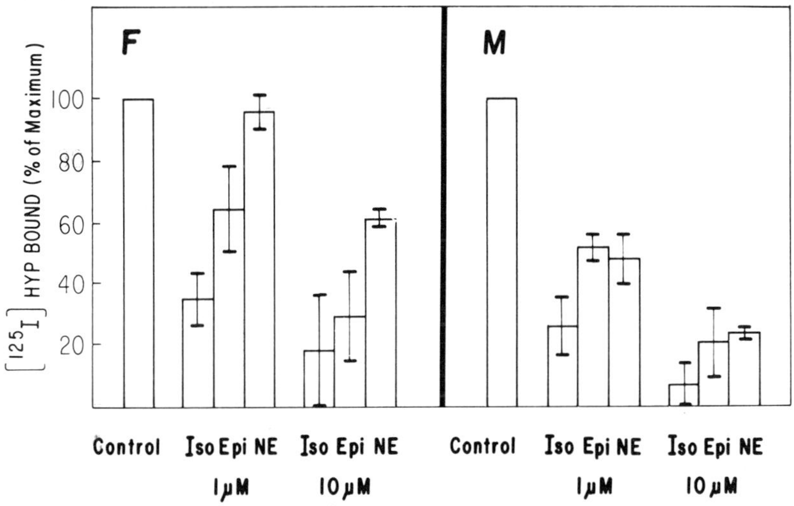

Figure 5–3. The relative potencies of beta-adrenergic agonists in inhibiting the specific binding of the iodonated beta-adrenergic antagonist [^{125}I]iodohydroxybenzylpindolol ([^{125}I]IHYP) to muscle [M] and fibroblast [F] cells obtained from primary cultures of neonatal rat ventricle. I-Isoproterenol [Iso], I-epinephrine [Epi], and I-norepinephrine [NE], each 1 μM or 10 μM, were tested as inhibitors of maximal specific [^{125}I]IHYP binding to F and M cells. The results shown are the mean ± SD of three experiments for which data points were obtained in triplicate. At each concentration, the differences between inhibition by epinephrine and norepinephrine were significant for F cells [$P<0.05$] but not for M cells. The data are consistent with the binding site on the M cells being of the beta$_1$ subtype and that on the F cells being of the beta$_2$ subtype. Reprinted from [55].

to young than old hearts (or vice versa), and thus must be carefully optimized and adjusted for each age studied.

It is of interest to note that progress has recently been reported in preparing acute dispersions of both rabbit and nodal tissue [56, 57] and cow [58] and dog [59, 60] Purkinje fibers. These preparations result in a relatively low yield of cells and therefore may not be suitable for biochemical or receptor binding studies. However, it is possible to obtain cells that are electrophysiologically comparable to intact tissue (figure 5–5), and these may prove useful in electrophysiological studies of developmental changes in the cardiac conducting system.

Long Term

There have been a few published reports of adult ventricle cells surviving in culture for periods of several days or more. The earliest was a study comparing adult and newborn rat heart cells. The adult cells were incubated for a maximum of only 24 hours, and no details of the dispersion were provided. It was found that in response to irradiation damage, the newborn but not the adult cells were capable of undergoing unscheduled DNA synthesis [61]. This suggests that maintaining adults cells in culture, where they would have to repair any enzyme damage and balance protein degradation with new synthesis, could be difficult. The first suc-

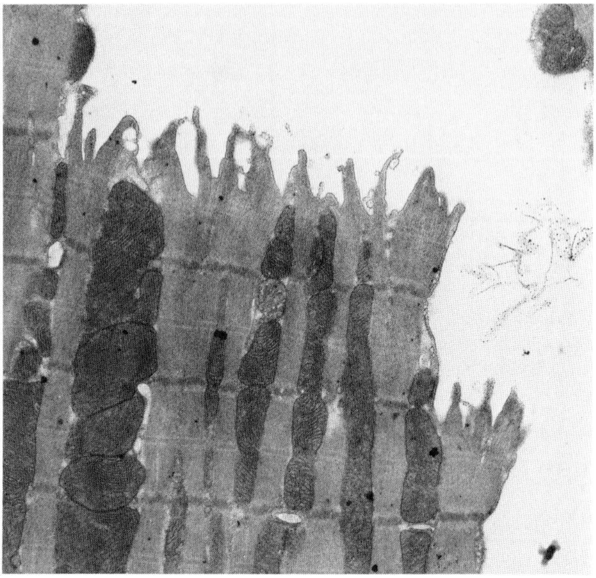

Figure 5–4. Electron micrograph of isolated adult dog ventricle cell illustrating that the separation of the cell from its neighbor has been achieved with preservation of the interdigitating fingers of cytoplasm that make up the intercalated disc. Even the terminal myofilaments retain their orderly register in spite of the separation. 12,170X. Reprinted from [44].

cessful report of long-term culture of adult cardiac cells was published in 1977 [62]. Adult rat heart cells were maintained for up to 60 days. However, only 20% of the cultures maintained any spontaneous activity for more than 14 days. Also, when rats greater than 30 days old were used, the yield was less than 1,000 cells per heart. Another laboratory has employed a dispersion method similar to the common acute dispersion protocols [63]. One culture was reported to maintain spontaneous activity for up to 118 days. Yield was not stated but is generally greater than 10^7 cells/heart by this type of dispersion protocol. Nag and Cheng [64] used a similar protocol and reported electron microscopic evidence which they interpreted as indicating that the adult cells dedifferentiated in culture to a newborn or fetal stage. In addition, they employed autoradiography to demonstrate (^{3}H)-thymidine incorporation into cell nuclei. Mitosis and cell division were also reported.

Common to all these studies are the findings that attachment is slow (48 to 72 hours), the cells undergo a transition from rod-shaped first to round then to flat and stellar, and the cells remain spontaneous over many days or weeks, although not all cells are initially spontaneous or maintain activity during the rounding process. The most recent study maintained the cells for only four days but provided detailed metabolic information [65]. It was found that only rod-shaped cells attached, the cells attached early, were stable and energetically well controlled during the four days in culture, and showed no evidence of either cell division or

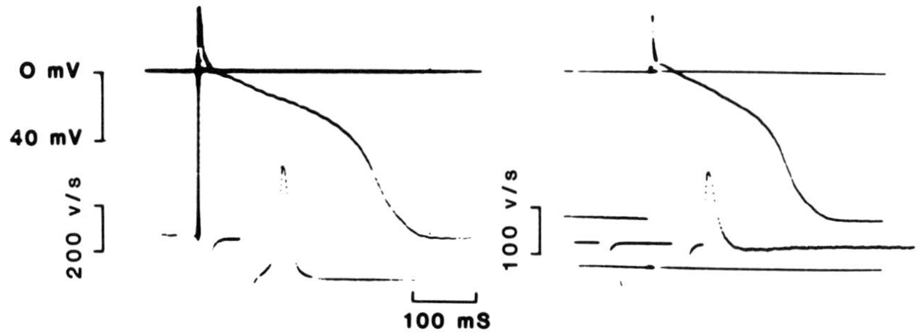

Figure 5–5. Comparison of transmembrane action potential recordings from an intact Purkinje fiber and a dispersed single Purkinje cell. In each case the cell was driven at a cycle length of 1000 mS. The upper traces represent the zero potential level, the middle traces are the action potential recordings, and the lower traces are the first-time derivatives of the action potentials displayed at a faster sweep speed. *a*. Record from intact canine Purkinje fiber. *b*. Record from single dispersed canine Purkinje cell.

dedifferentiation over that period. The cells were not spontaneous but were electrically excitable.

All these preparations are recent and still not fully characterized. Still, they are potentially useful for developmental studies where different age source tissue must be dispersed into single cells, allowed to recover, then measured with respect to some developmentally relevant parameter which cannot be easily or accurately measured in intact tissue. However, when it is desired to actually vary the in vitro environment in order to study the role of a putative developmental stimulus under defined conditions, newborn or fetal cultures may be more appropriate. This is because one would be starting with source tissue not previously exposed to the putative stimulus in vivo.

CELL CULTURE

Cell Lines

The difficulty with producing cell lines from cardiac tissue is that the cultures are likely to become overgrown by the more rapidly proliferating non-muscle cells present at the time of initial plating or the myocardial cells may lose the functional characteristics of muscle upon successive passages. A cell line was produced from a biopsy sample of the adult human atrial appendage in 1958 but was reported to be fibroblastic in appearance [66]. Later studies reported a low resting potential (-20 mV) and the absence of any dependence of calcium content or calcium flux on the sodium gradient [67]. Thus, this cell line lacks the functional hallmarks of a cardiac muscle cell. In addition, it has characteristics indicative of HeLa cell contamination [68]. An epithelial-like cell line was reported using the box turtle heart [69], and an endothelial cell line has been established from the tadpole heart [70]. There is an intriguing report of an embryonal carcinoma cell line which differentiates into contracting striated muscle cells upon exposure to dimethylsulphoxide [71]. Cells possessing both cardiac and skeletal muscle appearance were reported, but the basis for the identification was unclear. Still, this cell line may prove valuable when more fully characterized.

Primary Cultures

Primary cardiac cultures can be grown in several configurations, depending on experimental requirements. Because of the absence of diffusional barriers, cell culture is ideally suited for radioactive tracer studies of ionic fluxes across the cell membrane. In some of these studies the cells have been grown as a monolayer directly on scintillation glass to permit sensitive measurement of calcium exchange [72]. Other experiments have grown the cells as synthetic strands where the nonmuscle cells migrate to the periphery, simplifying the kinetic analysis [73]. Because of their cable-like geometry, strands also have been employed in studies of intercellular coupling and impulse propagation [74, 75]. In other electrophysiological studies the cells have been reaggregated into spheres [76]. Microelectrode impalement is easier in reaggregates than in monolayer cultures, and the electrotonic interactions are better defined [77]. However, for many developmental studies monolayer cultures are preferable to either strand or reaggregate configurations. Monolayers permit the most unencumbered access to the cell surface of all possible configurations. Individual cells are more easily visualized than with the thicker preparations. The cells are easily detached and redispersed into single cells, e.g., for receptor binding studies.

Whatever configuration is employed, central to the concept of using primary cardiac cultures as an in vitro model of the developing heart is the hypothesis that under appropriate conditions cardiac development can continue in vitro. At the very least, conditions are required wherein the cells remain fixed at the developmental stage of the source tissue, without reversion to a more primitive or dedifferentiated form. There has been much controversy and confusion in the literature, resulting in part from inconsistent word usage, over whether cardiac cells in culture do dedifferentiate. In 1959 Grobstein [78] defined two types of dedifferentiation. True dedifferentiation was characterized by a loss of (future) potential, that is, a reversion to an embryonic cell along an uncomplicated dedifferentiative path. In contrast, overt dedifferentiation was defined as the loss of some specialized functions of the cell. Romanoff [79] made a similar distinction when he stated that "dedifferentiation *in vitro* may not necessarily imply a degradation of biological value . . . but merely that the conditions of culture do not permit the different potencies of cells to remain in evidence." Although some cell types, e.g., vascular smooth muscle [80], do undergo true dedifferentiation in culture, there is no evidence that this occurs in the cardiac cell. Whether overt dedifferentiation occurs appears to depend on culture conditions and the function being used as an index of dedifferentiation. In addition, as will be made clear in the next section, many of the early observations of loss of cell function may have arisen from overgrowth of the cultures by fibroblast cells or enzyme destruction of specialized cell structures.

In Vitro *Dedifferentiation*

Although reports of cardiac cell dispersion and culture exist as early as 1934 [81], modern studies of cardiac cells in primary monolayer culture date from the papers by Cavanaugh [82] for the embryonic chick and Harary and Farley [83] for the newborn rat. These two species continue to be the most common sources of cardiac tissue for culture, although success has also been reported with tissue from mouse [84], rabbit [85, 86], guinea pig [87, 88], cat [89], dog [90, 91], and human [92, 93]. In general, the ventricle is cultured, but there have been some studies of the atrium [94, 95] and Purkinje system [96].

The hypothesis that cardiac cells in monolayer culture might dedifferentiate dates back at least to the early 1960s [83, 97] (although in vitro myocardial dedifferentiation was proposed earlier based on studies with explants [79]), and was based in that paper on the fact that cells stopped beating and apparently had decreased enzyme activities. The loss of beating in vitro has been reported by some laboratories [97, 98], but others find that cultures exhibit spontaneous activity for up to several months [99]. Further, it is not clear how to interpret loss of beating from a developmental viewpoint. It may represent overt dedifferentiation in that the ability to contract is lost. On the other hand, the in vivo ventricular cell is not normally spontaneous, and the disappearance of this feature in vitro may represent a return to "normalcy." This could result, for example, from repolarization of the cell membrane as enzyme damage is repaired in vitro. There have been several demonstrations that non-spontaneous cells may still possess the ability to contract when stimulated electrically or pharmacologically [98, 100].

With respect to enzyme activity, Kuramitsu and Harary reported a decrease in several enzyme levels within four to five days in vitro and, based on this and a loss of contractile function, suggested that a true dedifferentiation of heart cells may be taking place [97]. However, it has been demonstrated that fibroblast proliferation occurs in vitro and will dilute biochemical markers specific to the muscle cell [101, 102, 103]. In addition, a decrease in enzyme level by itself is not an indicator of either overt dedifferentiation or true dedifferentiation. It is necessary to demonstrate the relevance of the change to the in vivo situation. Even the demonstration of muscle cell division in culture is not evidence of dedifferentiation since myocardial cell division persists after birth in vivo in many species [104, 105].

This is not to say that the in vitro heart cell is identical to the in vivo one. Indeed, numerous biochemical indices can be unusual in culture. Frelin et al. [106] reported that glucose uptake waas high in culture, although Schroedl et al. [107] reported values consistent with intact tissue [108]. Frelin et al. also reported that palmitic acid oxidation decreased after one week but explicitly stated that the heart cells did not lose their specific metabolic pattern. There have been several reports that heart cells in culture shift from production of heart-type lactate dehydrogenase to muscle-type [109, 110], the inverse of what occurs during in vivo cardiac development [111]. However, this has been ascribed to overgrowth by non-muscle cells in culture [112]. Further, the experimental results depend on culture con-

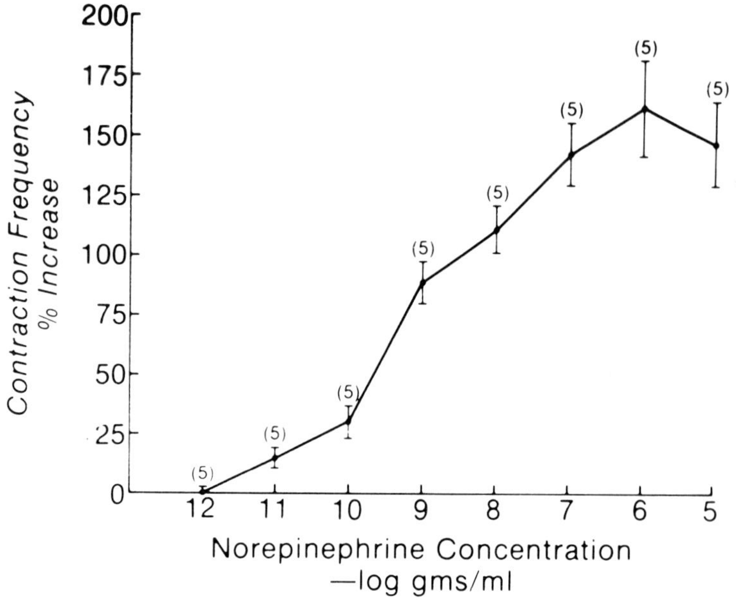

Figure 5–6. High sensitivity of spontaneously contracting isolated neonatal rat myocardial cells in culture, shown by the positive chronotropic effect of norepinephrine [ED_{50} = 2.3 nM]. The cells were exposed to 10 μl pulses of norepinephrine in the concentrations plotted, added at the inlet of a 300 μl superfusion chamber [37°C], and the peak response was recorded. The number of cells is indicated at each datum point. Each cell included was exposed to the neurotransmitter for 30 seconds for each of five or less doses and was always observed to return to baseline rate ± 20% in control solution. Intact source hearts [n = 3] exposed to norepinephrine superfusion had an ED_{50} of 100 nM [not shown]. Reprinted from [120].

ditions. Fujimoto and Harary [113] found that the addition of lipid to their culture media reversed the loss of enzyme activities earlier reported by Kuramitsu and Harary [97]. Freerskin et al. [114] reported an effect of exogenously added triiodothyronine (T3) on the activity of certain metabolic enzymes in culture. Electrophysiological characteristics may also vary with culture conditions, as will be discussed below.

Additional evidence for dedifferentiation in monolayer culture came from the pharmacological studies of Lehmkuhl and Sperelakis [115], in which it was reported that spontaneously contracting embryonic chick cardiac cells in monolayer culture were unresponsive to either neurotransmitters or the specific fast sodium channel blocker tetrodotoxin (TTX). The lack of response to autonomic agonists was consistent with dedifferentiation because it is known that this characteristic develops during fetal and early neonatal life and that the source tissue employed for culture was responsive in vivo [116]. However, in recent years numerous laboratories have been able to demonstrate responsiveness of monolayer cardiac cultures to autonomic agents, generally at levels of sensitivity even greater than that of the source tissue [117, 118, 119, 120, 121, 122] (figure 5–6). While the reason for the initial failure to observe such pharmacologic responsiveness is unclear, there are several possibilities. It has been shown that the conditions employed for the pharmacological testing can have a marked effect on the observed sensitivity [117]. In addition, the proteolytic enzymes used in cell dispersion may impair the responsiveness to some hormones [123, 124], so that if excess enzyme is employed or if subsequent culture conditions are inadequate for complete repair of initial enzyme damage then autonomic responsiveness could be impaired or absent.

The insensitivity of cardiac monolayer cultures to TTX, also reported in 1965 [115], took on increased significance when it was later discovered that the TTX insensitive action potentials with slow rates of rise that were seen in culture were also characteristic of the early embryonic heart [125]. While this could suggest dedifferentiation [126], more recent evidence supports the hypothesis stated above, that the loss of certain functional characteristics is predominantly due to inadequate repair of enzyme damage occurring during the initial cell dispersion protocol. Evidence for this alternative explanation comes in part from studies of spherical reaggregate cultures. Chick heart cells grown in this configuration redevelop TTX sensitivity [127, 128]. These redeveloped channels in reaggregate cultures can be destroyed or inactivated in vitro by exposure to proteolytic enzymes and recovery requires protein synthesis [127]. In addition, flux studies with the Na^+ ionophore veratridine suggest that a TTX sensitive Na^+ channel is present in chick monolayer myocardial cultures [129]. Finally, a recent electrophysiological study has succeeded in demonstrating functional channels with moderate TTX sensitivity in similar cultures [130]. Other laboratories have found that by adding a putative mRNA [131] or by growing the cells in delipidated medium [132] it is possible to demonstrate fast response action potentials in monolayer cultures of embryonic chick heart cells. Conditions have been developed in this laboratory under which both canine [91] and rat [133] monolayer cultures exhibit TTX sensitive action potentials with rapid rates of rise in standard culture medium. In the latter case it also has been demonstrated that these fast response action potentials are not lost with extended time in culture [134] (figure 5–7). Other laboratories have also reported fast response action potentials in rat cardiac monolayers, although neither TTX sensitivity nor time-dependence were examined [17, 135]. In addition, a careful examination of the ultrastructural evidence reveals that the level of myofibril organization in rat monolayer cultures, although initially quite low, increases rather than decreases with extended time in culture [136, 137]. After one week in vitro the ultrastructural appearance of cells from the neonatal rat ventricle is very similar to the intact source tissue [133].

Future Directions

Although the cell culture preparation appears well suited to cardiac developmental studies, the past difficulty in demonstrating autonomic responsiveness and other functional characteristics of the differentiated myocardial cell has slowed progress in this area. However, there have been a variety of reports which suggest the direction future investigations will take.

One area that has been the subject of many developmental studies is the role of the extracellular matrix in regulating cardiac cell characteristics (see [138]). This

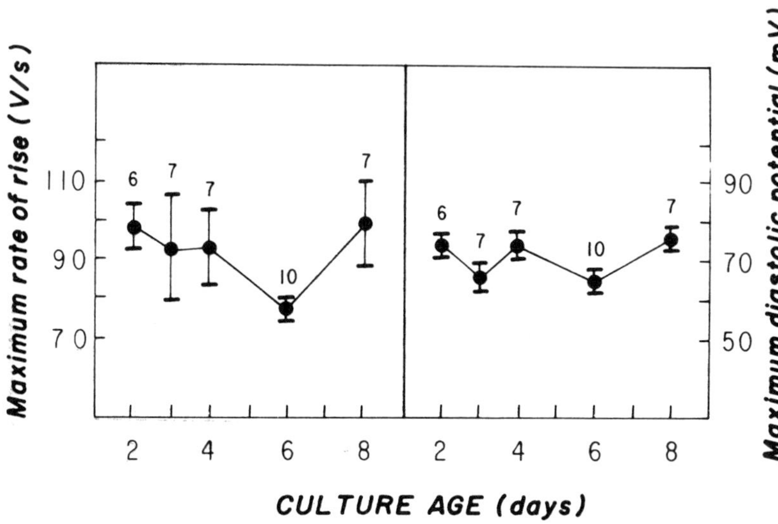

Figure 5–7. Effect of time in culture on maximum rate of rise and maximum diastolic potential of the action potential in spontaneously active neonatal rat ventricle monolayer cell cultures. The data were tested by regression analysis, and no statistically significant trend with culture age was found for either parameter. n values are indicated above the data points [means ± SE]. Reprinted from [134].

question has been explored in culture by using enzymes to selectively strip off components from the glycocalyx. These experiments suggest a role for the glycocalyx in regulating calcium, but not potassium exchange [139]. This in turn may lead to alterations in electrophysiological characteristics [140] and rhythmicity [141].

Another area of interest is the role of circulating hormones in cardiac development. For example, thyroid hormones have long been suspected of influencing adrenergic responsiveness and have also been shown to accelerate the development of the sympathetic nervous system [142]. In the human, maturation of function in the hypothalamic-pituitary-thyroid axis begins to develop in mid-gestation, and serum levels of thyroxine (T4) and subsequently T3 begin to rise at or after this time [143]. The rat shows a similar but delayed sequence of events, beginning the first week of life [144], demonstrating that with respect to the ontogenesis of thyroid function the neonatal rat is in an immature state and may therefore be a suitable culture source. Similarities between the cardiovascular manifestations of hyperthyroidism and enhanced sympathoadrenal activity led early researchers to look for enhanced catecholamine sensitivity in the hyperthyroid state [145]. Although some recent studies have failed to demonstrate enhanced catecholamine sensitivity of the heart in hyperthyroidism [146, 147], experiments on isolated fetal mouse hearts in organ culture demonstrated enhanced sensitivity to beta-adrenergic agonists after incubation with thyroid hormone [36]. In addition, numerous laboratories have reported increased density of beta-adrenergic receptors and decreased density of alpha-adrenergic receptors in hyperthyroidism [148, 149, 150, 151]. Given the efficacy of the beta-adrenergic blocker propranolol in ameliorating the cardiovascular symptoms of thyrotoxicosis [152, 153], it seems clear that at least one component of thyroid hormone action relates to endogenous catecholamines. However, it is equally clear that thyroid hormone must exert additional actions on the heart, either directly or indirectly. Maximally effective blocking doses of propranolol fail to reverse all the cardiac effects of thyroid hormone [154]. In addition, isolated papillary muscle of hyperthyroid cats exhibit enhanced myocardial contractility [155]. Finally, although neither cardiac sympathetic nervous activity [156] nor plasma catecholamine level [157, 158] are elevated in hyperthyroidism, a study on the newborn rat demonstrated that T3 injection accelerated by several days the time at which function could be demonstrated in both the adrenal medulla and the sympathetic nerves of the heart [142].

Thus, there are numerous reasons to believe that the ontogenesis of thyroid function can alter the development of the myocardial cell and cardiac adrenergic responsiveness, either directly or through an action on the developing sympathetic nervous system. However, these intact tissue studies have not been able to confirm a direct action of T3 or T4 on the myocardial cell or cardiac adrenergic receptors, as opposed to an action mediated by another aspect of the hyperthyroid state in vivo.

Cell cultures have been used to demonstrate direct effects of T3 on cardiac muscle cells. It has been shown that T3 added to a culture will cause an increase in the spontaneous rate of the cardiac cells [159]. In addition, exogenously added T3 causes an increase in the number of beta-adrenergic receptors in culture [160] and alters both the affinity and number of cardiac alpha-adrenergic receptors [161]. Thus, onset of thyroid function in the developing animal could play a direct role in cardiac cell maturation, and cell culture appears capable of helping to elucidate that role.

Finally, the question of the role of peripheral innervation in cardiac development can be addressed by the cell culture method. The time course of cardiac innervation has been extensively studied in vivo and found to more or less parallel changes in responsiveness to autonomic agonists [116]. However, a cause-and-effect relation has not been demonstraated, in part because in the intact developing animal many factors are changing at the time of cardiac innervation and one or more of them may contribute to the altered autonomic responsiveness. Cell culture may be useful in this respect because co-cultures of myocardial cells and peripheral neurons can be grown under conditions in which other medium components are either defined or unchanging.

Considerable progress has been made in recent years with respect to the production and maintenance of nerve cells in culture. Both explanted ganglia and isolated dispersed neurons can be maintained in vitro. The nerve cells send out processes that form intimate associations with the myocardial cells (figure 5–8). As with cardiac cultures, in vitro development and functional expression have proven more complex than originally anticipated. Predominantly adrenergic sympathetic ganglia cells, when grown in the presence of non-neuronal cells (including cardiac

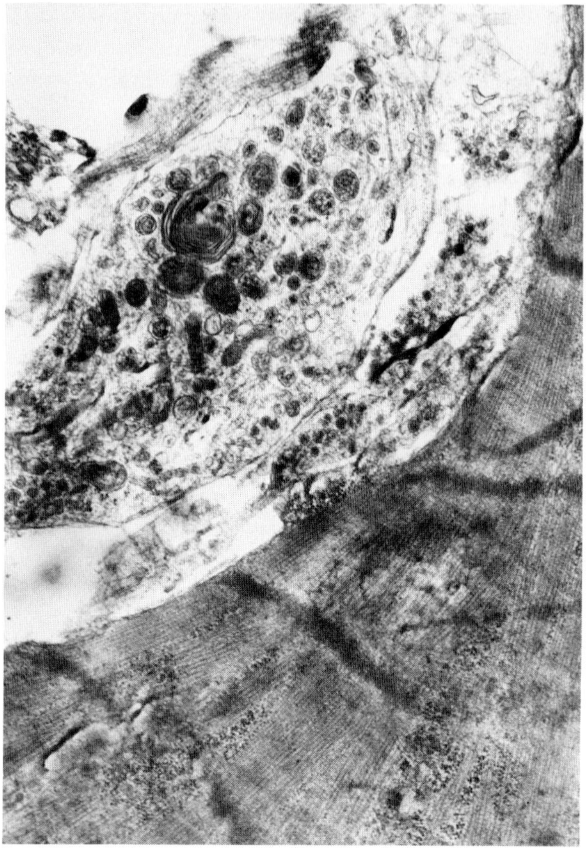

Figure 5–8. Electron micrograph of a close contact between a cardiac cell and neuron in culture, showing typical synaptic vessicle-like structures within the nerve cell and the high degree of myofibril organization of the muscle cell. 23,000X

myocytes and fibroblasts), become partially cholinergic. Individual neurons can at one time express both adrenergic and cholinergic function [162]. Although this plasticity of neuronal expression complicates the study of in vitro adrenergic innervation of heart cells, several laboratories have demonstrated functional nerve-muscle adrenergic junctions in co-cultures. In these experiments, functional adrenergic innervation was defined as a positive chronotropic response of the heart cell to pharmacological [163] or electrical [162] stimulation of the nerve cell which could be blocked by propranolol. The former method is quick and easy, but does not actually demonstrate functional coupling of individual nerve and muscle cells. Examples of both pharmacological and electrical stimulation are shown in figures 5–9 and 5–10. To date there have been no full reports of the effect of nerves in culture on any cardiac muscle cell characteristics, although abstracts have appeared [164, 165]. With the increased understanding that is now available on the charac-

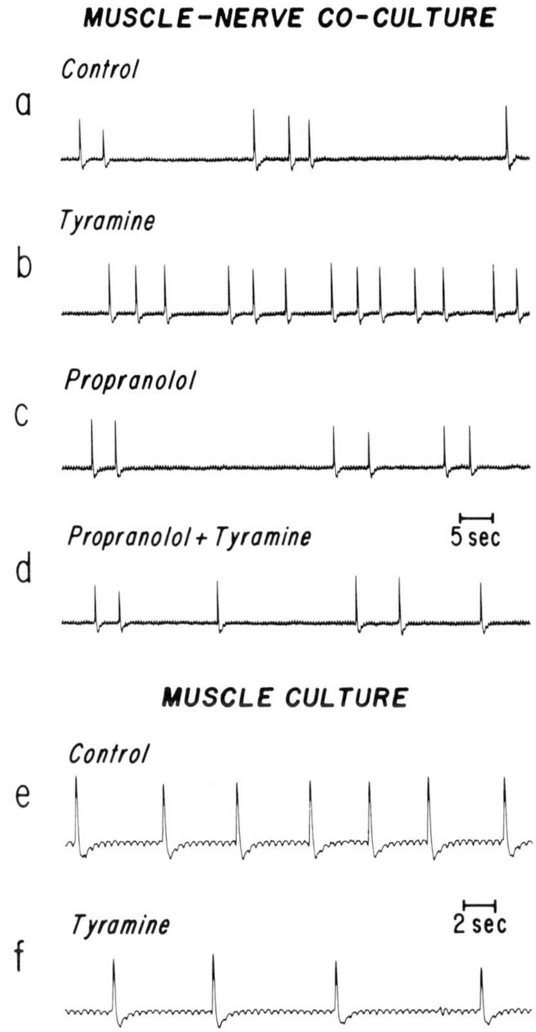

Figure 5–9. Effect of tyramine on spontaneous rate of muscle cells cultured in the presence and absence of sympathetic neurons. *a*. Opto-electrical recording of spontaneous beating of a neonatal rat cardiac muscle cell in culture for seven days, grown in the presence of sympathetic neurons for six days. *b*. Recording from same cell during superfusion with the indirectly acting amine tyramine [3×10^{-6}M], which causes an increase in spontaneous rate in the co-culture. *c*. Recording from the same cell after washing out the tyramine, during superfusion with 2×10^{-7}M propranolol. *d*. Recording from the same cell during superfusion with tyramine [3×10^{-6}M] in the continued presence of propranolol. After wash-out of the tyramine and propranolol this same cell still responded to fresh tyramine [not shown]. In other muscle cells in co-culture propranolol reduced but did not always completely eliminate the response to tyramine. *e*. Opto-electric recording of spontaneous beating of a rat cardiac muscle cell in culture for seven days, taken from the same litter as the cell shown in records (a–d), but not co-cultured in the presence of neurons. *f*. Recording from the same cell as in (e), showing a direct negative chronotropic effect of tyramine on muscle cells in culture, rather than the positive chronotropic effect observed in co-cultures.

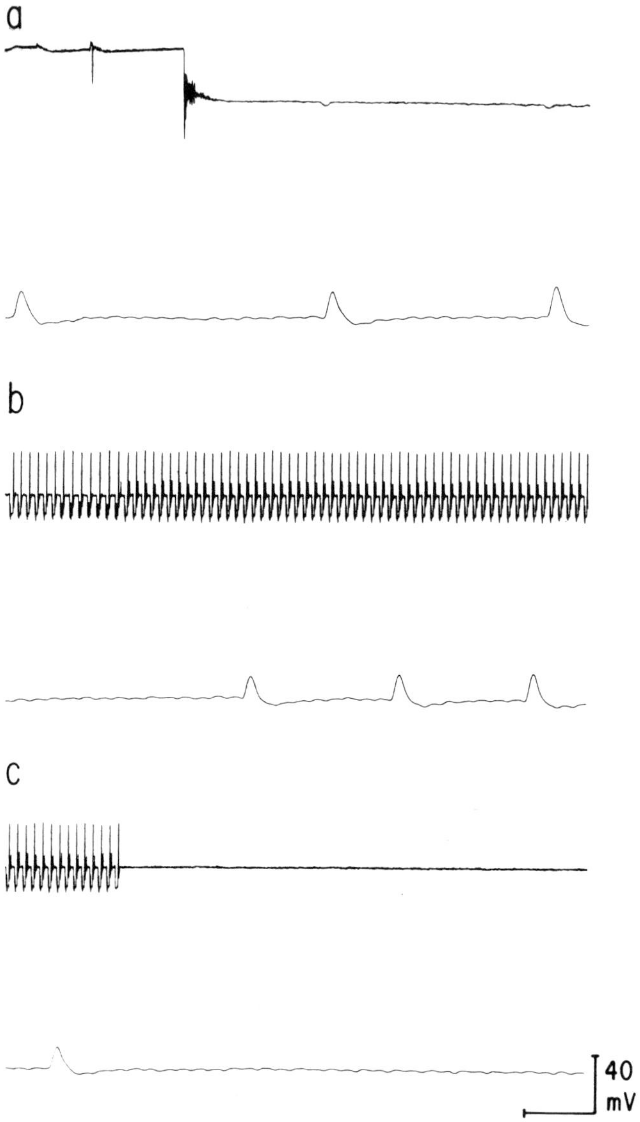

Figure 5–10. Effect of electrical stimulation of a neuron in culture on the spontaneous beating of a cardiac muscle cell which is innervated by it. In each panel the top trace is the intracellular recording from the soma of a single neuron, and the bottom trace is the opto-electric record of the spontaneous beating in a nearby muscle cell, not in contact with the nerve soma. *a*. Spontaneous rate of the muscle cell at the time the nerve cell is impaled. The muscle cell rate was not altered by the impalement of the nerve cell. *b*. The start of a period of rapid stimulation of the nerve cell by injection of hyperpolarizing current pulses through the recording electrode. After a brief lag the muscle cell begins to beat at a faster rate. *c*. The end of the period of nerve stimulation, which is followed by a period of quiescence of the muscle cell.

teristics of both nerve and myocardial cells in culture, preparation of co-cultures should become more common.

CONCLUSION

From the preceding sections it can be seen that the various model systems available for cardiac research have not been fully utilized in developmental studies. It is hoped that this chapter will stimulate increased activity in this area. Each of the preparations has unique advantages, as well as disadvantages, that will determine which model to use in any particular situation. The various transplant methods are not easily adapted to developmental studies but have provided some insight into the influence of innervation on cardiac function. Organ culture is useful when more precise control of the tissue's environment is necessary but when it is still desirable to preserve the three dimensional geometry of the heart. Adult cell dispersion, a relatively recent technique, is of benefit when experimental parameters are being measured that are not easily obtainable in intact tissue. Finally, cell culture provides the most complete control of experimental conditions, at the expense of a more technically complex and difficult preparation, considerably removed from the in vivo situation. For this reason, cell culture results cannot generally be directly compared to intact heart data. With cell culture, and in fact with all these preparations, one should compare two different in vitro conditions, as, for example, cardiac cultures grown in the presence and absence of nerve cells. In this way, the question of whether a particular developmental stimulus is capable of directly altering cardiac cellular characteristics can be unambiguously answered. Once such an effect is documented, the underlying mechanism is much more accessible for further elucidation using an in vitro preparation than with the intact animal. Of course, the results must always be related back to data from more intact systems to determine whether such an effect is of physiological significance.

REFERENCES

1. Harris, C.R.S. 1973. *The heart and the vascular system in ancient Greek medicine*. Oxford: Clarendon Press.
2. Paul, J. 1975. *Cell and tissue culture*. London: Churchill Livingstone.
3. Bertrand, M.E.; LaBlanche, J.M.; Tilmant, P.Y.; Ducloux, G.; Warembourg, H., Jr.; and Soots, G. 1981. Complete denervation of the heart (autotransplantation) for treatment of severe, refractory coronary spasm. *Am. J. Card.* 47:1375–1378.
4. Mason, J.W. 1980. Overdrive suppression in the transplanted heart: effect of the autonomic nervous system on human sinus node recovery. *Circulation* 62:688–696.
5. Romhilt, D.W.; Doyle, M.; Sagar, K.B.; Hastillo, A.; Wolfgang, T.C.; Lower, R.R.; and Hess, M.L. 1982. Prevalence and significance of arrhythmias in long-term survivors of cardiac transplantation. *Circulation* 66:1219–1222.
6. Lund, D.D.; Schmid, P.G.; Kelley, S.E.; Corry, R.J.; and Roskoski, R., Jr. 1978. Choline acetyltransferase activity in rat heart after transplantation. *Am. J. Physiol.* 235:H367–H371.
7. Peiss, C.N.; Cooper, T.; Willman, V.L.; and Randall, W.C. 1966. Circulatory responses to electrical and reflex activation of the nervous system after cardiac denervation. *Circ. Res.* 19:153–166.
8. Westfall, D.P., and Fleming, W.W. 1968. Sensitivity changes in the dog heart to norepinephrine, calcium and aminophylline resulting from pretreatment with reserpine. *J. Pharmacol. Exptl. Ther.* 159:98–106.

9. Burnstock, G.; Evans, B.; Gannon, B.J.; Heath, J.W.; and James, V. 1971. A new method of destroying adrenergic nerves in adult animals using guanethidine. *Br. J. Pharmacol.* 43:295–301.
10. Finch, L.; Haeusler, G.; and Thoenen, H. 1973. A comparison of the effects of chemical sympathectomy by 6-hydroxydopamine in newborn and adult rats. *Br. J. Pharmacol.* 47:249–260.
11. Johnson, E.M., Jr., and Macia, R.A. 1979. Unique resistance to guanethidine-induced chemical sympathectomy of spontaneously hypertensive rats: a resistance overcome by treatment with antibody to nerve growth factor. *Circ. Res.* 45:243–249.
12. Abbot, C.P.; DeWitt, C.W.; and Creech, O., Jr. 1965. The transplanted rat heart: histologic and electrocardiographic changes. *Transplantation* 3:432–445.
13. Corry, R.J., Kelley, S.E. 1974. Immunological enhancement of primarily vascularized rat heart xenotransplants in mice. *Transplantation* 18:503–507.
14. Zisbrod, Z.; Racelis D.; and Schanzer, H. 1981. Heterotopic heart transplantation in the rat. *Mt. Sinai J. Med.* 48:455–458.
15. Lee, S.; Macedo, A.R.; Curtis, G.P.; Lee, D.; and Orloff, M.J. 1982. A simplified model for heterotopic rat heart transplantation. *Transplantation* 33:438–442.
16. Lurie, K.G.; Billingham, M.E.; Masek, M.A.; Ginsburg, R.; Bristow, M.R.; Harrison, D.C.; and Reitz, B.A. 1982. Ultrastructural and functional studies on prolonged myocardial preservation in an experimental heart transplant model. *J. Thorac. Cardiovasc. Surg.* 84:122–129.
17. Phillips, M.I.; Hermsmeyer, K.; Landas, S.; Hadley, G.; and Gerndt, J. 1980. Culturing of non-brain cells in the brain ventricles. *Life Sci.* 27:2583–2588.
18. Gilmour, R.F. Jr., and Zipes, D.P. 1982. Electrophysiological response of vascularized hamster cardiac transplants to ischemia. *Circ. Res.* 50:599–609.
19. Olson, L., and Seiger, A. 1976. Beating intraocular hearts: light-controlled rate by autonomic innervation from host iris. *J. Neurobiology* 7:193–203.
20. Taylor, D.; Seiger, A.; Freedman, R.; Olson, L.; and Hoffer, B. 1978. Electrophysiological analysis of reinnervation of transplants in the anterior chamber of the eye by the autonomic ground plexus of the iris. *Proc. Natl. Acad. Sci.* 75:1009–1012.
21. Campbell, G.R.; Chamley-Campbell, J.; Short, N.; Robinson, R.B.; and Hermsmeyer, K. 1981. Effect of cross-transplantation on normotensive and spontaneously hypertensive rat arterial muscle membrane. *Hypertension* 3:534–543.
22. Abel, P.W., and Hermsmeyer, K. 1981. Sympathetic cross-innervation of SHR and genetic controls suggests a trophic influence on vascular muscle membranes. *Circ. Res.* 49:1311–1318.
23. Burrows, M.T. 1912. Rhythmische Kontraktionen der isolierter Herzmuskelzelle ausserhalb des Organismus. *Munch Med. Wochenschr* 59:1473–1475.
24. Wildenthal, K. 1971. Long term maintenance of spontaneously beating mouse hearts in organ culture. *J. Appl. Physiol.* 20:153–157.
25. Ingwall, J.S.; Roeske, W.R.; and Wildenthal, K. 1980. The fetal mouse heart in organ culture: maintenance of the differentiated state. *Methods in Cell Biology* 12A:167–186.
26. Wildenthal, K. 1972. Protein breakdown inhibited by insulin to improve heart culture. *Nature* 239:101–102.
27. Wildenthal, K.; Griffin, E.E.; and Ingwall, J.S. 1976. Hormonal control of cardiac protein and amino acid balance. *Circ. Res.* 38:1138–1144.
28. Ingwall, J.S., and Wildenthal, K. 1978. Fetal mouse hearts in organ culture: studies in cardiac metabolism. In *Recent advances in studies on cardiac structure and metabolism,* vol. 12, ed. T. Kobayashi, Y. Ito, and G. Rona, pp. 621–633. Baltimore: University Park Press.
29. Sperelakis, N., and Shigenobu, K. 1974. Organ-cultured chick embryonic hearts of various ages: Part I, Electrophysiology. *J. Mol. Cell. Cardiol.* 6:449–471.
30. Shigenobu, K., and Sperelakis, N. 1974. Failure of development of fast Na^+ channels during organ culture of young embryonic chick hearts. *Dev. Biol.* 39:326–330.
31. DeHaan, R.L.; McDonald, T.F.; and Sachs, H.G. 1976. Development of tetrodotoxin sensitivity of embryonic chick heart cells *in vitro*. In *Developmental and Physiological Correlates of Cardiac Muscle,* ed. M. Lieberman, and T. Sana, pp. 155–168. New York: Raven Press.
32. Wildenthal, K. 1973. Studies of foetal mouse hearts in organ culture: metabolic requirements for prolonged function *in nitro* and the influence of cardiac maturation on substrate utilization. *J. Mol. Cell. Cardiol.* 5:87–99.

33. Wildenthal, K. 1973. Maturation of responsiveness to cardioactive drugs: differential effects of acetylcholine, norepinephrine, theophylline, tyramine, glucagen, and dibutyryl cyclic AMP on atrial rate in hearts of fetal mice. *J. Clin. Invest.* 52:2250–2258.
34. Wildenthal, K.; Allen, D.O.; Karlsson, J.; Wakeland, J.R.; and Clark, C.M., Jr. 1976. Responsiveness to glucagon in fetal hearts: species variability and apparent disparities between changes in beating, adenylate cyclase activation, and cyclic AMP concentration. *J. Clin. Invest.* 57:551–558.
35. Chen, F.M.; Yamamura, H.I.; and Roeske, W.R. 1979. Ontogeny of mammalian myocardial beta-adrenergic receptors. *Eur. J. Pharm.* 58:255–264.
36. Wildenthal, K. 1971. Studies of fetal mouse hearts in organ culture: influence of prolonged exposure to triiodothyronine on cardiac responsiveness to isoproterenol, glucagon, theophylline, acetylcholine and dibutyryl cyclic 3′, 5′-adenosine monophosphate. *J. Pharmacol. Exptl. Ther.* 190:272–299.
37. Powell, T.; Terrar, D.A.; and Twist, V.W. 1980. Electrical properties of individual cells isolated from adult rat ventricular myocardium. *J. Physiol.* 302:131–153.
38. Dani, A.M.; Cittadini, A.; Flamini, G.; Festuccia, G.; and Terranova, T. 1977. Preparation and some properties of isolated beating myocytes from adult rabbit heart. *J. Mol. Cell. Cardiol.* 9:777–784.
39. Matsuda, H.; Noma, A.; Kurachi, Y.; and Irisawa, H. 1982. Transient depolarization and spontaneous voltage fluctuations in isolated single cells from guinea pig ventricles. *Circ. Res.* 51:142–151.
40. Belardinelli, L., and Isenberg, G. 1983. Actions of adenosine and isoproterenol on isolated mammalian ventricular myocytes. *Circ. Res.* 53:287–297.
41. Isenberg, G. 1982. Ca entry and contraction as studied in isolated bovine ventricular myocytes. *Z. Naturforsch* 37:502–512.
42. Silver, L.H.; Hemwall, E.L.; Marino, T.A.; and Houser, S.R. 1983. Isolation and morphology of calcium-tolerant feline ventricular cells. *Am. J. Physiol.* 245:H891–H896.
43. Spanier, A.M., and Weglicki, W.B. 1982. Ca^{2+}-tolerant adult canine myocytes: preparation and response to anoxia/acidosis. *Am. J. Physiol.* 243:H448–H455.
44. Hewett, K.; Legato, M.J.; Danilo, P., Jr.; and Robinson, R.B. 1983. Isolated myocytes from the adult canine left ventricle: calcium tolerance electrophysiology and ultrastructure. *Am. J. Physiol.* 245:H830–H839.
45. Dow, J.W.; Harding, N.G.; and Powell, T. 1981. Isolated cardiac myocytes: I. Preparation of adult myocytes and their homology with the intact tissue. *Cardiovas. Res.* 15:483–514.
46. Dow, J.W.; Harding, N.G.; and Powell, T. 1981. Isolated cardiac myocytes: II. Functional aspects of mature cells. *Cardiovas. Res.* 15:549–579.
47. Bernard, C., and Gargouil, Y.M. 1970. Acquisitions successives, chez l'embryon de Rat, des permeabilities specifiques de la membrane myocardique. *C.R. Acad. Sc. Paris, Ser. D* 270:1495–1498.
48. Cavoto, F.V., and Kelliher, G. 1973. Age-induced changes in the transmembrane action potential of cardiac muscle. *Fed. Procs.* 32:811.
49. Langer, G.A.; Brady, A.J.; Tan, S.T.; and Serena, S.A. 1975. Correlation of the glycoside response, the force staircase and the action potential configuration in the neonatal rat heart. *Circ. Res.* 36:744–752.
50. Danilo, P.; Reder, R.; Mill, J.; and Petrie, R. 1979. Developmental changes in cellular electrophysiologic characteristics and catecholamine content of fetal hearts. *Circulation* 60:41–50.
51. Nehr, E., and Sakmann, B. 1976. Single channel currents recorded from membrane of denervated frog muscle fibers. *Nature* 260:799–802.
52. Hamill, O.P.; Marty, A.; Neher, E.; Sakmann, B.; and Sigworth, F.J. 1981. Improved patch-clamp techniques for high-resolution current recording from cells and cell-free membrane patches. *Pflugers Arch.* 391:85–100.
53. Reuter, H.; Stevens, C.F.; Tsien, R.W.; and Yellen, G. 1982. Properties of single calcium channels in cardiac cell culture. *Nature* 297:501–504.
54. Lee, K.S.; Weeks, T.A.; Kao, R.L.; Akaike, N.; and Brown, A.M. 1979. Sodium current in single heart muscle cells. *Nature* 278:269–271.
55. Lau, Y.H.; Robinson, R.B.; Rosen, M.R.; and Bilezikian, J.P. 1980. Subclassification of beta-adrenergic receptors in cultured rat cardiac myoblasts and fibroblasts. *Circ. Res.* 47:41–48.

56. Taniguchi, J.; Kokubun, S.; Noma, A.; and Irisawa, H. 1981. Spontaneously active cells isolated from the sino-atrial and atrio-ventricular nodes of the rabbit heart. *Jap. J. Physiol.* 31:547–558.
57. Masson-Pevet, M.; Jongsma, H.J.; Bleeker, W.K.; Tsjernina, L.; van Ginneken, A.C.G.; Treijtel, B.W.; and Bouman, L.N. 1982. Intact isolated sinus node cells from the adult rabbit heart. *J. Mol. Cell. Cardiol.* 14:295–299.
58. Callewaert, G.; Carmeliet, E.; Van der Heyden, G.; and Vereecke, SJ. 1982. The pacemaker current in a single cell preparation of bovine cardiac Purkinje fibers. *J. Physiol.* 326:66P.
59. Sheets, M.F.; January, C.T.; and Fozzard, H.A. 1981. Method for isolation of single canine Purkinje cells. *Fed. Procs.* 41:1504.
60. Robinson, R.B., and Hewett, K. 1984. Electrophysiological characteristics of isolated canine Purkinje cells. *Biophys. J.* 45:54a.
61. Lampidis, T.J., and Schaiberger, G.E. 1975. Age-related loss of DNA repair synthesis in isolated rat myocardial cells. *Exptl. Cell. Res.* 96:412–416.
62. Jacobson, S.L. 1977. Culture of spontaneously contracting myocardial cells from adult rats. *Cell Structure and Function* 2:1–9.
63. Claycomb, W.C., and Palazzo, M.C. 1980. Culture of the terminally differentiated adult cardiac muscle cell: a light and scanning electron microscope study. *Dev. Bio.* 80:466–482.
64. Nag, A.C., and Cheng, M. 1981. Adult mammalian cardiac muscle cells in culture. *Tissue and Cell* 13:515–523.
65. Piper, H.M.; Probst, I.; Schwartz, P.; Hutter, F.J.; and Spieckermann, P.G. 1982. Culturing of calcium stable adult cardiac myocytes. J. Mol. and Cell. Cardiol. 14:397–412.
66. Girardi, A.J.; Warren, J.; Goldman, C.; and Jeffries, B. 1958. Growth and CF antigenicity of measles virus in cells deriving from human heart. *Procs. Soc. Exptl. Biol. Med.* 98:18–22.
67. Lamb, J.F., and McCall, D. 1972. Effect of prolonged ouabain treatment on Na, K, Cl, and Ca concentration and fluxes in cultured human cells. *J. Physiol.* 225:599–617.
68. Lavappa, K.S. 1978. Survey of ATCC stocks of human cell lines for HeLa contamination. *In Vitro* 14:469–475.
69. Clark, H.F., and Karzon, D.T. 1967. Terrapene heart (TH-1), a continuous cell line from the heart of the box turtle Terrapene caroline. *Exptl. Cell. Res.* 48:263–268.
70. Schlage, W.K., and Bereiter-Hahn, J. 1983. A microscopic perfusion respirometer for continuous respiration measurement of cultured cells during microscopic observation. *Microsc. Acta.* 87:19–34.
71. McBurney, M.W.; Jones-Villeneuve, E.M.V.; Edwards, M.K.S.; and Anderson, P.J. 1982. Control of muscle and neuronal differentiation in a cultured embryonal carcinoma cell line. *Nature* 299:165–167.
72. Burt, J.N., and Langer, G.A. 1982. Ca^{++} distribution after Na^{+} pump inhibition in cultured neonatal rat myocardial cells. *Circ. Res.* 51:543–550.
73. Horres, C.R., and Lieberman, M. 1977. Compartmental analysis of potassium efflux from growth-oriented heart cells. *J. Mem. Bio.* 34:331–350.
74. Lieberman, M.; Sawanobori, T.; Kootsey, J.H.; and Johnson, E.A. 1975. A synthetic strand of cardiac muscle: its passive electrical properties. *J. Gen. Physiol.* 65:527–550.
75. Sachs, F. 1976. Electrophysiological properties of tissue cultured heart cells grown in a linear array. *J. Memb. Biol.* 28:373–399.
76. Moscona, A. 1961. Rotation-mediated histogenic aggregation of dissociated cells. *Exp. Cell Res.* 22:455–475.
77. Nathan, R.D., and DeHaan, R.L. 1979. Voltage clamp analysis of embryonic heart cell aggregates. *J. Gen. Physiol.* 73:175–198.
78. Grobstein, C. 1959. Differentiation of vertebrate cells. In *The cell,* ed. J. Brachet and A. E. Mirsky, pp. 437–496. New York: Academic Press.
79. Romanoff, A.L. 1960. *The avian embryo: structural and functional development.* New York: MacMillan Co.
80. Chamley-Campbell, J.; Campbell, G.R.; and Ross, R. 1979. The smooth muscle cell in culture. *Physiol. Rev.* 59:1–61.
81. Hogg, B.M.; Goss, C.; and Cole, K.S. 1934. Potentials in embryo rat heart muscle cultures. *Proc. Soc. Exptl. Biol. Med.* 32:304–308.
82. Cavanaugh, M.W. 1955. Pulsation, migration and division in dissociated chick embryo heart cells *in vitro*. *J. Exptl. Zool.* 128:573–585.

83. Harary, I., and Farley, B. 1960. *In vitro* studies of single beating heart cells. *Science* 131:1674–1675.
84. Lane, M.L.; Sastre, A.; and Salpeter, M.M. 1976. Innervation of heart cells in culture by an endogenous source of cholinergic neurons. *Proc. Natl. Acad. Sci.* 73:4506–4510.
85. Fioretti, A.; Ghione, M.; Pozzoli, E.; and Lambertenghi-Deliliers, G. 1976. Autoantibodies and spleen cell-mediated cytotoxicity in adriamycin-induced myocardiopathy in rabbits. *Cancer Res.* 36:1462–1469.
86. Thompson, A., and Halbert, S.P. 1971. The cardiac auto-immune system: III. Studies on the cytotoxicity of heart antibodies for pulsating rabbit and rat heart cells in tissue culture. *Int. Arch. Allergy Applied Immun.* 40:274–286.
87. Auclair, M.C. 1976. Activite electrique des cells en culture de coeur de cobaye, comparison avec le rat. *CR des Seances de la Soc. Biol. et de Fil.* 170:543–548.
88. Bonventre, P.F., and Imhoff, J.G. 1967. Studies on the mode of action of diptheria toxin: II. Protein synthesis in primary heart cell cultures. *J. Exptl. Med.* 126:1079–1086.
89. Kaumann, A.J., and Wittmann, R. 1975. Apparent equilibrium constant between beta-adrenoceptors and a competitive antagonist on a cultured pacemaker cell of mammalian heart. *Naunyn-Schmied Arch. Exptl. Path. Pharmacol.* 287:23–32.
90. Andreasen, T.; Castles, J.J.; Saito, W.Y.; Chacko, K.; Fenner, C.; Mason, D.T.; and Wikman-Coffelt, J. 1975. The behavior of fetal canine cells in culture: synthesis and phosphorylation of myosin. *Dev. Biol.* 47:366–375.
91. Robinson, R.B.; Reder, R.F.; Danilo, P., Jr. In press. Preparation and electrophysiological characterization of cardiac cell cultures derived from the fetal canine ventricle. *Dev. Pharmacol. Ther.*
92. Bernik, M.D., and Kwaan, H.C. 1971. Inhibitors of fibrinolysis in human tissues in culture. *Am. J. Physiol.* 221:916–921.
93. Halbert, S.P.; Bruderer, R.; and Thompson, A. 1973. Growth of dissociated beating human heart cells in tissue culture. *Life Sci.* 36:969–975.
94. Norwood, C.R.; Castaneda, A.R.; and Norwood, W.J. 1980. Heterogeneity of rat cardiac cells of defined origin in single cell culture. *J. Mol. Cell. Cardiol.* 12:201–210.
95. Saetersdal, T.; Odegarden, S.; Rotevatn, S.; and Engedal, H. 1980. Atrial specific granules of the human auricle in embryogenesis, tissue culture, and hypertrophy. *Cell. Tissue Res.* 209:345–351.
96. Canale, E.; Campbell, J.H.; and Campbell, G.R. 1983. Cardiac Purkinje cells in culture. *J. Mol. Cell. Cardiol.* 15:197–206.
97. Kuramitsu, H., and Harary, I. 1964. Studies *in vitro* on single beating rat heart cells. III. Enzyme changes and loss of specific function in culture. *Bioch. Bioph. Acta.* 86:65–73.
98. Schanne, O.F. 1971. Factors involved in the loss of spontaneous contractile and electrical activity in clusters of cultured cardiac cells. *Can. J. Physiol. Pharmacol.* 50:523–532.
99. Mark, G.E., and Strasser, F.F. 1966. Pacemaker activity and mitosis in cultures of newborn rat heart ventricle cells. *Exptl. Cell. Res.* 44:217–233.
100. Goshima, K. 1974. Initiation of beating in quiescent myocardial cells by norepinephrine, by contact with beating cells and by electrical stimulation of adjacent FL cells. *Exptl. Cell. Res.* 84:223–234.
101. Polinger, I.S. 1973. Growth and DNA synthesis in embryonic chick heart cells *in vivo* and *in vitro*. *Exptl. Cell. Res.* 76:253–262.
102. Chacko, S. 1973. DNA synthesis, mitosis and differentiation in cardiac myogenesis. *Dev. Biol.* 35:1–18.
103. Bogenmann, E., and Eppenberger, H.M. 1980. DNA synthesis and polyploidization of chicken heart muscle cells in mass cultures. *J. Mol. Cell. Cardiol.* 12:17–27.
104. Leblond, C.P.; Messier, B.; and Kopriva, B. 1959. Thymidine-H^3 as a tool for the investigation of the renewal of cell populations. *Lab. Invest.* 8:296–308.
105. Petersen, R.O., and Baserga, R. 1965. Nucleic acid and protein synthesis in cardiac muscle of growing and adult mice. *Exptl. Cell. Res.* 40:340–352.
106. Frelin, C.; Pinson, A.; Athias, P.; Surville, J.M.; and Padieu, P. 1979. Glucose and palmitate metabolism of beating rat heart cells in culture. *Path. Biol.* 27:45–50.
107. Schroedl, N.A.; Hartzell, C.R.; Ross, P.D.; and McCarl, R.L. 1982. Glucose metabolism, insulin effects, and developmental age of cultured neonatal rat heart cells. *J. Cell. Physiol.* 113:231–239.

108. Hoerter, J.A., and Opie, L.H. 1978. Perinatal changes in glycolytic function in response to hypoxia in the incubated or perfused rat heart. *Biol. Neonate* 33:144–161.
109. Cahn, R.D. 1967. Changes in enzyme patterns in functional heart, cartilage and pigmented retina cells in culture. In *Factors influencing myocardial contractility,* ed. R.D. Tanz, F. Kavaler, and J. Roberts, pp. 293–299. New York: Academic Press.
110. Karsten, U.; Kossler, A.; Janiszewski, E.; and Wollenberger, A. 1973. Influence of variations in pericellular oxygen tension on individual cell growth, muscle-characteristic proteins, and lactate dehydrogenase isoenzyme pattern in cultures of beating rat heart cells. *In Vitro* 9:139–146.
111. Styka, P.E., and Penney, D.G. 1977. The perinatal rat: body weight, hematocrit, and regional changes in heart weight, and lactate dehydrogenase isozyme composition and activity. *Growth* 41:325–336.
112. Van der Laarse, A.; Hollaar, L.; Kokshoorn, L.J.M.; and Witteveen, S.A.G.J. 1979. The activity of cardio-specific isoenzymes of creatine phosphokinase and lactate dehydrogenase in monolayer cultures of neonatal rat heart cells. *J. Mol. Cell. Cardiol.* 11:501–510.
113. Fujimoto, A., and Harary, I. 1965. The effect of lipids on enzyme levels in beating rat heart cells. *Bioch. Bioph. Res. Comm.* 20:456–461.
114. Freerskin, D.L.; Schroedl, N.A.; and Hartzell, C.R. 1983. Triiodothyronine depresses the NAD-linked glycerol-3-phosphate dehydrogenase activity of cultured neonatal rat heart cells. *Fed. Procs.* 42:325.
115. Sperelakis, N., and Lehmkuhl, D. 1964. Insensitivity of cultured chick heart cells to autonomic agents and tetrodotoxin. *Am. J. Physiol.* 209:693–698.
116. Pappano, A.J. 1977. Ontogenetic development of autonomic neuroeffector transmission and transmitter reactivity in embryonic and fetal hearts. *Pharmacol. Rev.* 29:3–33.
117. Hermsmeyer, K., and Robinson, R.B. 1977. High sensitivity of cultured cardiac muscle cells to autonomic agents. *Am. J. Physiol.* 233:C172–C179.
118. Karsten, U.; Wallukat, G.; and Wollenberger, A. 1977. Alpha- and beta-adrenergic receptor stimulation in cultures of beating rat heart cells and its effect on automaticity and cyclic AMP levels. *Acta. Biol. Med. Germ.* 36:k63–k69.
119. Lane, M.-A.; Sastre, A.; Law, M.; and Salpeter, M.M. 1977. Cholinergic and adrenergic receptors on mouse cardiocytes *in vitro*. *Dev. Biol.* 57:254–269.
120. Marvin, W.J., Jr.; Robinson, R.B.; and Hermsmeyer, K. 1979. Correlation of function and morphology of neonatal rat and embryonic chick cultured cardiac and vascular muscle cells. *Circ. Res.* 45:528–540.
121. Kupfer, L.E.; Robinson, R.B.; and Bilezikian, J.P. 1982. Identification of alpha$_1$-adrenergic receptors in cultured rat myocardial cells with a new iodinated alpha$_1$-adrenergic antagonist, [^{125}I]IBE 2254. *Circ. Res.* 51:250–254.
122. Hermsmeyer, K.; Mason, R.; Griffen, S.H.; and Becker, P. 1982. Rat cardiac muscle single cell automaticity responses to alpha- and beta-adrenergic agonists and antagonists. *Circ. Res.* 51:532–537.
123. Kono, T. 1969. Destruction and restoration of the insulin effector system of isolated fat cells. *J. Biol. Chem.* 244:5777–5784.
124. Rodbell, M.; Birnbaumer, L.; and Pohl, S.L. 1970. Adenyl cyclase in fat cells. III. Stimulation by secretin and the effects of trypsin on the receptors for lipolytic hormones. *J. Biol. Chem.* 245:718–722.
125. Shigenobu, K., and Sperelakis, N. 1971. Development of sensitivity to tetrodotoxin of chick embryonic hearts with age. *J. Mol. Cell. Cardiol.* 3:271–286.
126. Sperelakis, N.; Shigenobu, K.; McLean, M.J. 1976. Membrane cation channels: changes in developing hearts, in cell culture and in organ culture. In *Developmental and physiological correlates of cardiac muscle,* ed. M. Lieberman and T. Sano, pp. 209–234. New York: Raven Press.
127. Sachs, J.G.; McDonald, T.F.; and DeHaan, R.E. 1973. Tetrodotoxin sensitivity of cultured embryonic heart cells depends on cell interactions. *J. Cell. Biol.* 56:255–258.
128. McLean, M.J., and Sperelakis, N. 1976. Retention of fully differentiated electrophysiological properties of chick embryonic cells in culture. *Dev. Biol.* 50:134–141.
129. Galper, J.B., and Catterall, W.A. 1978. Development changes in the sensitivity of embryonic heart cells to tetrodotoxin and D600. *Dev. Biol.* 65:216–227.
130. Lompre, A.M.; Pogglioli, J.; and Vassort, G. 1979. Maintenance of fast Na^+ channels during primary cultures of embryonic chick heart cells. *J. Mol. Cell. Cardiol.* 11:813–825.

131. McLean, M.J.; Renaud, J.-F.; Sperelakis, N.; and Niu, M.C. 1976. Messenger RNA induction of fast sodium ion channels in cultured cardiac myoblasts. *Science* 191:297–299.
132. Renaud, J.F.; Scanu, A.M.; Kazazoglou, T.; Lombet, A.; Romey, G.; and Lazdunski M. 1982. Normal serum and lipoprotein-deficient serum give different expressions of excitability, corresponding to different stages of differentiation, in chick cardiac cells in culture. *Proc. Natl. Acad. Sci.* 79:7768–7772.
133. Robinson, R.B., and Legato, M.J. 1980. Maintained differentiation in rat cardiac monolayer cultures: tetrodotoxin sensitivity and ultrastructure. *J. Mol. Cell. Cardiol.* 12:493–498.
134. Robinson, R.B. 1982. Action potential charcteristics of rat cardiac cells do not change with time in culture. *J. Mol. Cell. Cardiol.* 14:367–370.
135. Athias, P.; Prelin, C.; Groz, B.; Dumas, J.P.; Klepping, J.; and Padieu, P. 1979. Myocardial electrophysiology: intracellular studies on heart cell cultures from newborn rats. *Pathologie Biologis (Paris)* 27:13–19.
136. Legato, M.J. 1972. Ultrastructural characteristics of the rat ventricular cell grown in tissue culture with special reference to sarcomerogenesis. *J. Mol. Cell. Cardiol.* 4:299–317.
137. Perissel, B.; Charbonne, F.; Moalic, J.M.; and Malet, P. 1980. Initial stages of trypsinized cell culture of cardiac myoblasts: ultrastructural data. *J. Mol. Cell. Cardiol.* 12:63–75.
138. Manasek, F.J. 1976. The extracellular matrix of the early embryonic heart. In *Developmental and physiological correlates of cardiac muscle,* ed. M. Lieberman and T. Sano, pp. 1–20. New York: Raven Press.
139. Langer, G.A.; Frank, J.S.; Nudd, L.M.; and Seraydarian, K. 1976. Sialic acid: effect of removal on calcium exchangeability of cultured heart cells. *Science* 193:1013–1015.
140. Nathan, R.D.; Fund, S.J.; Stocco, D.M.; Barron, E.A.; and Markwald, R.R. 1980. Sialic acid: regulation of electrogenesis in cultured heart cells. *Am. J. Physiol.* 239:C197–C207.
141. Barron, E.A.; Markwald, R.R.; and Nathan, R.D. 1982. Localization of sialic acid at the surface of embryonic myocardial cells. *J. Mol. Cell. Cardiol.* 14:381–395.
142. Lau, C., and Slotkin, T.A. 1979. Accelerated development of rat sympathetic neurotransmission caused by neonatal triiodothyronine administration. *J. Pharmacol. Exptl. Ther.* 208:485–490.
143. Evans, T.C.; Kretzchmer, R.M.; Hodges, R.E.; and Song, C.W. 1967. Radioiodine uptake studies of the human fetal thyroid. *J. Nucl. Med.* 8:157–165.
144. Walker, P.; Dubois, J.D.; and Dussault, J.H. 1980. Free thyroid hormone concentrations during postnatal development in the rat. *Ped. Res.* 14:247–249.
145. Brewster, W.R., Jr.; Issacs, J.T.; Osgood, P.F.; and King, T.J. 1956. The hemodynamic and metabolic interrelationships in the activity of epinephrine, norepinephrine, and the thyroid hormones. *Circulation* 13:1–20.
146. Cairoli, V.J., and Crout, J.R. 1967. Role of the autonomic nervous system in the resting tachycardia of experimental hyperthyroidism. *J. Pharmacol. Exptl. Ther.* 158:55–65.
147. Levey, G.S.; Skelton, C.L.; and Epstein, S.E. 1969. Influence of hyperthyroidism on the effects of norepinephrine on myocardial adenyl cyclase activity and contractile state. *Endocrinology* 85:1004–1009.
148. Williams, L.T.; Lefkowitz, R.J.; Watanabe, A.M.; Hathaway, D.R.; and Besch, J.R., Jr. 1977. Thyroid hormone regulation of beta-adrenergic receptor number. *J. Biol. Chem.* 252:2787–2789.
149. Ciaraldi, T.P., and Marinetti, G.V. 1977. Hormone action at the membrane level: VIII. Adrenergic receptors in rat heart and adipocytes and their modulation by thyroxine. *Bioch. Bioph. Acta.* 541:334–346.
150. Chang, H.Y. and Kunos G. 1981. Short-term effects of tri-iodothyronine on rat heart adrenoceptors. *Biochem. Biophys. Res. Comun.* 100:313–320.
151. McConnaughey, M.M.; Jones, L.R.; Watanabe, A.M.; Besch, H.R., Jr.; Williams, L.T.; and Lefkowitz, R.J. 1979. Thyroxine and propylthiouracil effects on alpha- and beta-adrenergic receptor number, ATPase activities and sialic acid content of rat cardiac membrane vessicles. *J. Cardiovasc. Pharm.* 1:609–623.
152. Shanks, R.G.; Hadden, D.R.; Lowe, D.C.; McDevitt, D.G.; and Montgomery, D.A.D. 1969. Controlled trial of propranolol in thyrotoxicosis. *Lancet* No. 7603:993–998.
153. Sterling, K., and Hoffenberg, R. 1971. Beta blocking agents and antithyroid drugs as adjuncts to radioiodine therapy. *Semin. Nucl. Med.* 1:422–431.
154. McDevitt, D.G.; Shanks, R.G.; Hadden, D.R.; Montgomery, D.A.D.; and Weaver, J.A.

1968. The role of the thyroid in the control of heart rate. *Lancet* No. 7550:998–1000.
155. Buccino, R.A.; Spann, J.F. Jr.; Pool, P.E.; Sonneblick, E.H.; and Braunwald, E. 1967. Influence of thyroid state on the intrinsic contractile properties and energy stores of the myocardium. *J. Clin. Invest.* 46:1669–1682.
156. Prange, A.J. Jr.; Meek, J.L.; and Lipton, M.A. 1970. Catecholamines; diminished rate of synthesis in rat brain and heart after thyroxine pretreatment. *Life Sci.* 9:901.
157. Coulombe, P.; Dussault, J.H.; and Walker, P. 1976. Plasma catecholamine concentrations in hyperthyroidism and hypothyroidism. *Metabolism* 25:973–979.
158. Coulombe, P.; Dussault, J.H.; Letarte, J.; and Simard, S.J. 1976. Catecholamine metabolism in thyroid disease: I. Epinephrine secretion rate in hyperthyroidism and hypothyroidism. *J. Clin. Endocrinol. Metab.* 42:125–131.
159. Wollenberger, A. 1965. Action of triiodothyronine on the rhythmicity of single isolated heart cells cultured *in vitro*. In *Current topics in thyroid research,* ed. C. Cassano and M. Andreoli, pp. 377–381. New York: Academic Press.
160. Tsai, J.S., and Chen, A. 1978. Effect of L-triiodothyronine on (-) [^{3}H]-dihydroalprenolol binding and cyclic AMP response to (-) adrenaline in cultured heart cells. *Nature* 275:138–140.
161. Kupfer, L.E., Robinson, R.B. and Bilizekian, J.P. 1984. The regulation by l-triiodothyronine of α and β adrenergic receptors in cultured rat heart cells. *Fed. Proc.* 43:871.
162. Furshpan, E.J.; MacLeisch, P.R.; O'Lague, P.H.; and Potter, D.D. 1976. Chemical transmission between rat sympathetic neurons and cardiac myocytes developing in microcultures: evidence for cholinergic, adrenergic and dual-function neurons. *Proc. Natl. Acad. Sci.* 73:4225–4229.
163. King, K.L.; Boder, G.N.; Williams, D.C.; and Harley, R.J. 1978. Chronotropic effect of tyramine on rat heart cells cultured with sympathetic neurons. *Eur. J. Pharm.* 51:331–335.
164. Marvin, W.J., Jr.; Chittick, V.L.; and Hermsmeyer, R.K. 1980. Desensitization of developing mammalian myocardium during adrenergic innervation. *Circulation* 62:III–24.
165. Marvin, W.J., Jr.; Atkins, D.L.; Chittick, V.L.; and Hermsmeyer, R.K. 1981. Densensitization of developing mammalian myocardium during cholinergic innervation. *Circulation* 64:IV–287.

6. CALCIUM EXCHANGE IN THE DEVELOPING MYOCARDIUM

GLENN A. LANGER AND JAY M. JARMAKANI

CHARACTERISTICS OF CALCIUM EXCHANGE IN CULTURED MYOCARDIAL CELLS

It might be expected that cells derived from hearts at the embryonic or neonatal stage of development would have quite different characteristics when compared to cells in the fully developed adult organ. This possibility is, however, not a reason to abandon study of cells in culture especially since, with appropriate techniques, physiological characteristics can be better defined and quantified than in the whole adult organ. It will be an advantage to have characteristics of cultured embryonic and neonatal cells in hand as the relatively new model of the enzymatically dissected adult myocyte develops and is exploited. The cultured cell model has been particularly useful in analysis of exchange and compartmentation of calcium ions (Ca^{2+}) in the myocardium, and that will be the focus of this section. When the data exist, results from culture will be compared with those of the adult myocardium.

It is appropriate to discuss Ca exchange from two aspects: (1) characteristics of Ca binding; and (2) transsarcolemmal movement and distribution.

Calcium Binding

The development of the scintillation disk-flow cell technique for on-line measurement of ^{45}Ca exchange in cultured cells from neonatal rat heart [1, 2] and the iso-

Supported by USPHS grants HL28539-01 and HL 24148, American Heart Association, Greater Los Angeles Affiliate grants 744G1-2 and 506 IG9, and the Castera Foundation.

M.J. Legato (ed.) THE DEVELOPING HEART, pp. 95–111.

lation of sarcolemmal membrane by "gas dissection" [3] has permitted kinetic identification and characterization of sarcolemmal-bound Ca^{2+} in the cultured preparation.

When the cultured cells from neonatal rat heart are perfused with HEPES-buffered solution, all of the exchangeable Ca^{2+} is rapidly exchangeable [2]. The Ca^{2+} exchanges as rapidly as the perfusate can be exchanged in the flow cell system and is, therefore, perfusion-limited with a $t^1\!/_2 = 10$ to 20 seconds. This rapidly exchangeable fraction represents about 42% of the total cellular Ca^{2+} with the remainder being very slowly exchangeable — i.e., little evidence of exchange over the course of one to two hours. Barry and Smith [4] also found a rapidly exchangeable ($t^1\!/_2$ about 10 secs) component in cultured chick embryo cells. They also described a slower compartment but with a much faster exchange ($t^1\!/_2$ about 10 min) rate than in the rat cells. The composition of the perfusate and the species were, however, different, and these factors could account for the difference in the slower compartment (see below). Sedlis et al. [5] found, in agreement with our studies [2], in cells derived from neonatal rat, no difference in the amount of ^{45}Ca exchanged within two minutes as compared to six hours.

The locus of the rapidly exchangeable Ca^{2+} was investigated using the lanthanum (La^{3+}) ion as a probe in both whole cells [6] and in the sarcolemma from these cells prepared by gas dissection [7]. La^{3+} is a potent displacer of Ca^{2+} and, in intact cells, does not penetrate intracellularly. It binds and is limited to the sarcolemma-glycocalyx complex at the cell surface. La^{3+} applied to whole cells in one series displaced 3.3 mmoles Ca^{2+}/kg dry wt cell. Ca^{2+} displacement by La^{3+} from sarcolemmal membranes was 168 nmoles/mg membrane protein. There is 19.1 mg membrane protein/g dry wt cells and therefore sarcolemmal-bound Ca^{2+} accounts for 3.2 mmoles Ca^{2+}/kg dry wt of original cells or virtually all of the La-displaceable Ca^{2+}. Of the rapidly exchangeable fraction of Ca^{2+}, the La displaceable component represents approximately 80% [6]. These experiments then provide conclusive evidence that at least 80% of the rapidly exchangeable fraction of cellular Ca^{2+} is bound to the sarcolemmal glycocalyx complex. This represents about 0.6 mmoles Ca^{2+}/kg wet wt cells.

Application of La^{3+} produces immediate uncoupling of excitation from contraction in cultured cells as well as in adult cardiac tissue. Divalent cations demonstrate a range of uncoupling potency proportional to their ability to displace Ca^{2+} (7, 8) from sarcolemmal binding sites in the order of $Cd^{2+} > Mn^{2+} \sim Co^{2+} > Mg^{2+}$. This is in descending order of the crystal ionic radii (0.97, 0.80, 0.72, 0.65 A) and increasingly removed from the crystal radius of Ca^{2+} at 0.99 A. Therefore, within the sequence of cations of equal charge, the ability to uncouple excitation from contraction and displace Ca from the sarcolemma is dependent upon the size of the non-hydrated ion relative to Ca^{2+}.

The studies presented above indicate that displacement of Ca^{2+} bound to sarcolemmal sites prevents contraction. Is there a change in the amount of bound Ca secondary to interventions that usually produce a positive inotropic state? At present three interventions are associated with increased amounts of sarcolemmal-bound Ca^{2+} in cultured cells:

1. Increased Stimulation Rate — Increase in beating rate from 60 to 145/min increases exchangeable Ca^{2+} by 65%. All of this additional Ca^{2+} was localized to the rapidly exchangeable cellular component of which 90% was La-displaceable [2].

2. Ouabain Application — Inhibition of the Na-pump with secondary stimulation of $Na^+ - Ca^{2+}$ exchange increased rapidly exchangeable Ca by 11%, nearly 80% of which was La^{3+}-displaceable [6]. Depression of the pump with low potassium perfusion produced the same result. Therefore, increased $Na^+ - Ca^{2+}$ exchange resulted in increased sarcolemmal Ca^{2+} binding.

3. Low $[Na]_o$ Perfusion — A decrease of extracellular $[Na]_o$ of 100 mM in HEPES perfusate increases Ca^{2+} binding to sarcolemmal sites by more than 100% [9]. This component of Ca^{2+} can be clearly identified apart from intracellular Ca^{2+}.

All of the preceding positively inotropic interventions are, then, associated with increased sarcolemmal Ca^{2+} binding. This provides additional evidence for the importance of this component of cellular Ca^{2+} in the support and control of contractile force.

The molecular sites to which the Ca^{2+} is bound are of interest since the nature of the binding should provide further insight into the mechanism of Ca^{2+} movement from the binding sites across the sarcolemma. As indicated previously, La^{3+} displaces Ca^{2+} from sarcolemma prepared from cultured cells by gas dissection. La^{3+} is non-specific in that it will displace Ca^{2+} from any charged site on the cellular surface — whether sugar, protein, or lipid. A recent study [10] in sarcolemma isolated from rabbit ventricle suggests, however, that more than 80% of sarcolemmal Ca^{2+} is bound to phospholipid. On the basis of evidence that the amphiphilic, cationic peptidolipid Polymyxin B (PXB) selectively binds to negatively charged membrane phospholipid [11], we used PXB as a probe for Ca^{2+}-binding sites in sarcolemma prepared by gas dissection from cultured cells [12]. Five of PXB's ten amino acids are positively charged 2,4 diaminobutyric acid groups. It is proposed that the nonpolar residues of PXB penetrate the hydrophobic domain of the membrane with the polar groups situated at the membrane/aqueous interface, where they interact with negatively charged phospholipid and displace Ca^{2+}. In the recent study [12] total Ca^{2+} displaced from the gas-dissected membranes by La^{3+} at maximal displacing concentration (1 mM) was 172 nmoles/mg membrane protein. At 0.1 mM, PXB displaced 42% of the total displaced by La^{3+}. Pretreatment of the membranes with neuraminidase and protease to remove an anionic sugar (sialic acid) and possible protein-binding sites lead to a decrease in La^{3+}-displaceable Ca^{2+} but to an *increase* in the fraction displaced by 0.1 mM PXB from 42% to 54%. Membranes were also treated with phospholipase D (PLD) before application of PXB. PLD cleaves phospholipid at the phosphodiester linkage at C_3 and converts neutral (or negatively charged) phospholipids to phosphatidic acid.

Therefore it would be expected to increase the proportion of anionic phospholipid and the amount of Ca^{2+} bound to the sarcolemma. This was found to be the case. PLD pretreatment significantly increased the La^{3+} displaceable Ca by 32% to 227 nmoles/mg membrane protein. *All* of this phospholipid specific increment (55 nmoles) was displaced by 0.1 mM PXB. These results and the fact that higher concentrations of PXB (up to 5 mM) displace amounts of Ca nearly equivalent to La^{3+}-displacement indicate that the major proportion of sarcolemmal Ca is bound to phospholipid.

It is likely that the anionic phospholipids (e.g., phosphatidylserine and phosphatidylinositol) would be responsible for a significant component of the binding. 0.1 mM PXB displaced 72 nmoles Ca^{2+}/mg protein. If it is assumed that 2 mol of phospholipid bind 1 mol Ca^{2+} then 144 nmoles of anionic phospholipid are participating in Ca binding under the conditions used (1 mM $[Ca^{2+}]_o$, 0.1 mM PXB). Total phospholipid was measured at 760 nmol/mg protein in the study [12] and also by Tibbits et al. [13] in sarcolemmal vesicles from the adult rat. Therefore, a minimum of 19% of the membrane phospholipid would be involved in Ca^{2+} binding in the presence of 1 mM $[Ca]_o$.

In summary, there is strong evidence in the cultured and neonatal heart that sarcolemmal-bound Ca plays a significant role in the control of contractile force and that the specific class of molecules involved may be the phospholipids. The next section will review the movement of Ca^{2+} into the cell and some of the factors that determine its distribution.

Transsarcolemmal Movement and Distribution

The previous section indicated that in cultured cells, in HEPES-buffered solution, that some 42% of the total cellular Ca^{2+} was very rapidly exchangeable. The remainder of the Ca^{2+} exchanged very slowly (t½ of at least several hours) and, therefore, could not be involved in the beat-to-beat control of contractile force. Eighty percent of the rapidly exchangeable fraction is La^{3+}-displaceable which localizes it to the sarcolemmal-glycocalyx complex. The localization of the remaining 20% is unknown. Under HEPES-buffered perfusion conditions a fraction of the rapidly exchangeable Ca^{2+} component must enter the cell and interact with the myofilaments to produce force and then return to the rapidly exchangeable sites without distribution to other more slowly exchangeable cellular sites.

An increase of heart rate does not introduce a more slowly exchangeable component [2]. A 65% increase in exchangeable Ca^{2+} secondary to increased contractile frequency was localized entirely to the rapidly exchangeable fraction. A more recent study [9], again with cells in HEPES-buffered perfusate, examined the response of ^{45}Ca uptake to lowering $[Na]_o$. Lowering $[Na]_o$ to 33 mM produced a 113% gain in rapidly exchangeable Ca^{2+} which was completely La^{3+}-displaceable. Therefore, under the stimulus of low $[Na]_o$, rapidly exchangeable Ca^{2+} localized to the sarcolemma-glycocalyx complex and to the "screening layer" is more than doubled. This experiment emphasizes the large capacity of the cellular surface sites — up to 7.1 mmoles Ca^{2+}/kg dry wt cells or more than 1.8 mmoles/

kg wet wt tissue. None of this Ca^{2+}, under HEPES-buffered perfusion conditions, exchanges with more slowly exchangeable cellular Ca^{2+} pools — despite the fact that $[Na]_o$ reduction is markedly positively inotropic and a component of the added fraction must reach the myofilaments. As with increased frequency the Ca^{2+} must cycle between sarcolemma-glycocalyx and myofilaments without mixing with more slowly exchangeable cellular sites.

The cultured neonatal rat cell provides kinetic evidence for marked compartmentalization of Ca^{2+}. Further studies provide evidence that the compartmentalization can be predictably manipulated. Substitution of phosphate for HEPES in the perfusate at pH less than 7.2 produces an immediate linear increase in Ca uptake at a rate of approximately 2.5×10^{-5} moles/kg dry wt/min [14]. This added component is relatively slowly exchangeable with $t\frac{1}{2} \sim 40$ minutes. Addition of inhibitors of mitochondrial respiration (e.g., Warfarin sodium or Antimycin A) completely prevented the increased Ca uptake induced by the phosphate. The mechanism of the phosphate effect comes from studies of isolated mitochondria [15]. The extent of Ca^{2+} uptake by respiring mitochondria is limited by a rising pH in the mitochondrial matrix that occurs as Ca enters in exchange for matrix H^+. If no source of protons is available, the Ca uptake by the mitochondria will cease. Perfusion of the cultured cells with HEPES-buffered solution containing chloride salts provides no source of a permeant anion capable of proton donation. The addition of $H_2PO_4^-$ to the perfusate provides such a donor in that it exchanges matrix hydroxide ions on the phosphate-hydroxide antiporter. This reaction is equivalent to the entry of H_3PO_4 followed by a loss of a proton to the alkaline matrix. The resulting excess anion provides the milieu for accumulation of Ca as the phosphate salt. The reaction is specific in its requirement of an anion capable of proton donation such as phosphate. Nitrate, for example, does not produce the increased slow component of uptake. The cellular organelle responsible for the enhanced uptake with phosphate addition at pH < 7.2 is clearly the mitochondrion because specific inhibition of its respiration and, therefore, electron transport and H^+ ejection, completely blocks the effect of PO_4 addition on Ca uptake. As would be expected, the Ca^{2+} that exchanges with the mitochondrial compartment is not La^{3+}-displaceable [14].

As pH is raised above 7.2 in the presence of phosphate, further significant increase of ^{45}Ca uptake is produced. This further increase cannot be blocked by inhibition by mitochondrial respiration and, therefore, is due to activation of still another component of exchange. As pH is increased $HPO_4^{=}$ and $PO_4^{\equiv}$ increase in concentration, and further study indicated that the augmented Ca^{2+} uptake correlated best with the increase in $PO_4^{\equiv}$ [16]. It has been demonstrated in the red blood cell [17] that $PO_4^{\equiv}$ inhibits the Na-K pump, probably through product inhibition. A similar inhibition has been shown in adult rabbit heart [18]. Therefore, it would seem that as elevation of pH increases the level of $PO_4^{\equiv}$ the Na-K pump is increasingly inhibited. This, as proposed for Na-K pump inhibition by cardiac glycosides, would increase $[Na]_i$ and increase Ca^{2+} uptake via the Na-Ca exchange system. This additional transsarcolemmal Ca^{2+} flux is added to, but distinct from, the mitochondrial component of exchange. Further studies give an indication of the intracellular system which is involved with this additional component of uptake.

Caffeine was used as an inhibitor of Ca^{2+} accumulation by the sarcoplasmic reticulum (SR) [19]. Under conditions in which all exchangeable Ca^{2+} was rapidly exchangeable (HEPES perfusion) or in which a mitochondrial component had been specifically added (phosphate, pH < 7.2) 10 mM caffeine had no effect on ^{45}Ca uptake pattern in cultured cells. If, however, transsarcolemmal Ca^{2+} flux was increased by addition of phosphate at pH $= 7.35$, 10 mM caffeine consistently produced a marked further increase in ^{45}Ca uptake. Since caffeine does not reduce the ^{45}Ca efflux rate from a previously labelled culture, the effect of the drug is to cause retention of a component of influxing Ca and to cause Ca to accumulate somewhere within the cell. A complex sequence of influx-efflux studies [19] is consistent with the following effect of caffeine: if it is assumed that the primary action of caffeine is to inhibit Ca^{2+} uptake by the SR, results indicate the once Ca is pumped into the SR the drug does not inhibit its efflux from the cell. However, that Ca^{2+} which is retained within the cell because caffeine inhibits SR entry cannot exit (or exits only very slowly) unless caffeine is removed.

Caffeine produced Ca^{2+} retention within the cultured cell only under the condition in which Ca^{2+} uptake had been increased by, presumably, augmentation of Na-Ca exchange. This suggests that the increased Ca influx associated with phosphate perfusion at pH $= 7.35$ increases intracellular Ca^{2+} to a level which activates Ca^{2+} uptake by the SR. Under this condition caffeine's inhibiting effect on SR Ca^{2+} uptake becomes manifest as a retention of Ca^{2+} within the cell. The perfusion conditions (HEPES or phosphate, pH < 7.2) in which caffeine has no effect apparently are not associated with high enough Ca^{2+} flux to produce levels of Ca which significantly activate the SR. If the SR is not activated, inhibition by caffeine would have little effect on Ca^{2+} kinetics.

As discussed in the section on Ca^{2+} binding, low $[Na]_o$ perfusion in HEPES perfusate causes a large increment in cellular Ca^{2+} — all of which is localized to sites at the sarcolemma — glycocalyx [9] or in the "screening layer." Low $[Na]_o$ perfusion produces essentially the same effect on Ca^{2+} distribution when mitochondrial exchange is activated by perfusion with phosphate at pH < 7.2. If, however, Ca^{2+} flux is augmented with phosphate at pH $= 7.35$, low $[Na]_o$ perfusion now causes a marked increase in the slower component of Ca^{2+} exchange as well as the increased binding at the sarcolemmal-glycocalyx sites. It would seem that activation of Na-Ca exchange in the presence of increased surface binding causes Ca^{2+} to be distributed to a cellular compartment — which, in light of the caffeine studies summarized above, could involve the SR.

In summary, three distinct compartments of Ca^{2+} can be kinetically identified in the cultured cell system: (1) a rapidly exchangeable ($t\frac{1}{2}$ of seconds) fraction at least 80% of which is localized to the sarcolemma-glycocalyx; (2) a slowly exchangeable fraction ($t\frac{1}{2}$ ~40 min) localized to the mitochondria; (3) an additional slowly exchangeable fraction associated with conditions of increased Ca^{2+} uptake. This component may involve activation of exchange through the SR.

CALCIUM EXCHANGE IN THE FETAL AND NEONATAL HEART

Studies of developmental changes in myocardial contractility have utilized three different methods: (1) in vivo experiments in the intact animal; (2) the papillary muscle preparation; or (3) arterially perfused whole heart (Langendorff) or septal preparation. Although in vivo experiments might be more physiological than the other two methods, it is impossible to control the variables (perfusion rate, heart rate, perfusate content, and others) which affect cardiac contractility. Therefore, most investigators have used the other two methods. The superfused papillary muscle preparation is very useful for contractility studies. However, slow diffusion and the small size of the muscle hinder metabolic and ionic exchange studies. We used the arterially perfused whole heart (Langendorff) or septal preparation for the following reasons: the perfusion rate and perfusate content can be controlled; mechanical function and tissue high energy phosphate are stable over five-hour periods; it is possible to change perfusate within seconds; and physiological temperature and loading conditions are maintained. One can perturb the preparation and monitor changes in contractility simultaneously with changes in ionic exchange and metabolic events.

Differences in methodology, developmental stage, and species variation have contributed to controversial results in the literature. In this section, we will describe calcium exchange in the neonatal and fetal heart and discuss various developmental changes that affect Ca^{2+} exchange.

An Overview of the Developing Heart

In the developing heart, the physical structure and biochemical composition of various organelles are changing continuously, and this causes constant change in cellular ionic content, membrane depolarization, and enzyme activities. These chronological changes vary from one species to another, and there is significant interspecies difference in the function of various organelles. For these reasons, one might obtain controversial or inconsistent results at various stages of development in each species and at similar stages in different species. Some of the developmental changes in organelle structure that have a direct effect on Ca exchange and contractility include: (1) the high cellular surface/volume ratio in the fetus and newborn [20, 21, 22]; (2) the absence of T-tubules and diadic junctions at birth [20, 21, 22]; (3) the decreased quantity of myofibrils [23] and myofibrillar Ca-ATPase activity in the neonate [24, 25, 26]; (4) developmental changes in sarcoplasmic reticulum (SR) Ca binding and release [25, 27], sarcolemmal Na^+-K^+-ATPase, and the quantity of mitochondrial cellular content and Ca^{2+} uptake [25].

Below we will discuss: (1) Ca^{2+} and contractility; (2) sarcolemmal Ca^{2+} flux; and (3) SR and mitochondrial Ca^{2+} binding and release.

Ca^{2+} and Contractility

Extracellular Ca^{2+} is essential for maintenance of contraction in the newborn and adult heart but not in skeletal muscle. The amount of Ca^{2+} that enters the cardiac cell via the slow inward current is not sufficient to produce contraction [19], and two other sources of contractile Ca have been considered. The first is Ca^{2+}-stimulated Ca^{2+} release from the SR [28, 29, 30], and the second is Ca^{2+} influx via $[Na]_i$-$[Ca]_o$ exchange [31, 32]. The contribution of each mechanism to contractile Ca^{2+} remains controversial. Ca^{2+} influx and that released from the SR increases myoplasmic Ca^{2+} which activates the contractile protein (actin and myosin) leading to contraction. Relaxation occurs when myoplasmic Ca^{2+} concentration is reduced to resting level by SR sequestration, transsarcolemmal Ca^{2+} efflux (sarcolemmal Ca-ATPase), or via $[Ca^{2+}]_i$-$[Na^+]_o$ exchange (probably during phase 4 of the action potential). Again, the relative role which each of these processes plays is unknown.

The maximal force of contraction is dependent on the quantity and organization of myofibrils, myofibrillar Ca^{2+}-ATPase, and contractile Ca^{2+} (myoplasmic Ca concentration during systole). In a given muscle, the absolute increase in contractility depends on the amount of contractile Ca^{2+}, and the relative increase (% of baseline) depends on initial level of myoplasmic Ca^{2+}.

Ultrastructural studies showed that the myofibrils are disorganized, and the quantity in the newborn is less than in the adult [20, 21, 22, 23]. Myofibrillar Ca-ATPase activity (V_{max}) increases with age, but the sensitivity (K_m) to Ca^{2+} is similar in all age groups [25]. The reasons that myofibrillar ATPase activity in the fetus and newborn is lower than in the adult is not known. Lompre et al. [33] showed that myosin isoenzyme V_1 (which has high enzyme activity) in the newborn rabbit was greater than in the adult. Our preliminary studies [25] showed that myosin ATPase activity in the myofibrillar preparation (measured at high ionic strength) [34] in the fetus and newborn rabbit was greater than in the adult rabbit. This finding is in agreement with those of Lompre et al. [33]. Measurement of myofibrillar ATPase at low ionic strength is more physiological because it represents actin-activated myosin ATPase in the presence of troponin and tropomyosin [34]. We found that this myofibrillar ATPase in the fetus and newborn was lower than in the adult [25]. The reason for the lower ATPase in the developing heart remains unknown. The decreased quantity and disorganization of myofibrils in addition to the lower ATPase activity in the developing heart explains the lower contractile force in the fetus and newborn [35, 36, 37, 38, 39, 40, 41].

Transsarcolemmal Ca^{2+} Flux

It is generally accepted that Ca^{2+} influx occurs via the slow inward Ca^{2+} current and Na^+-Ca^{2+} exchange. Ca^{2+} efflux is determined by Na^+-Ca^{2+} exchange, sarcolemmal Mg^{2+}-Ca^{2+}-ATPase, and the SR via diadic junction with T-tubules. We will discuss developmental changes in Ca^{2+} influx, Ca^{2+} efflux, and net Ca^{2+} exchange.

Rolph et al. [26] showed that Na^+-K^+-ATPase activity increased in the fetus, and Marsh et al. [42] demonstrated that the activity of this enzyme in the newborn is greater than in the adult. Our data [43] are in agreement with these findings. This has a direct effect on myoplasmic ionic concentration, Na^+-Ca^{2+} exchange, and the toxicity as well as positive inotropism of cardiac glycosides. Furthermore, the magnitude and duration of the action potential in the newborn are different from those in the adult [55]. This difference will probably modify the slow inward Ca^{2+} current and Na^{+}-Ca^{2+} exchange.

Several investigators have shown that the force of myocardial contraction in the newborn is less than in the adult [35, 36, 37, 38, 39, 40, 41]. Because of interspecies difference in development and organelle function, it is important to characterize organelle function and correlate these results with the function of the whole heart. To define the developmental changes in sarcolemmal function, a number of studies have been completed by us and others.

Philipson et al. [45] and Bers and Langer [8] showed good correlation between sarcolemmal Ca^{2+} binding and +dT/dt (max). This suggested that Ca^{2+} bound to the sarcolemma has a direct relationship to contractile Ca^{2+}. Since the contractile force in the neonate is less than in the adult, it is possible that this difference might be due to a decreased sarcolemmal Ca^{2+} binding in the neonate. In a previous study [41] we examined this possibility and found no difference between the newborn and adult rabbit. This, however, does not rule out a difference in Ca^{2+} influx for the reasons discussed above.

We studied the relationship between extracellular Ca^{2+} ($[Ca^{2+}]_o$) and contractility in the fetus, newborn, and adult rabbit [25, 41]. These studies showed that maximal contractility was achieved at $[Ca^{2+}]_o$ of 15 mM in the fetal and adult hearts and at 30 mM in the newborn. Increasing $[Ca]_o$ from physiological concentration to 15 mM caused greater increase in contractility in the newborn than in the adult. Examination of the relationship between +dT/dt (max) and $[Ca]_o$ to determine the K_m value and V_{max} showed that the K_m value in the newborn (3.02 mM) was significantly greater than that in the adult (1.47 mM) and fetus (1.40 mM). The V_{max} in the adult was significantly greater than in the fetus and newborn. One might argue that raising $[Ca]_o$ may shorten the duration of the Ca^{2+} inward current (46) and may not result in maximal contractile force (maximal myoplasmic Ca achieved). It should be noted that further increase in $[Ca^{2+}]_o$ (more than that causing maximal contraction) produced cardiac contracture. In a previous study [39] we showed that contracture is caused by an increase in myoplasmic Ca^{2+} and/or a decrease in tissue ATP. Since tissue ATP was not different from control in high Ca^{2+} experiments, contracture must be due to a significant increase in myoplasmic Ca^{2+}. Of interest, the effect of $[Ca]_o$ on contractility in the fetus is unlike that of the newborn and more similar to that of the adult. At physiological $[Ca]_o$ of 1.5 mM [25], +dT/dt max as a percentage of maximal value was 55% in the fetus, 52% in the adult, and 33% in the newborn. Since the sensitivity of myofibrillar Ca^{2+}-ATPase to Ca^{2+} is similar in all age groups [25], this difference in contractility in the three age groups suggests a difference in myoplasmic Ca^{2+} concentration. A determination of total cellular Ca^{2+} at physiological Ca^{2+} concentration showed that cellular Ca in the newborn (5.54 ± 0.71 mmol/kg dry weight) was significantly less than in the adult (9.92 ± 0.69 mmol/kg dry weight) [41, 47]. A direct measurement of myoplasmic Ca is necessary to confirm this observation.

We also determined the effect of barium (Ba^{2+}) on myocardial contracture [41]. Ba^{2+} influx is similar to Ca influx [48], but Ba is not sequestered by the SR [49]. Ba^{2+} caused a similar degree of contracture in the newborn and adult [41]. Assuming that the sensitivity of contractile proteins to Ba^{2+} is similar in the newborn and adult, the relative increase in the resting tension can be used as an index of tissue Ba^{2+} uptake. The Ba^{2+} dose-response curve was similar in the newborn and adult in the presence or absence of Ca^{2+}. The time to one-half maximal contracture in the newborn was not significantly different from that in the adult [41]. These data suggest that the rate of Ba^{2+} influx is similar in the two age groups. This also suggests that Ca^{2+} influx is similar in all age groups, and the difference in myoplasmic Ca^{2+} is due to a difference in the amount of Ca^{2+} attributable to the SR.

A change in $[Na]_o$ concentration has an effect on sarcolemmal Ca^{2+} binding, Ca^{2+} influx, and Na^+-Ca^{2+} exchange [31, 32]. An increase in $[Na^+]_o$ from 142 mM to 200 mM caused a decrease in $+dT/dt$ (max), and this decrease in the newborn was significantly greater than in the adult [50]. This initial decrease was followed by gradual recovery which was dependent on the stimulation rate. This recovery can be explained by the increase in intracellular ($[Na^+]_o$) and Na^+-Ca^{2+} exchange [32]. The initial decrease in contractility was due to Na^+ competition with Ca^{2+} on external membrane binding sites, and the effect of this competition in the newborn is greater than in the adult. The subsequent recovery is due to an increased intracellular Na^+ and stimulation of Na^+-Ca^{2+} exchange. The increased effect of $[Na^+]_o$ in the newborn suggests that the newborn is more dependent on extracellular Ca^{2+} than the adult.

Perfusion with solution that contained La^{3+} in low concentration causes displacement of Ca^{2+} from the sarcolemma [51, 52] but La^{3+} does not enter the cell [53]. La^{3+} caused a depression in myocardial contractility, and this decrease in the fetus and newborn [25, 54] was significantly greater than in the adult. Reder et al. [55] showed that the neonatal Purkinje fibers were less sensitive to verapamil than the adult fibers. Rosen et al. [44] reported that ouabain-induced changes in the action potential of Purkinje fibers in the newborn were less than in the adult, and this was explained by the difference in Na^+-K^+-ATPase activity and cellular ionic concentration in the two age groups.

In the newborn and adult rabbit muscles perfused with cardiac glycosides, we [56] found that cardiac glycosides induced a similar decrease in cellular K in the newborn and adult, and this was associated with similar increase in tissue Ca^{2+}. However, mechanical toxicity was reached in the adult sooner than in the newborn. A decrease in $[Ca^{2+}]_o$ delayed the onset of cardiac toxicity in the adult [56] and fetus [43].

From the experiments described above, we can conclude that for a given $[Ca]_o$ concentration Ca^{2+} influx in the neonate is different from that in the adult. This difference varies with the species and the stage of development in each species. In general, the newborn is more dependent on extracellular calcium and myoplasmic Ca^{2+} concentration in the newborn tends to be less than in the adult. The effect of

Ca paradox on the newborn is significantly less than that in the adult [70, 71]. This suggests developmental changes in sarcolemmal structure.

The various studies on the effect of various inotropic agents in the newborn as compared with the adult show conflicting results [43, 47, 56, 57, 58, 59]. These differences may be explained by the differences in species and experimental preparation. In our experiments [43, 47, 56] we found that the positive inotropism of isoproterenol, paired electrical stimulation, hyperosmolarity, and cardiac glycosides in the newborn is greater than that in the adult. However, when the increase in contractility is calculated per gram muscle mass, the inotropism in both age groups was similar. These results are consistent with lower myoplasmic Ca^{2+} in the newborn. It is also interesting that these inotropic agents produced positive inotropism in the newborn muscles perfused with high $[Ca]_o$ but not in the adult muscles perfused with the same Ca^{2+} concentration [47]. These data provide additional evidence that myoplasmic Ca^{2+} concentration in the newborn is less than in the adult, and partially explain the delayed mechanical toxicity of cardiac glycosides in the newborn [56].

Ca^{2+} Binding and Release by Sarcoplasmic Reticulum and Mitochondria

Nayler et al. [27] showed that Ca^{2+} binding by the SR increased with age, but there was no increase in the quantity of SR per gram muscle. We [25] found that the rate of Ca^{2+} uptake by SR was similar in the fetus, newborn, and adult rabbit. Ca^{2+} binding at pCa = 4 was also similar in the three age groups. The amount of SR per gram muscle in the fetus was significantly less than in the newborn, and the newborn value was significantly less than in the adult [25]. Therefore, relative SR Ca^{2+} uptake/g muscle in the fetus and newborn were 17% and 58% of the adult value, respectively. Thus our data and those of Nayler et al. [27] showed an increase in SR Ca^{2+} uptake (per gram muscle) with age. Fabiato and Fabiato [29] showed that the threshold of Ca^{2+} stimulated Ca^{2+} release from the SR in the newborn was greater than in the adult. For these reasons, one might expect that for a given amount of Ca^{2+} influx, the amount of Ca^{2+} released form the SR in the newborn is less than in the adult. This suggests that the sum of Ca^{2+} influx and Ca^{2+} released from the SR in the newborn is less than in the adult and is consistent with the previous proposition that myoplasmic Ca^{2+} in the newborn is less than in the adult.

Caffeine stimulates SR Ca release and subsequently inhibits SR Ca^{2+} uptake [60, 61, 62, 63]. For this reason, caffeine has been used as a probe of SR function. In preliminary experiments [64] we perfused newborn and adult hearts with Kreb's-Henseleit solution containing caffeine. Caffeine induced an initial positive inotropism that was significantly greater in the adult than in the newborn. Jourdon et al. [65] also showed that the caffeine effect on the contractile activity in the newborn rat myocardium differs from those previously described in the adult. These data indicate that caffeine effect is different in the two age groups. More detailed studies are needed to define the role of the SR in Ca^{2+} release and Ca^{2+} sequestration in the developing heart, but indications are that SR contribution to contraction depends on a relation between SR capacity and Ca^{2+} load as was shown to be the case in the tissue culture experiments reviewed in the first section of this chapter.

We [25] have shown that mitochondrial content (per gram muscle) in the fetus was significantly less than in the newborn, and the newborn values were significantly less than in the adult. Mitochondrial Ca^{2+} uptake is zero at pCa more than six (physiological range) in all age groups. This finding is consistent with the high in vitro K_m value (33 μM) for mitochondrial Ca^{2+} uptake [66]. At pCa less than six, the rates of mitochondrial Ca^{2+} uptake in the fetus and newborn is greater than in the adult, and mitochondrial Ca^{2+} uptake/g muscle in the newborn is 1.4 times the values in the fetus and adult [25]. Mitochondria probably do not play a role in Ca^{2+} uptake and release under physiological conditions at any developmental stage, but might play a role under pathological conditions that cause Ca^{2+} overload.

In summary, the SR and not the mitochondria might contribute to contractile calcium, and the amount of Ca^{2+} released from the SR (per gram muscle) in the newborn is less than in the adult.

Ca Efflux

The mitochondria bind with Ca when myoplasmic Ca reaches high level and have no role in the maintenance of steady state myoplasmic Ca^{2+} or Ca^{2+} efflux. The increased myoplasmic Ca^{2+} during systole stimulates SR Ca^{2+}-ATPase and results in Ca^{2+} sequestration by the SR [67] and subsequent reduction of myoplasmic Ca^{2+}, thus leading to cardiac relaxation. Although the SR controls myoplasmic Ca on a beat-by-beat basis, net change in cellular Ca^{2+} is dependent on the balance between Ca influx and Ca efflux. Ca efflux occurs via: (1) $[Ca^{2+}]_i$-$[Na^+]_o$ exchange (probably stage 4 of the action potential); (2) Mg^{2+} sensitive sarcolemmal Ca^{2+}-ATPase [68]; and (3) sarcolemmal Ca^{2+} extrusion via the SR-diadic junctions with the T-tubules. The exact role of each mechanism has not been defined. In our experiments, perfusion with lanthanum [54], manganese [54], and caffeine [64] caused a significant increase in the resting tension in the newborn but not in the adult. Because tissue ATP concentration in these muscles is not depressed, contracture must be caused by an increased myoplasmic Ca^{2+} [39]. Since lanthanum displaces Ca^{2+} bound to the sarcolemma and blocks Ca^{2+} influx and efflux [51, 52] but does not enter the cell [53], the increase in the resting tension is probably due to a decrease in Ca^{2+} efflux.

Jourdon et al. [65] suggested that caffeine stimulated Ca^{2+} extrusion via the Na^+-Ca^{2+} exchange. Since caffeine inhibits Sr Ca^{2+} uptake, and contracture occurred when Na^+-Ca^{2+} exchange was inhibited (by low Na^+), it is conceivable that Na^+-Ca^{2+} exchange plays a significant role in Ca^{2+} efflux.

It has been suggested that cytosolic calcium might be extruded to the extracellular space by the SR via diadic junction with the T-tubules. Our studies suggest that resting myoplasmic Ca^{2+} in the fetus was higher than in the newborn, and high $[Ca^{2+}]_o$ caused significant increase in the resting tension and one-half relaxation time in the fetus but not in the newborn [25]. Furthermore, in vitro SR Ca^{2+} binding in the fetus was approximately 30% of that in the newborn [25]. Thus, it is possible that the SR might be responsible for Ca efflux via the diadic junctions and that this activity increases with age.

The Mg^{2+} sensitive Ca^{2+}-ATPase in the sarcolemma is capable of Ca extrusion from the cell [68]. The in vitro (Ca^{2+}) K_m value for this enzyme is 0.3 μM [68] and therefore it is very sensitive to physiological Ca^{2+} concentrations. Although this enzyme might play a significant role in Ca^{2+} efflux, quantitative data in the developing heart are lacking.

THE ROLE OF CALCIUM IN CONTROL OF MEMBRANE PERMEABILITY

The original studies of Zimmerman and Hulsmann [69] characterized a phenomenon which has been called the "Ca paradox." Many studies since have documented that following a period of Ca-free perfusion return of Ca to the perfusate produces marked Ca entry, potassium (K) loss, loss of myoglobin and intracellular enzymes (e.g., creatine phosphokinase), and marked contracture. Associated with the Ca-depletion there occur major structural changes in the sarcolemma-glycocalyx [70]. These consist of a peeling of the external lamina component of the glycocalyx away from the rest of the membrane and a reorientation of intramembrane particles within the bilayer. These structural defects seem to be consistently related with the Ca paradox phenomenon.

Of interest, given the focus on development in this volume, is the age-dependence of the paradox. Chizzonite and Zak [71] showed that rat hearts below the age of six days were insensitive to Ca depletion. By 15 days the hearts developed sensitivity to Ca depletion as exhibited by contracture and release of intracellular enzymes upon return to Ca. A recent ultrastructural study [72] showed virtually no peeling of the external lamina with Ca depletion in hearts from three-day-old rats. By 20 days of age there was extensive peeling and even more peeling in adult hearts. La^{3+} was used as a cytochemical probe for evaluation of membrane permeability. It was totally excluded from the cytoplasm of three-day-old hearts after both Ca depletion and repletion, whereas 20-day-old hearts had extensive La deposits in the cytoplasm secondary to Ca depletion. Release of creatine phosphokinase upon Ca repletion was negligible at three days but marked in 22-day-old hearts. The study points to an increasing dependency of the developing sarcolemma on the presence of Ca for maintenance of structural integrity and selective permeability of the cell. Specifically, the increased sensitivity of external lamina peeling to Ca depletion with increasing age would seem to be correlated with the increased Ca paradox response of older hearts.

CONCLUSIONS

The studies in *cultured cells* indicate an important role for Ca^{2+} bound at the sarcolemma in the control of myocardial force development. At this time there is support for the proposal that anionic phospholipid within the sarcolemma is the site of a major component of the Ca^{2+} binding. Ca^{2+} moves across the sarcolemma and into the cell via two routes: the Ca^{2+} channel and the Na^+-Ca^{2+} exchanger. The distribution of Ca^{2+} intracellularly is critically dependent upon the constituents of the perfusion solution and the magnitude of Ca^{2+} influx. The presence of a proton donor (e.g., phosphate, bicarbonate) promotes mitochondrial Ca^{2+} exchange and SR activation occurs upon an increase of Ca^{2+} influx.

In *fetal and neonatal heart* myofibrils are disorganized and fewer in number as compared to the adult. The myofibrillar Ca-ATPase V_{max} increases with age but its sensitivity, K_m, to Ca is similar in all age groups. Sarcolemmal Ca^{2+} binding is not different in the newborn and adult, nor is the rate of Ba^{2+} transsarcolemmal influx (Ba^{2+} used as an analog for sarcolemmal Ca movement). However, total cellular Ca^{2+} is significantly less in the newborn which may, therefore, be attributable to a lower Ca^{2+} content of the SR in the newborn. "Mechanical toxicity" (Ca dependent contracture) upon cardiac glycoside administration occurs more rapidly in the adult consistent with the higher cellular Ca^{2+} which occurs with age.

Finally, the newborn heart is remarkably resistant to the "Ca paradox." Hearts below the age of six days show little structural defect of the sarcolemma upon Ca depletion and little evidence of Ca^{2+} overload or enzyme release upon Ca repletion. By postnatal day 15, hearts develop sensitivity to Ca depletion as exhibited by contracture and release of intracellular enzymes upon return to Ca^{2+}-containing perfusate.

REFERENCES

1. Langer, G.A., and Frank, J.S. 1972. Lanthanum in heart cell culture: effect on calcium exchange correlated with its localization. *J. Cell. Biol.* 54:441–455.
2. Langer, G.A.; Frank, J.S.; and Nudd, L.M. 1979. Correlation of calcium exchange, structure, and function in myocardial tissue culture. *Am. J. Physiol.* 237:H239–H246.
3. Langer, G.A.; Frank, J.S.; and Philipson, K.D. 1978. Preparation of sarcolemmal membrane from myocardial tissue culture monolayer by high-velocity gas dissection. *Science* 200:1388–1391.
4. Barry, W.H., and Smith, T.W. 1982. Mechanisms of transmembrane calcium movement in cultured chick embryo ventricular cells. *J. Physiol.* London 325:243–260.
5. Sedlis, S.P.; Corr, P.B.; Sobel, B.E., and Ahumada, G.G. 1983. Lysophosphatidylcholine potentiates Ca^{2+} accumulation in rat cardiac myocytes. *Am. J. Physiol.* 244:H32–H38.
6. Burt, J.M., and Langer, G.A. 1982. Ca^{2+} distribution after Na^+ pump inhibition in cultured neonatal rat myocardial cells. *Circ. Res.* 51:543–550.
7. Langer, G.A., and Nudd, L.M. 1983. Effect of cations, phospholipases and neuraminidase on calcium binding to gas dissected membranes from cultured cardiac cells. *Circ. Res.* 53:482–490.
8. Bers, D.M., and Langer, G.A. 1979. Uncoupling cation effects on cardiac contractility and sarcolemmal Ca^{2+} binding. *Am. J. Physiol.* 237:H332–H341.
9. Langer, G.A., and Nudd, L.M. In press. Ca compartmentalization in cardiac tissue culture: the effects of $[Na]_o$ depletion. *J. Mol. Cell Cardiol.*
10. Philipson, K.D.; Bers, D.M.; and Nishimoto, A.Y. 1980. The role of phospholipids in the Ca^{2+} binding of isolated cardiac sarcolemma. *J. Mol. Cell. Cardiol.* 12:1159–1173.
11. Teuber, M., and Miller, I.R. 1977. Selective binding of Polymyxin B to negatively charged lipid monolayers. *Biochim. Biophys. Acta.* 467:280–289.
12. Burt, J.M., and Langer, G.A. 1983. Ca^{2+}-displacement by Polymyxin B from sarcolemmal isolated by gas dissection from cultured neonatal rat myocardial cells. *Biochim. Biophys. Acta.* 729:44–52.

13. Tibbits, G.; Sasaki, M.; Ikeda, M.; Shimada, K.; Tsuruhura, T.; and Nagatomo, T. 1981. Characterization of rat myocardial sarcolemma. *J. Mol. Cell. Cardiol.* 13:1051–1061.
14. Langer, G.A., and Nudd, L.M. 1980. Addition and kinetic characterization of mitochondrial calcium in myocardial tissue culture. *Am. J. Physiol.* 239:H769–H774.
15. Lehninger, A.L. 1974. Role of phosphate and other proton-donating anions in respiration-coupled transport of Ca^{2+} by mitochondria. *Proc. Nat. Acad. Sci. USA* 71:1520–1524.
16. Ponce-Hornos, J.E.; Langer, G.A.; and Nudd, L.M. 1982. Inorganic phosphate: its effects on Ca exchange and compartmentalization in cultured heart cells. *J. Mol. Cell. Cardiol.* 14:41–51.
17. Garay, R.P., and Garrahan, P.J. 1975. The interaction of adenosinetriphosphate and inorganic phosphate with the sodium pump in red cells. *J. Physiol.* 249:51–67.
18. Ponce-Hornos, J.E., and Langer, G.A. 1982. Effects of inorganic phosphate on ion exchange, energy state and contraction in mammalian heart. *Am. J. Physiol.* 242:H79–H88.
19. Langer, G.A.; Rich, T.L.; and Nudd, L.M. 1983. Calcium compartmentalization in cultured and adult myocardium: activation of a caffeine-sensitive component. *J. Mol. Cell. Cardiol.* 15:459–474.
20. Page, E., and Buecker, J.L. 1981. Development of diadic junctional complexes between sarcoplasmic reticulum and plasmalemma in rabbit left ventricular myocardial cells. *Circ. Res.* 48:519–522.
21. Hoerter, J.; Mazet, F.; and Vassort, G. 1981. Perinatal growth of the rabbit cardiac cell: possible implications for the mechanism of relaxation. *J. Mol. Cell. Cardiol.* 13:725–740.
22. Legato, M. 1979. Cellular mechanisms of normal growth in the mammalian heart. II. A quantitative and qualitative comparison between the right and left ventricular myocytes in the dog from birth to five months of age. *Circ. Res.* 44:263–279.
23. Maylie, J.G. 1982. Excitation-contraction coupling in neonatal and adult myocardium of cat. *Am. J. Physiol.* 242:H834–H843.
24. Baldwin, K.M.; Cooke, D.A.; and Cheadle, W.G. 1977. Enzyme alterations in neonatal heart muscle during development. *J. Mol. Cell. Cardiol.* 9:651–660.
25. Nakanishi, T., and Jarmakani, J.M. 1984. Developmental changes in myocardial mechanical function and subcellular organelles. *Am. J. Physiol.* 246 (*Heart Circ. Physiol.* 15):H615–H625.
26. Rolph, T.P.; Jones, C.T.; and Parry, D. 1982. Ultrastructural and enzymatic development of fetal guinea pig heart. *Am. J. Physiol.* 243:H87–H93.
27. Nayler, W.G., and Fassold, E. 1977. Calcium accumulation and ATPase activity of cardiac sarcoplasmic reticulum before and after birth. *Cardiovasc. Res.* 11:231–237.
28. Fabiato, A., and Fabiato, F. 1975. Contractions induced by a calcium-triggered release of calcium from the sarcoplasmic reticulum of single skinned cardiac cells. *J. Physiol. Lon.* 249:469–495.
29. Fabiato, A., and Fabiato, F. 1978. Calcium-induced release of calcium from the sarcoplasmic reticulum of skinned cells from adult human, dog, cat, rabbit, rat and frog hearts and from fetal and newborn rat ventricles. *Ann. N.Y. Acad. Sci.* 307:491–522.
30. Fabiato, A., and Fabiato, F. 1979. Calcium and cardiac excitation-contraction coupling. *Ann. Rev. Physiol.* 41:473–484.
31. Mullins, L.J. 1979. The generation of electric currents in cardiac fibers by Na/Ca exchange. *Am. J. Physiol.* 236:C103–C110.
32. Langer, G.A. 1982. Sodium-calcium exchange in the heart. *Ann. Rev. Physiol.* 44:435–449.
33. Lompre, A.M.; Mercadier, J.J.; Wisnewsky, C.; Bouveret, P.; Pantaloni, C.; D'Albis, A.; and Schwartz, K. 1981. Species and age-dependent changes in the relative amounts of cardiac myosin in isozymes in mammals. *Dev. Biol.* 84:286–290.
34. Litten, R.Z.; Martin, B.J.; Howe, E.R.; Alpert, N.R.; and Solaro, R.J. 1981. Phosphorylation and adenosine triphosphatase activity of myofibrils from thyrotoxic rabbit heart. *Circ. Res.* 48:498–510.
35. Boerth, R.C. 1972. Postnatal development of myocardial function in the cat. *Circulation* 46, Suppl. II: 11–36.
36. Davies, P.; Dewar, J.; Tynan, M.; and Ward, R. 1975. Postnatal development changes in the length-tension relationship of cat papillary muscles. *J. Physiol.* 253:95–102.
37. Friedman, W.F. 1972/3. The intrinsic physiologic properties of the developing heart. *Prog. Cardiovasc. Dis.* 15:87–111.
38. Hopkins, S.F.; McCutcheon, E.P.; Wekstein, D.R. 1973. Postnatal changes in rat ventricular function. *Circ. Res.* 32:685–691.

39. Jarmakani, J.M.; Nagatomo, T.; and Langer, G.A. 1978. The effect of calcium and high-energy phosphate compounds on myocardial contracture in the newborn and adult rabbit. *J. Mol. Cell. Cardiol.* 10:1017–1029.
40. Jarmakani, J.M.; Nakazawa, M.; Nagatomo, T.; Langer, G.A. 1978. Effect of hypoxia on mechanical function in the neonatal mammalian heart. *Am. J. Physiol.* 235:H469–H474.
41. Jarmakani, J.M.; Nakanishi, T.; George, B.L.; and Bers, D. Effect of extracellular calcium on myocardial mechanical function in the neonatal rabbit. *Dev. Pharm. Ther.* 5:1–13.
42. Marsh, A.J.; Lloyd, B.L.; and Taylor, R.R. 1981. Age dependence of myocardial Na^+,K^+-ATPase activity and digitalis intoxication in the dog and guinea pig. *Circ. Res.* 48:329–333.
43. Nakanishi, T.; Shimizu, T.; Vemura, S.; and Jarmakani, J.M. 1984. Ouabain effect on myocardial mechanical function and sodium pump in the fetus. *Am. J. Physiol.* 246 (*Heart Circ. Physiol.* 15):H213–H221.
44. Rosen, M.R.; Hordof, A.J.; Hodes, A.B.; Verosky, M.; and Vulliemoz, Y. 1975. Ouabain-induced changes in electrophysiologic properties of neonatal, young and adult canine cardiac Purkinje fibers. *J. Pharmacol. Exp. Ther.* 194:253–263.
45. Philipson, K.D.; Bers, D.M.; Langer, G.A. 1981. The role of sarcolemmal-bound Ca^{2+} in myocardial contractility. In *The regulation of muscle contraction,* ed. A.D. Grinnell and M.A.B. Brazier, pp. 215–223. New York: Academic Press.
46. Colatsky, T.J., and Hogan, P.M. 1980. Effects of external calcium, calcium channel-blocking agents, and stimulation frequency on cycle length-dependent changes in canine cardiac action potential duration. *Circ. Res.* 46:543–552.
47. Nishioka, K.; Nakanishi, T.; George, B.L.; and Jarmakani, J.M. 1981. The effect of calcium on the inotropy of catecholamine and paired electrical stimulation in the newborn and adult myocardium. *J. Mol. Cell. Cardiol.* 13:511–520.
48. Kohlhardt, M.; Haastert, H.P.; and Krause, H. 1973. Evidence of nonspecificity of the Ca channel in mammalian myocardial fiber membranes. *Pflugers Arch. Ges. Physiol.* 342:125–136.
49. Vanderkoor, J.M., and Martonosi A. 1971. Sarcoplasmic reticulum. XIII Changes in the fluorescence of 8-amino-I-naphthalene sulfonate during Ca^{2+} transport. *Archs. Biochem. Biophys.* 144:99–106.
50. Nakanishi, T., and Jarmakani, J.M. 1981. Effect of extracellular sodium on mechanical function in the newborn rabbit. *Dev. Pharmacol. Ther.* 2:188–200.
51. Bers, D.M., and Langer, G.A. 1979. Uncoupling cation effects on cardiac contractility and sarcolemmal Ca^{2+} binding. *Am. J. Physiol.* 237:H332–H341.
52. Langer, G.A. 1976. Events at the cardiac sarcolemma: localization and movement of contractile-dependent calcium. *Fed. Proc.* 35:1274–1278.
53. Frank, J.S.; Langer, G.A.; Nudd, L.M.; and Seraydarian, K. 1977. The myocardial cell surface, its histochemistry, and the effect of sialic acid and calcium removal on its structure and cellular ionic exchange. *Circ. Res.* 41:702–714.
54. George, B.L., and Jarmakani, J.M. 1983. The effects of lanthanum and manganese on excitation-contraction coupling in the newborn rabbit heart. *Dev. Pharmacol. Ther.* 6:33–44.
55. Reder, R.F.; Miura, D.S.; Danilo, P., Jr.; and Rosen, M.R. 1981. The electrophysiologic properties of normal neonatal and adult canine cardiac Purkinje fibers. *Circ. Res.* 48:658–668.
56. Nakanishi, T., and Jarmakani, J.M. 1981. The effect of acetyl strophanthidin on myocardial function and potassium and calcium exchange in the newborn rabbit. *Am. J. Physiol.* 241:H637–H645.
57. Park, M.K.; Sheridan, P.H.; Morgan, W.W.; and Beck, No. 1980. Comparative inotropic response of newborn and adult rabbit papillary muscles to isoproterenol and calcium. *Dev. Pharmacol. Ther.* 1:70–82.
58. Park, I.; Michael, L.H.; and Driscoll, D.J. 1982. Comparative response of the developing canine myocardium to inotropic agents. *Am. J. Physiol.* 242:H13–H18.
59. Hougen, T.J., and Friedman, W.F. 1982. Age-related effects of digoxin on myocardial contractility and Na-K pump in sheep. *Am. J. Physiol.* 243:H517–H522.
60. Blayney, L.; Thomas, H.; Muir, J.; and Henderson, A. 1978. Action of caffeine on calcium transport by isolated fractions of myofibrils, mitochondria, and sarcoplasmic reticulum from rabbit heart. *Circ. Res.* 43:520–526.
61. Blinks, J.R.; Olson, C.B.; Jewel, B.R.; and Braveny, P. 1972. Influence of caffeine and other methylxanthines on mechanical properties of isolated mammalian heart muscle. *Circ. Res.* 30:367–393.

62. Goto, Mo.; Abe, Y.; and Ohba, M. 1975. A comparative study of the effects of caffeine on the contractility of frog and canine myocardium under voltage clamp. In *Developmental and physiological correlates of cardiac muscle,* ed. M. Lieberman and T. Sano, pp. 275–286. New York: Raven Press.
63. Shine, K.I., and Langer, G.A. 1971. Caffeine effects upon contraction and calcium exchange in rabbit myocardium. *J. Mol. Cell. Cardiol.* 3:255–270.
64. George, B.L.; Shimizu, T.; and Jarmakani, J.M. In press. Caffeine effect on excitation-contraction coupling in neonatal rabbit hearts. *Devel. Pharmacol. Ther.*
65. Jourdon, P.; Auclair, M.C.; and Lechat, P. 1981. Caffeine effects on mechanical activity in newborn rat myocardium. *J. Mol. Cell. Cardiol.* 13:861–865.
66. Langer, G.A.; Frank, J.S.; and Philipson, K.D. 1982. Ultrastructure and calcium exchange of the sarcolemma, sarcoplasmic reticulum and mitochondria of the myocardium. *Pharm. Ther.* 16:331–376.
67. Jones, L.R., and Besch, H.R. 1979. Calcium handling by cardiac sarcoplasmic reticulum. *Tex. Rep. Biol. Med.* 39:19–35.
68. Caroni, P., and Carafoli, E. 1980. An ATP-dependent Ca^{2+} pumping system in dog heart sarcolemma. *Nature* 283:765–767.
69. Zimmerman, A.N.E., and Hulsmann, W.C. 1966. Paradoxical influence of calcium ions in the permeability of the cell membranes of the isolated rat. *Nature* 211:646–647.
70. Frank, J.S.; Rich, T.L.; Beydler, S.; and Kreman, M. 1982. Calcium depletion in rabbit myocardium. Ultrastructure of the sarcolemma and correlation with the calcium paradox. *Circ. Res.* 51:117–130.
71. Chizzonite, R.A., and Zak, R. 1981. Calcium induced cell death: susceptibility of cardiac myocyte is age-dependent. *Science* 213:1508–1511.
72. Frank, J.S., and Rich, T.L. 1983. Age-dependent changes in the ultrastructure of the neonatal rat heart after Ca depletion and repletion. *Am. J. Physiol.* 245:H343–H353.

7. REGULATION OF CONTRACTILITY IN DEVELOPING HEART

Z.J. PENEFSKY

The striking mechanical property of muscle fibers is contractility. It is observed in myocytes before the heart acquires its final form at birth, and although mammalian heart fibers at birth are not fully differentiated, the organ is nevertheless contractile. The developing heart responds to electrical stimuli. Adrenergic agents and calcium ions cause inotropy. It is not fully appreciated exactly how a developing heart can respond to electrical and chemical challenges despite its immature morphological structure.

This chapter describes known morphological differences between developing and mature myocardium and discusses the functional correlates of these differences. It will be seen that calcium (and cytosolic calcium concentration) are responsible for the transition of resting to active myocardium and that the regulatory complex troponin C-troponin I is intimately involved in the fine control of contraction. The two components of the complex interact in a feedback system that determines the contractile state of the myocardium. The marked differences in the regulation of contractility in developing versus adult myocardium may be due to the apparent immaturity of troponin I in the former.

This chapter also emphasizes contractile events at the cellular level. Interested readers may wish to consult recent reviews and other studies on the newborn [1], comparison of parameters between species [2], development of adrenergic responses [3], biochemical [4] and mechanical [5] aspects of development, and the role of calcium in contractility. Studies of developing human heart (12 to 24 weeks

M.J. Legato (ed.) THE DEVELOPING HEART, pp. 113–148. *Martinus Nijhoff Publishing. Boston/The Hague/Dordrecht/Lancaster.*

of gestation) deal with the response to catecholamine and acetylcholine [6, 7, 8], action potentials and the active state [9], and ultrastructural aspects [10].

MORPHOLOGICAL STRUCTURE OF MYOFILAMENTS: THE SARCOMERE

The sarcomere is the basic unit of the contractile machinery. It is readily recognized because of its characteristic repeated pattern. Each sarcomere is delimited at each end by structural proteins collectively called the Z band. Parallel strands of actin filaments (also called "thin filaments") insert into the Z band and extend into the sarcomere on each side of the Z band for a distance of 1 micron. The "thick filaments," composed primarily of myosin aggregates, span the central portion of the sarcomere for a length of 1.5 microns (see figure 7–1). Shortening of the sarcomeres is responsible for the systolic force of the heart that pushes blood from the right and the left ventricles into the pulmonary and systemic circulations, respectively.

Electron micrographs indicate that a significant fraction of mammalian myocytes are undifferentiated at birth. During the early deposition of the myofilaments, a band of protein material (M line), located in the center of the thick filaments, is absent from the sarcomere (figure 7–1). At this stage, the sarcomeres are not well organized. Following deposition of the M line protein in the central portion of the thick filaments, however, the structure of the sarcomeres approaches that characteristic of the adult. Although the function of the M line protein is not known, it may serve to keep the sarcomeric filaments in register.

C protein, which was first identified and characterized in skeletal muscle, has also been observed in the heart. The function of C protein is unknown, nor is it known whether it is present in developing heart.

MOLECULAR STRUCTURE OF CONTRACTILE ELEMENTS

Thin Filaments

These filaments are composed of several different proteins.

Actin Filaments

Actin filaments are formed in an ATP-dependent aggregation of globular actin monomers [5]. Product ADP appears to be protein-bound and shielded from solvent water since it is not accessible for further interaction. Two strands of actin polymers form a coiled coil with a half pitch of seven actin monomers. Each actin monomer contains a site that interacts with myosin (see below). Actin has been observed as a component of a variety of cells, many of them non-contractile. The structure of these is similar to that of the thin filaments in muscle. There is no evidence for structural differences between developing and adult myocardial actin.

Troponin-Tropomyosin Complex

In the groove made by the actin coils is situated an elongated protein complex that spans all seven actin monomers (see above). The complex has been dissociated into a protein called tropomyosin, which makes the backbone of the complex, and

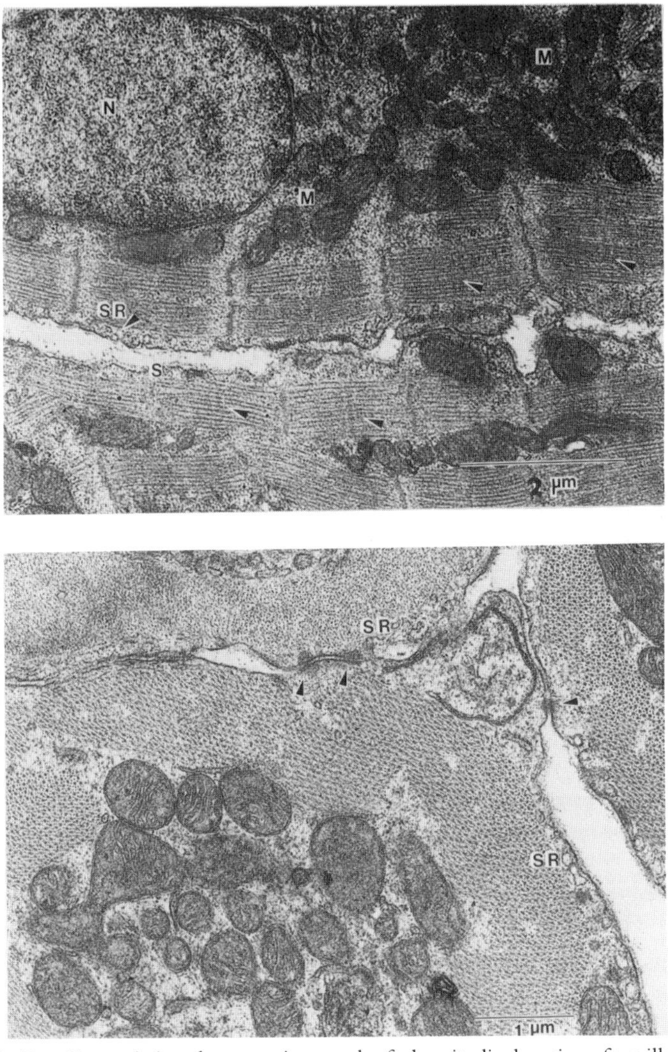

Figure 7-1. Top. Transmission electron micrograph of a longitudinal section of papillary muscle from a four-day-old piglet. Basic structural elements including mitochondria (M), cell nuclei (N), sarcoplasmic reticulum (SR), sarcolemma (S), and sarcomeres are evident. Note the difference in level of development of the M-band in adjacent myocytes (arrows). T tubules are not evident at this stage of development. Bottom. Transmission electron micrograph of a transverse section of an atrial trabecula from a piglet aged 24 hours. Junctional complexes (arrows) appear well developed. Subsarcolemmal sarcoplasmic reticulum (SR) is present. (Photographs courtesy of Dr. Thomas F. Robinson and Lee Cohen-Gould.)

three sub-units called troponins. These proteins have been the subject of intensive investigations in a number of laboratories [11, 12, 13, 14, 15, 16]. Because of the important role of the sub-units in the regulation of myocardial contractility [17], they will be described in detail.

SUB-UNITS OF TROPONIN

Troponin has been dissocated into three sub-units, each with uniquely different properties: (1) troponin C (Tn C) binds calcium; (2) troponin I (Tn I) inhibits the interaction between actin and myosin during rest periods between contractions; (3) troponin T (Tn T) serves to attach Tn C-Tn I complex to the backbone (tropomyosin).

PROPERTIES OF THE TROPONIN COMPLEX

Troponin C

Troponin C extracted from adult myocardium binds three molecules of calcium per molecule of complex [12]. Two different types of calcium binding sites have been identified: two high-affinity binding sites and one of low affinity. The high-affinity sites ($3 \times 10^8\ M^{-1}$) and one low-affinity site ($2 \times 10^6\ M^{-1}$). Magnesium competes with calcium at the high-affinity site. Milimolar concentrations of magnesium reduce the high-affinity binding constants to $2 \times 10^7\ M^{-1}$ [13]. On the other hand, the low-affinity site is calcium-specific and not altered in the presence of magnesium [17, 18, 19]. A computer simulation study supports the suggestion that because of its calcium specificity and low affinity, only the low-affinity site can have a regulatory function. The two high-affinity sites on cardiac troponin are similar to those observed on skeletal muscle troponin. While cardiac muscle troponin has one low-affinity site, two such sites are found on the protein in skeletal muscle.

Developmental changes in the properties of the troponin complex in mammalian myocardium have only recently been studied. Troponin C isolated from fetal rat heart appears to have the same calcium binding properties as adult troponin C [20]. In rat myocardium, there seems to be a progressively decreasing calcium sensitivity as the newborn heart develops [21]. It is not clear whether the decline in calcium sensitivity in rat heart is due to developmental changes in the properties of troponin C. It has been estimated that the low-affinity site of cardiac troponin provides a calcium binding capacity sufficient to regulate calcium concentration during various contractile states of the heart [15].

TROPONIN I

Troponin I inhibits the interaction between actin and myosin molecules. The sub-unit is phosphorylated (on serine 20) by a cAMP-dependent protein kinase [11]. A direct relationship was established between the concentration of beta adrenergic agents and the level of phosphorylation of troponin I. An inverse relation also has been shown between the phosphorylation of troponin I and calcium sensitivity of troponin C [17]. This paradoxical role of adrenergic agents (discussed below) on the Tn C-Tn I-complex acts as a feedback mechanism for the fine tuning and regulation of myocardial contractility [22].

Recent experiments in Solaro's laboratory [20] indicate that there is essentially no adult Tn I in the heart of the 9-day old rat (figure 7–2). In the 19-day-old rat, however, acrylamide gels indicate that two proteins migrate in the region of adult

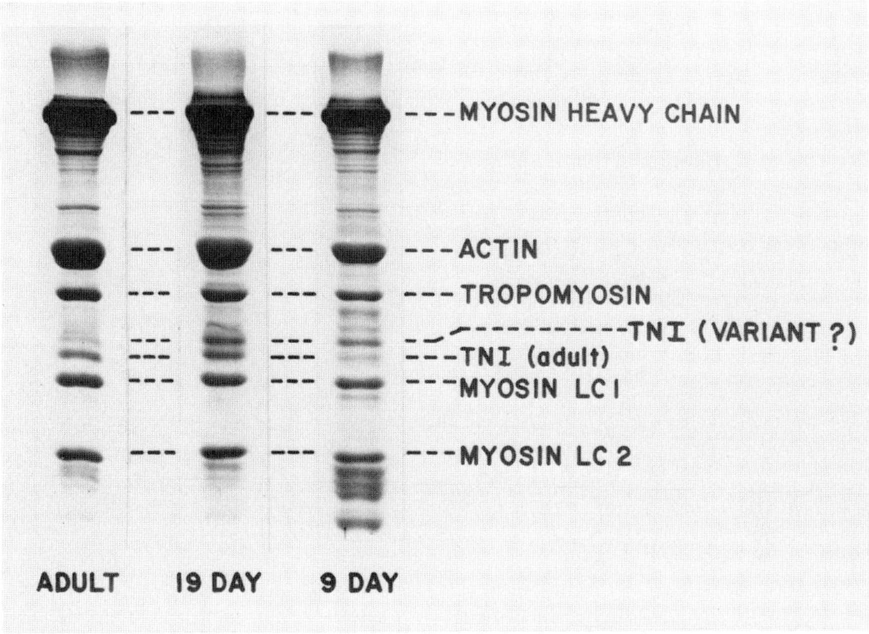

Figure 7–2. Sodium dodecylsulphate-polyacrylamide gel electrophoresis of cardiac myofibrils from adult, 19-day-old, and 9-day-old rats. There is essentially no adult Tn I in the 9-day-old rat heart, while in the 19-day-old heart there is a doublet, one band of which is adult Tn I. (Photograph courtesy of Dr. R. John Solaro.)

Tn I. One of the proteins in the doublet is in fact Tn I (see figure 7–2). Evidence for charge differences in the Tn I proteins from adult and nine-day-old rat hearts was obtained from acrylamide gel electrophoresis under acid conditions without sodium dodecylsulphate [20].

TROPONIN T

The third sub-unit of troponin, troponin T, may serve as a link between the tropomyosin backbone and the troponin I-troponin C complex. Although Tn I and Tn C bind very tightly to each other [23], the Tn I-Tn C complex dissociates in the presence of EGTA [13, 14, 16]. Since treatment of the tropomyosin-troponin complex with EGTA removes only Tn C, it may be that Tn C is the last component in the linear sequence of molecules as follows: Tropomyosin-Tn T-Tn I-Tn C (see figure 7–3).

New studies [20] reveal a difference in the molecular structure of fetal versus adult dog myocardial troponin T. Functional consequences of this difference, if any, are not known.

THICK FILAMENTS: MYOSIN

The thick filaments are formed by aggregation of myosin molecules [5]. Unlike actin, myosin is an oblong, assymetric molecule with a globular head and a long

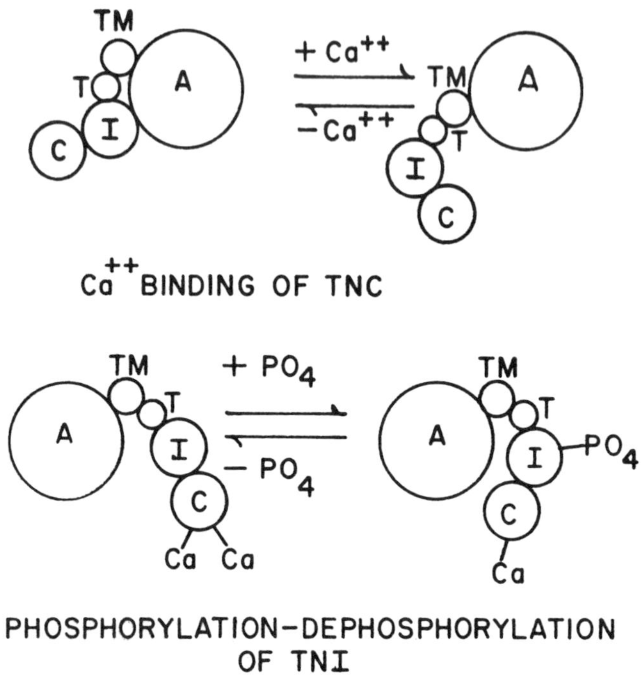

Figure 7–3. A schematic representation of the regulatory proteins. The troponin-tropomyosin system T:Tn T; I: Tn I; C:Tn C; A:actin. Top: Effects of calcium binding to Tn C. At rest, Tn I covers the active site on the actin molecule preventing interaction between actin and myosin. When Tn C binds calcium, the entire assembly shifts and the active site on the actin molecule is exposed for interaction with myosin. Bottom: representation of the effects of phosphorylation of Tn I on the calcium binding affinity of Tn C. Phosphorylated Tn I decreases the affinity of Tn C for calcium and restores the position of Tn I to the resting (relaxed) state.

alpha helical tail (figure 7–4). The tails of two myosin molecules interact to form a coiled coil, leaving the globular heads free [24]. The ATPase activity of myosin is an integral feature of the role of this protein in the mechanism of muscle contraction.

Sub-units of Myosin

Trypsinolysis of myosin results in two components: light meromyosin (LMM), comprising most of the long chain coil, and heavy meromyosin (HMM), containing a short length of the chain and the globular head. ATPase activity is in the HMM sub-unit. Further exposure of HMM to papain yields two smaller sub-units: S1, containing the globular head, and S2, the short straight chain [5]. ATPase activity is in S1. Further dissociation of S1 produces smaller fragments called the light-chains. Two pairs of such light chains have been identified: LC1 and LC2. The LC2 proteins are said to serve as hinges holding S1 and S2 sub-units together. The ATPase activity is not observed in the LC sub-units.

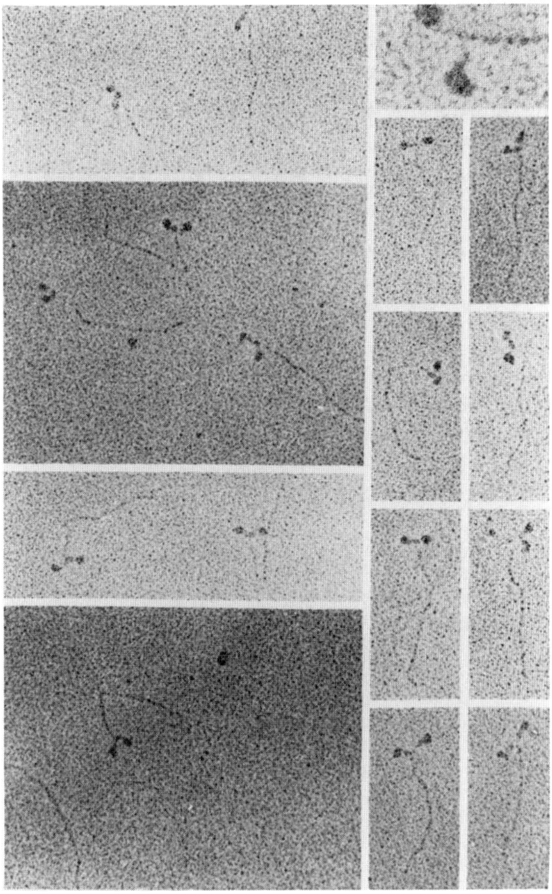

Figure 7–4. Myosin molecules isolated from skeletal muscle. Double strands making a coiled coil with two globular heads may be seen (Courtsey of Dr. S. Margossian, with permission from *J. Biol Chem.*)

LC1 (molecular weight 25,500), and LC2 (molecular weight 20,000) are found in mammalian myocardium of many mammals. Equal amounts of the two subunits are found in S1 isolated from fetal, neonatal, and adult myocardium and in neonatal myocytes in tissue culture. A third light chain molecule (molecular weight 24,500) is found in reduced amounts in fetal rat heart and in tissue culture. LC3 is about 15% of LC1 in these tissues. The ATPase activity of myosin reflects, in part, the LC composition of the molecule, particularly its responsiveness to calcium, potassium, and actin. The potassium-activated ATPase of myosin increases soon after birth and is associated with the disappearance of LC3 [25].

Cardiac myosin also binds calcium. Myosin is similar to the other calcium binding proteins described above in that it contains both high-affinity and low-affinity binding sites. The high-affinity calcium sites may also bind magnesium at physiological cytosolic concentrations of this ion (mM). The low-affinity sites are specific for calcium. Magnesium binding may be important for the integrity of the molecule, whereas the low-affinity sites may have a regulatory role in contraction. Beta adrenergic agonists influence the phosphorylation of LC [26]. However, phosphorylation can be induced either by a calcium-calmodulin-dependent protein kinase [27] or by a cAMP-dependent protein kinase [28, 29]. Epinephrine enhances the phosphorylation of LC, presumably by stimulating an adenylate-cyclase-cAMP system [30]. There is no evidence that phosphorylation of LC affects contractility.

DEVELOPMENT OF MYOSIN ISOENZYMES

Myosin exhibits an ATPase activity that is influenced by physiological or pathological states. These altered states of ATPase activity may be due to the presence in myosin of different isoenzymes. Indeed, three isoenzymes of myosin (V1, V2, and V3) with progressively decreasing ATPase activity were identified in mammalian cardiac myosin. The amounts present depended on the condition of the animal [31, 32, 33]. Fetal and neonatal myocardial ATPase activity was low. The activity increased after birth and was maximal in the mature heart. Concomitant studies have shown that myosin isoenzymes are totally of the slow type (V3) in fetal and neonatal myocardium. The properties of myosin change soon after birth, reflecting increases in the amount of V1 (the most active ATPase) and disappearance of V3 (the slowest ATPase). In the adult, the fast V1 is the predominant isoenzyme with little or none of the intermediate form (V2). It is not known if V2 is a distinct isoenzyme form or an intermediate stage in the formation of V1 from V3.

A concomitant of aging is a shift from a predominance of type V1 into type V3 isomyosin in the myocardium [31].

Myosin Strands

During differentiation of a myocardial fiber, a number of myosin strands are believed to aggregate tail to tail to form the initial backbone of the thick filaments [5]. Further aggregation of myosin at regular intervals along the length of the initial backbone produces the thick filaments. While the coiled portion of myosin in each row adheres to those molecules below, the free globular head and the S2 portion form the cross-bridges. Organization is poor in newly differentiated filaments. Appearance of M line protein during development helps to order the thick filaments (see figure 7–1). Orderly arrangement of the thick and thin filaments provides the spatial positioning of the two elements necessary for appropriate interaction and force generation.

Other Calcium Binding Proteins

Two additional and important calcium binding proteins, parvalbumin and calmodulin, are located in the cytosol. These proteins also have both high- and low-affinity sites for the metal. While parvalbumin may be involved in calcium sequestration in the cytosol, calmodulin participates in the calcium-dependent phosphorylation of a number of contractile proteins.

Normal systole in the heart is thought to be mediated by a calmodulin-dependent phosphorylation of the sarcoplasmic reticulum [34]. Calmodulin may be of importance in abnormal states such as anoxia when the cytosol becomes overloaded with calcium. In such a situation, it has been suggested, calmodulin may function as a backup system to relieve the heart of an excess calcium load [35]. Calmodulin also is involved in the phosphorylation of sarcolemmal proteins [36]. Some investigators suggest that calmodulin may be involved in beat to beat removal of calcium from the cytosol [37]. No information is available regarding developmental changes in the structure of calmodulin. However, a high concentration of calmodulin was reported in fetal rat and in murine heart. Two weeks after birth the level of calmodulin in heart homogenates was reduced to about 50% of the fetal stage [38]. A reason for, or a consequence of, the reduction in calmodulin level is not known.

The Sarcoplasmic Reticulum

The sarcoplasmic reticular system (SR) in the adult myocardium removes calcium from the cytosol during rest and stores it. Following stimulation of the heart, the SR delivers calcium to the contractile apparatus. Most, if not all, of the calcium utilized during contraction is believed to be stored in the elaborate SR system. Branches of these tubules are seen in the region of the transverse tubules (see below) and near sub-sarcolemmal vesicles. Such structures are called the sub-sarcolemmal cisternae, in contrast to the closed end of the SR tubules near the transverse tubules (T tubules) which are called terminal cisternae.

In the adult myocardium, at the level of the Z lines where the sarcolemmal membrane invaginates into the cytosol, the apposition of the terminal cisternae and the T tubule produces a diadic structure. The structure and function of this diadic function was recently illustrated in electron micrographs prepared from isolated junctions [39, 40]. The SR portion of the diad is believed to be the site of release of stored calcium ("activator calcium"). The sarcolemmal portion of the diad is the site of release of "trigger calcium," which activates the release of the ions from the SR. Highly purified preparations of sarcolemmal and sarcoplasmic reticular membranes bind tightly to each other and reconstitute diadic structures as shown in figure 7–5 [40]. The ability to form diadic structures in vitro has been related to the physiological affinity of the two systems, which couple excitation to contraction. Coupling of the two membranes has been studied morphologically in vivo [41]. Membrane surface area appears to increase four-fold just after birth. The surface-to-volume ratio is much greater in fetal myocytes than it is after the appearance of the T tubules [42]. The diadic structures are the counterpart of the triads seen in skeletal muscle [43]. Both structures appear to fulfill the same function.

In newborn mammalian hearts, the SR system is scant and immature [5]. The SR appears to develop in parallel with the contractile elements. There is little in-

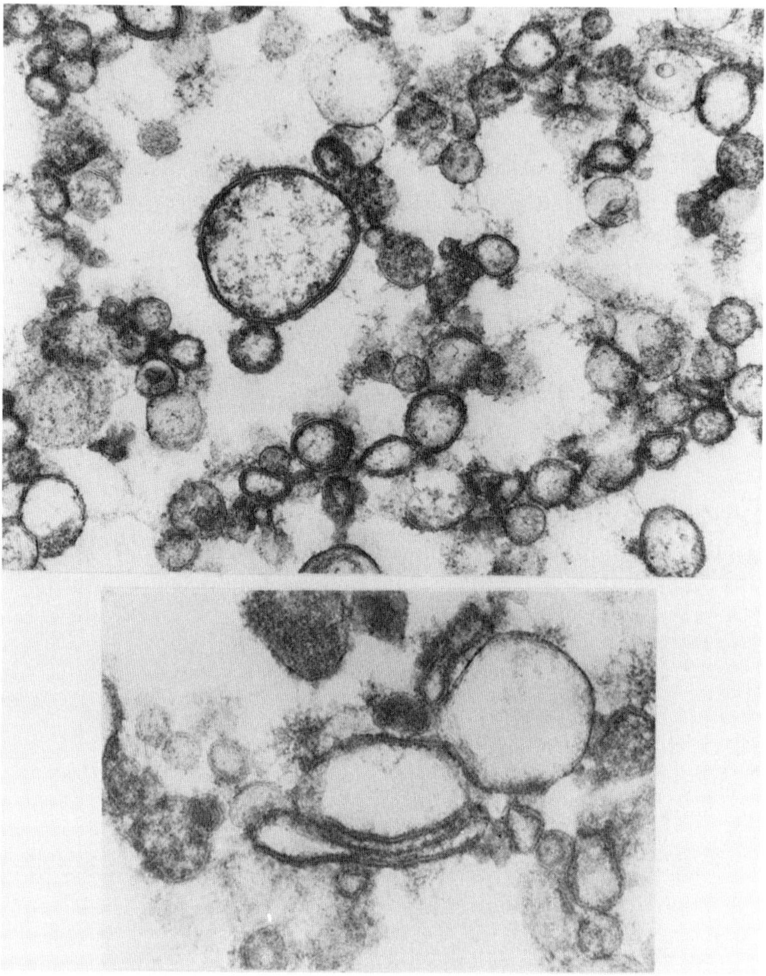

Figure 7–5. Electron micrograph of purified sarcolemmal membrane vesicles (top). A reconstituted diadic junction from skeletal muscle T tubular protein and cardiac sarcolemma (bottom). (Courtesy of Dr. N. Brandt.)

formation regarding the properties of immature SR except for electron micrograph studies of fetal and newborn myocardial tissues and embryonic chick heart. In the latter [44], there is a change in the composition and a decline in the ratio of phospholipid to protein as the SR develops. Concomitantly, an increase in the amount of transport ATPase is seen in the isolated SR vesicles. The relevance of these findings to mammals is not known. A postnatal increase in calcium accumulation and ATPase activity of the SR system was observed in guinea pig and rabbit SR [45]. Physiological studies do reveal, however, that in the rat there is a considerable dependence of the contractile system on calcium in early life. As the newborn rat develops, the calcium dependence diminishes [46, 47].

In the mature heart, the elaborate branches of the SR tubules are presumed to be the sites of sequestration of calcium from the cytosol. It has been shown that SR membranes can accumulate a maximum of 19 mmol of calcium. The concentration of calcium in the cytosol is estimated to be 100 nM at rest, and 10 uM during activity [48]. Virtually pure preparations of SR membrane can now be obtained [49], and these have been characterized [50]. A calcium-stimulated ATPase is present in the SR. The ATPase is an ATP-dependent calcium pump that removes calcium from the cytosol during relaxation. Calcium accumulation by the SR is stimulated by cAMP-dependent protein kinase [51, 52, 53]. A 22,000-dalton protein (phospholamban), is phosphorylated by the kinase and in turn activates the calcium pump [54, 55, 56]. Phospholamban is efficiently phosphorylated by either a cAMP-dependent protein kinase [57, 58, 59, 60] or by a calcium-calmodulin activated protein kinase [34, 35, 37]. The action of the two systems above is additive.

The closed ends of the SR system, the terminal cisternae, are presumed to be the sites at which activator calcium is released. These structures are located close to the T system (see below). Depolarization of the T system following stimulation and the release of trigger calcium from the sarcolemma results in release of stored calcium from the terminal cisternae into the cytosol. It is the release of this activator calcium that links excitation to contraction. A highly active cAMP-dependent protein kinase is present in fetal rat heart. The activity of the enzyme was lower in ten-day old heart. In contrast, phosphorylase and phosphorylase kinase activities were very low in fetal rat heart and increased postnatally.

The fact that calcium is sequestered from the cytosol by the calcium pump during relaxation and released by the terminal cisternae before contraction places certain restrictions on the characteristic properties of the system. These may be listed as follows: (1) the calcium pump must reside entirely or almost entirely in the tubular section of the SR, near the myofibers; (2) during rest calcium must be stored in the terminal cisternae so that it can be quickly released following stimulation; (3) there must be a mechanism for the transfer of calcium from one compartment of the SR system to the other in order to keep the ions circulating and ready for release. It is obvious that the failure of one or more of these systems will result in the failure of contractility. The mechanical behavior of developing rat heart provides ample evidence that the SR is functionally immature early in life. The myocardia of other immature mammals [8, 62] fail to respond positively to an increase in the frequency of stimulation. In such preparations (fetal human and newborn piglets, respectively) the calcium ion concentration appear to be the limiting factor in the ability of the heart to respond to stress. In developing rat myocardium, the negative force-frequency response is not reversed even in the presence of high concentrations of calcium [47]. It is possible that the limiting factor in this species is either the delivery of calcium from the sequestering region to the storage sites or an insufficient calcium supply.

Mitochondria

Mitochondria are seen in a variety of sizes and shapes in developing myocardium (figure 7–1). Various stages of division also are apparent. Since late in gestation and postnatally oxidative metabolism is vigorous, the mitochondrial membranes are well endowed with appropriate enzymes of oxidative phosphorylation. Adult heart mitochondria exhibit a high calcium pump activity [63, 64] that formerly was assigned a role in calcium sequestration during relaxation on a beat-to-beat basis. This viewpoint is no longer accepted. Possible calcium pump activity in the mitochondria of developing heart has not been reported.

THE SARCOLEMMA

The morphology of developing sarcolemma contrasts markedly with that of the adult. An example of such differences is shown in the electron micrograph of figure 7–1. Perhaps the most striking feature of developing sarcolemma is the absence of the transverse tubular system (T system). There is, in addition, a paucity of intercalated discs and related junctional structures [5]. The overall appearance of mammalian cardiac myocytes at birth is thus reminiscent of adult myocardial tissue in lower vertebrates. Development of the T system in mammals does not start until some weeks after birth, when the first signs of invaginations of the sarcolemma appear at the level of the Z lines. These developmental changes occur in parallel with the development of the SR system [42]. The sequence of development of the SR and T systems and the juxtaposition of the two membranes in a regular pattern indicate a close functional relationship.

The close morphological relationship of the SR and T systems is reflected in observations that the two membranes appear in the same fraction in purification procedures (figure 7–5). Diadic junctions are present even in highly purified sarcolemmal preparations, indicating the close affinity of the two membranes [40, 41, 42, 43, 44].

The outer surface of the sarcolemmal membrane is supported by a basement membrane and glycocalyx [5]. The structural protein of the glycocalyx contains sialic acid, a polyanionic molecule with large numbers of negatively charged sites to which calcium ions are presumed to be bound. It was recently shown that the glycocalyx is more tightly bound to the sarcolemma of newborn rat than it is in adult heart [65, 66]. Calcium deprivation and re-introduction failed to elicit a calcium paradox in newborn rat myocardium and was without effect on glycocalyx. In contrast, the same procedure applied to adult heart led to calcium paradox and to separation of the glycocalyx from the sarcolemmal basement membrane. These observations confirm an early report that within the first two postnatal weeks, rat myocardium could withstand both calcium deprivation and return to normal calcium concentration without visible signs of damage [66].

Sarcolemmal Channels

Sodium Channels

Much of our understanding of developing sodium channels is derived from studies of chick embryonic heart [67]. The channels are believed to be absent during early differentiation of myocardial fibers. They are known to be voltage-dependent and closed at rest. Upon stimulation of the membrane and depolarization to the threshold, the channels (gates) open. Sodium ion then flows inward along electrochemical gradients. Tetrodotoxin, a specific inhibitor of sodium channels, is a useful agent with which to study the role of sodium in the depolarization of muscle membranes as well as the properties of the channels.

Calcium Channels

These channels, also voltage-dependent, appear to be present early in the development of the myocytes. Unlike sodium channels which are very fast, current passes slowly through calcium channels. Calcium as well as sodium ions are thought to pass through the slow channels. The channels are tetrodotoxin-insensitive and are blocked by manganese. Calcium channels appear to be concentrated along the T tubules of the adult sarcolemma. This fact suggests that in fetal and newborn mammalian sarcolemma, when the T system is absent, calcium channels may be evenly dispersed along the sarcolemma or may be concentrated opposite the Z lines where the T tubules ultimately develop. Recently, it was suggested that calcium channels may be activated by phosphorylation of a specific phosphate receptor associated with the channel. Two proteins were identified in rat myocardium. Protein A (MW 36,000) was phosphorylated in the presence of 32-P alone, and protein B (MW 27,000) was phosphorylated only in the presence of cAMP [68].

A 23,000 dalton protein "calciductin" has been identified [69] in isolated sarcolemmal vesicles. It has been suggested that caciductin fulfills the same role in sarcolemmal membrane as phospholamban does in the sarcoplasmic reticulum. That is, calciductin is phosphorylated by a cAMP-dependent protein kinase or by calcium-calmodulin stimulation of an endogenous protein kinase. The two effects appear to be independent and additive. This implies a site-specific interaction of the kinases with calciductin (figure 7–6).

Other lines of evidence also indicate that activation of calcium channels is dependent on phosphorylation. Both cAMP and its analogue, 8-bromo-cAMP, are effective in increasing calcium flux through the channels [70].

Developmental changes in calcium channels were indicated in studies of action potentials recorded from developing [71] and fetal [72] sarcolemma, or cells in tissue culture [73], as well as from voltage clamping of myocytes [74] in rat myocardium. It is clearly established that there are developmental changes in calcium channels in the rat. Additional observations on changes in the contour of action potentials in developing Sprague-Dawley rats [75], dog [76], and piglet [77] myocardia also lend support to the possibility of differentiation of myocardial calcium channels during the course of development. It is interesting that in aging rat myocardium, the contour of the action potentials reverts towards that of newborn rat myocardium [78]. That is, a plateau is evident in the action potential recorded from the aged animal. The appearance of the plateau in the aged animal coincides with the reappearance of sensitivity of the contractile apparatus to calcium [79]. Thus the plateau of the action potential and sensitivity to calcium, characteristic

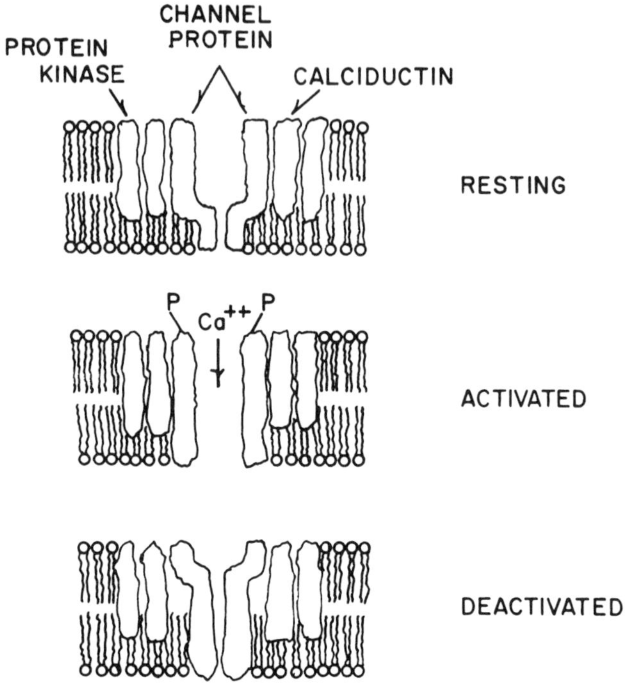

Figure 7–6. A schematic representation of calcium channels of the sarcolemmal membrane. Channel protein is surrounded by calciductin. Protein kinase mediates phosphorylation of calciductin either by stimulation via cAMP, or via calcium-calmodulin. The resting channel at the top is closed to calcium. Middle: following phosphorylation, the channel opens to permit calcium to pass through it. Bottom: following activation, the channel is deactivated and does not pass any calcium unless properly stimulated. The transition from the deactivated state to the resting state may be the rate-limiting step in calcium transport.

properties of newborn myocardium, are absent in the mature animal and regained in the aged animal [80].

Sarcolemmal Pumps

Sodium-Potassium Pump

A sodium-potassium activated ATPase, known as "sodium pump," has been isolated from the sarcolemmal fraction of the myocardium. The ATPase shares the properties of sodium pumps prepared from other tissues [81]. Like other sodium-potassium pumps, the role of the ATPase is to exchange sodium within the cell with potassium in the extracellular space [82]. The ATPase is assumed to span the sarcolemmal membrane, sending projections out from both surfaces. On the cytosolic side, the ATPase binds sodium and has little or no affinity for potassium. On the external side of sarcolemma, the pump has high affinity for potassium and little or no affinity for sodium. The two ions are exchanged across the sarcolemmal membrane in an energy-dependent fashion. For this reason, the process is called active transport. Like many energy-dependent systems, the pump stops in the absence of oxygen. Thus in anoxia, ionic equilibrium occurs via diffusion and the normal balance is lost.

The most potent inhibitor of the sodium-potassium pump, ouabain, increases the intracellular sodium concentration. The effect of the glucoside is to cause accumulation of intracellular sodium. The latter exchanges with extracellular calcium and promotes inotropy [83].

The exchange of sodium for potassium under normal physiological conditions may not be a strict 1:1 ratio. Other charged ions capable of crossing the membrane may exchange with sodium or potassium, and the overall neutrality of the pump is maintained even though the pump as such may be electrogenic.

There are no available reports on sodium pump activity of the neonatal sarcolemma. There is a report that ouabain is more effective on newborn than on adult rat myocardium [83]. More recent studies show that for the first eight to ten postnatal weeks ouabain is still effective. However, when the myocardium is exposed to high concentrations of the glycoside, the resulting inotropy is accompanied by a gradual increase in resting tension. If the exposure to the glycoside is prolonged, the myocardium will go into contracture. The contracture is slowly reversible [84]. Such a phenomenon is shown in figure 7–7.

Calcium Pump

The possibility that a calcium pump is present in the sarcolemmal membrane was initially greeted with skepticism. Much of the interest in calcium pumping in muscle was devoted to characterizing the SR, and it was assumed that an apparent calcium pump activity in sarcolemmal preparations was due to contamination by SR vesicles. However, as new techniques for the isolation and purification of the sarcolemmal membranes were developed, it became clear that there was an endogenous calcium pump in the sarcolemma [85, 86, 87]. A protein fraction with calcium-activated ATPase activity was found. A low molecular weight (150,000 dalton) protein was identified in the sarcolemmal membrane proteins that was phosphorylated either by a cAMP-dependent or by a calcium-calmodulin-dependent protein kinase [88]. Agents such as timolol stimulate the ATPase activity of the latter protein [89].

The ATPase activity of the sarcolemmal pump protein is influenced by its degree of phosphorylation, and there is evidence that the activity of the pump is regulated by phosphorylation-dephosphorylation of this pump protein [90]. It is now considered that a portion of the calcium that enters the cytosol during excitation is removed by the pumping action of this ATPase. Although there is thus far no available study on the development of a calcium pump in the sarcolemma, it should be remembered that the SR is immature if not rudimentary in the fetal and newborn mammal and that a sarcolemmal calcium pump may play a critical role in the maintenance of the cytosolic calcium balance in early life.

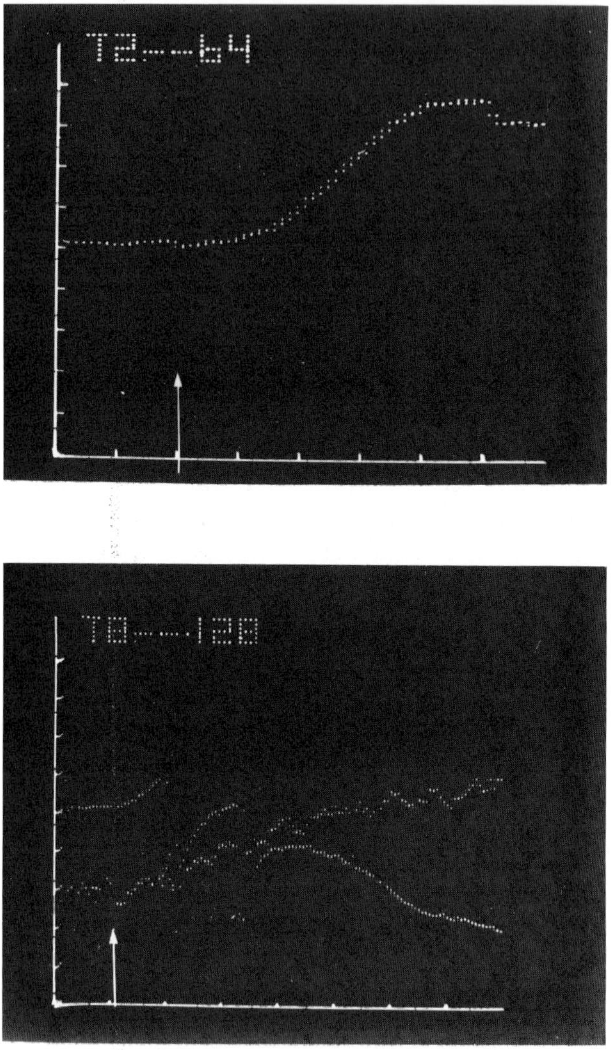

Figure 7–7. Comparison of the effects of Ouabain on developing and mature rat hearts. Top: 15-day-old preparation. Ouabain was added at arrow. Peak tension (T_2) gradually increased and levelled off. Bottom: resting (T_0) upper curve, and contractile tension (T_2, lower curve) recorded beat by beat from a 2-month-old rat heart preparation following addition of Ouabain (arrow). Ouabain induced spontaneous activity that caused initial fluctuations in resting and active tension. The contractile tension reached a peak and then decreased gradually. Meanwhile resting tension gradually increased with each consecutive contraction (upper curve). Calibration: ordinate; zero point at the mid-point (where the control record starts). Each division above or below the 0-point is ± 20% change respectively from the control value. Abcissa: Each division contains 8 consecutive contractions on the top record and 16 contractions in the bottom record. Total number of contractions and the parameters are marked on the records. (For purpose of comparison, T_2 was intentionally recorded 40% below the 0-point).

Sarcolemmal Na/Ca Exchange

The central importance of calcium in E-C coupling in myocardium is now clearly established. The interrelationships between calcium and sodium also are of vital importance. It was pointed out above that sodium entering the fiber during depolarization is not necessarily exchanged 1:1 for the potassium which leaves the fiber during repolarization. However, unless the balance between Na and K is controlled, charge distribution across the sarcolemma will be disrupted. The existence of a Na/Ca exchange mechanism for maintaining ionic balance was postulated from initial observations that when myocardial tissue was overloaded with sodium and then exposed to calcium, a major portion of the calcium crossed the sarcolemma in exchange for sodium within the cytosol [91, 92, 93, 94].

The nature and direction of the Na/Ca exchange has been the subject of debate. It is known that digitalis glycosides inhibit the Na-K pump. Myocardial tissues treated with digitalis become Na-loaded. It is precisely the exchange of this high internal Na with extracellular Ca that constitutes the basis for the inotropic action of the glycosides [95, 96].

A reverse Na/Ca exchange can be demonstrated under conditions of calcium overload in myocardial fibers. Removal of calcium from the fibers in exchange for sodium can prevent the contracture which would otherwise result. The Na/Ca exchange system is stimulated by magnesium and ATP, and it is thought that a cAMP-dependent as well as a Ca-calmodulin-dependent protein kinase can stimulate the exchange. Thus the exchange may be regulated by phosphorylation-dephosphorylation of the system [96][.

Myocardial tissues, Na-loaded by treatment with toxic doses of digitalis glycosides, will, when exposed to a high concentration of calcium, exchange intracellular Na for extracellular Ca. The exchange does not require ATP [97]. It was suggested that the electrochemical potential of high calcium outside the membrane was utilized to move intracellular Na out of the cell against its concentration gradient.

Comparable Na/Ca exchanges studies have not been carried out in fetal and neonatal myocardium. However, there is evidence that in these tissues an even greater exchange takes place between Na and Ca than in adult myocardium [98]. Fetal and newborn heart tissues are less similar to mammalian adults than they are to the myocardium of lower vertebrates such as the frog. They lack a T system, have a high surface-to-volume ratio, and a scant SR system. It is possible that these characteristics make the Na/Ca exchange a mechanism for control of imbalance in developing mammalian heart, as well as in the frog.

The functions of the three sarcolemmal membrane systems involved in calcium transfer may be summarized as follows: (1) specific calcium channels exclusively bring calcium into the cytosol; (2) calcium pumps remove calcium from the cytosol; (3) a Na/Ca exchange system, operating in both directions, serves to maintain physiological levels of calcium in the cytosol. The calcium pump is an ATPase. The calcium channels and Na/Ca exchange systems are regulated by phosphorylation and dephosphorylation of specific protein kinases under the control of cAMP-dependent or Ca-calmodulin-dependent protein kinases. The Na/Ca exchange system appears to be of greater functional significance in developing heart than in adult mammalian heart.

Sarcolemmal Receptors

Cathecholamine Receptors

The maturation of beta receptors in developing heart has attracted much interest in recent years. In fetal rat heart, beta receptors are identifiable on the 13th day of gestation [99]. They bind norepinephrine and are inhibited by propranolol. Stimulation of the receptors increased adenylate cyclase activity. However, at this stage of development, catecholamines are not inotropic. Adrenergic receptors appear on the tenth day of gestation followed about one and one-half days later by cholinergic elements [100]. In piglet heart, the in vivo response to catecholamines is present after birth [101]. Beta adrenergic response has been reported even from myocytes in tissue culture from rat heart [102] and from embyonic chicken heart [103]. The response to adrenergic agonists persists in myocardia until old age, when it diminishes [104]. Adrenergic agents have important effects on the mechanical responses of myocardium: (1) they stimulate the release of calcium from the sarcolemma; (2) they abbreviate time to peak tension; (3) they increase the rate of calcium uptake by the SR.

1. Stimulation of sarcolemmal beta receptors is considered to be connected to the stimulation of adenylate cyclase and synthesis of cyclic AMP. All fetal and neonatal hearts studied exhibit an increase in the activity of adenylate cyclase [105, 106, 107, 108, 109, 110] with development. The resulting cAMP stimulates the activity of a cAMP-dependent protein kinase which in turn can phosphorylate calciductin, the putative calcium channel protein. In its phosphorylated form, calciductin is believed to enhance the entry of calcium into the cytosol [111]. Direct microinjection of the catalytic sub-unit of protein kinase into guinea pig myocardium [112] is quickly followed by a prolongation of the plateau of the action potential and an increase in the level of the plateau voltage. Concomitant with changes in action potential, peak contractile tension increases. The velocities of contraction and of relaxation also increase significantly. Voltage clamp studies confirm that micro-injection of the catalytic sub-unit of protein kinase significantly increases the amplitude of the slow inward current. The voltage-current relationships however are not changed from the control state. This implies that effects of the injected subunit are expressed via an increase in the number of calcium channels activated rather than an alteration in the characteristics of the channels.

In newborn heart, the response to catecholamines persists for a longer time [5] than it does in the adult. It appears that the reason for the difference in persistence is that nerve endings in the adult take up circulating catecholamines whereas in the newborn, immature nerve endings are less effective in uptake and perhaps also in metabolic inactivation. While the beta receptors remain active early in postnatal life, at least in rat myocardium, the activity of adenylate cyclase diminishes after birth. The activity of adenylate cyclase appears to reach a nadir on the 14th postnatal day and gradually increases toward the mature level thereafter [113].

2. With regard to abbreviation of the time to peak tension (TPT): whereas in the adult myocardium, treatment with epinephrine shortens the time to peak tension, in the newborn myocardium this abbreviation is not observed [77]. It was mentioned above that in the fetal rat heart, catecholamines did not induce inotropy. It may be that the lack of abbreviation of contraction time is a reflection of an intermediate (i.e., immature) stage of development of the system responding to catecholamines. The factor participating in the abbreviation of contraction time may be an early onset of relaxation.

3. Catecholamines also increase the rate of calcium uptake by the sarcoplasmic reticulum. It has been suggested that the abbreviation of the time to peak tension is due to an early onset of relaxation that may result from increase in the rate of sequestration of calcium [51]. Since effects of catecholamines on fetal or neonatal preparations of the SR have not as yet been reported, it is not clear how the SR from the developing hearts may compare with that of the adult.

The known differences in the response of developing and adult hearts to catecholamines are open to the following interpretations. Catecholamines as a whole are effective in increasing the excitability of the myocardium. In the newborn heart, these agents do not alter the effective and relative refractory periods of the heart. In contrast, in the adult heart, catecholamines shorten the effective refractory period as well as the relative refractory period. As shown in figure 7–8, the early period (P1) as well as the late period (P2) of contraction remained constant following addition of epinephrine to an immature myocardium. Meanwhile peak contractile tension (T2) increased gradually with each beat. However, in the mature myocardium, epinephrine addition shortened only P2 and did so within a few beats. Thus TPT decreased. T2 again increased gradually with each beat.

There was an additional difference in the response of neonatal and mature myocardium to catecholamines. Following the inotropic response of mature myocardium to catecholamines, spontaneous activity was invariably observed. Spontaneous activity of this kind was not seen in neonatal myocardium. These differences in response may well be due to developmental changes in the excitability of the two types of tissues.

HORMONAL CONTROL OF CONTRACTILITY

Specific effects of hormones on either the myocardium or the cardiovascular system have until recently attracted little attention. There are, however, in vitro effects of hormones on contractility that may be demonstrated, and evidence is accumulating for basic hormonal regulation of these tissues. As yet systematic comparative studies of the effects of hormones on myocardial development have not been carried out.

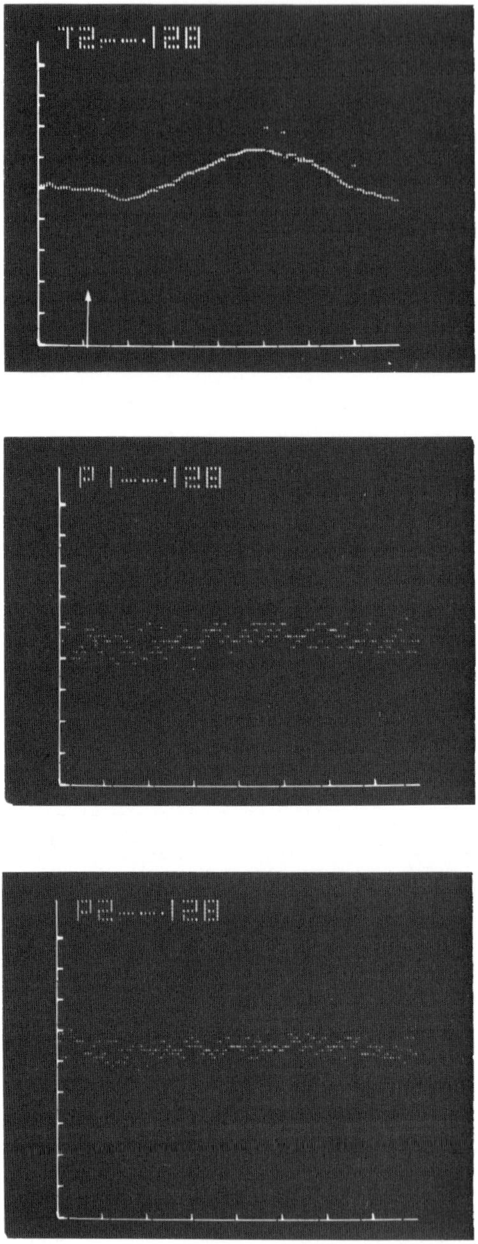

Figure 7–8. Effects of epinephrine on the contractile tension and duration of contraction. Top: beat-to-beat increase in peak tension in response to addition of epinephrine. The gradual decline followed washout of the catecholamine from the tissue bath. Middle record: early phase of contraction (P1: time from the beginning of contraction to dp/dt max. Bottom: later phase of contraction (P2: time between P1 and TPT). Although epinephrine increased contractile tension, the duration of contraction did not change. The abbreviation of contraction time, a hallmark of the effect of catecholamines on contraction time, was absent.

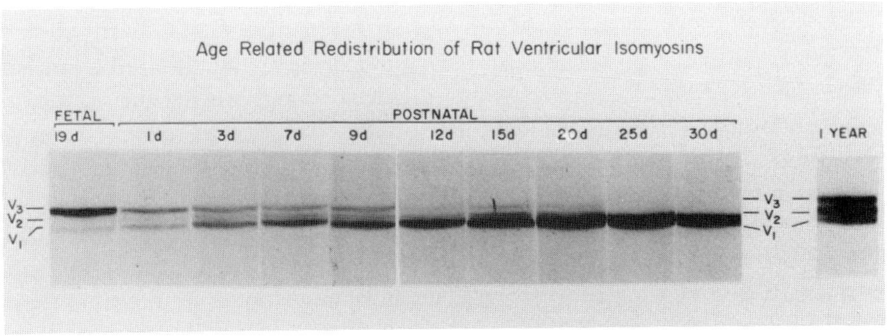

Figure 7–9. The distribution of myosin isoenzymes in fetal, developing, and mature rat heart. At the 19th gestation day, myosin isoenzyme was predominantly V_3. A conversion to V_1 is observed on the first postnatal day. The amount of V_1 increased steadily, and on the 30th postnatal day the isoenzyme was exclusively V_1. In a year-old heart, a significant amount of V_3 is observed. (Photograph courtsey of Dr. R Chizzonite.)

Pancreatic Hormones

Glucagon is inotropic in adult myocardium. The effect is due to stimulation of the adenylate cyclase-cyclic AMP system and may be indirect. There is no information on the role of this hormone in developing myocardium.

Thyroid Hormone

The effects of hyperthyroidism on the mechanical responses of the heart have long been recognized. This clinical condition causes hypertrophy of the myocardium. It was thought earlier that this form of hypertrophy was due to an overall stimulation of the tissues with associated demands for higher oxygen and substrate supply. More recently, however, with the recognition of a multiplicity of myosin isoenzymes [31, 32, 33] and the recognition that hyperthyroidism is correlated with high myosin ATPase activity, it is becoming evident that thyroid hormone may control the distribution of myosin isoenzymes. Within 72 hours after treatment of newborn animals with thyroid hormone, myosin isoenzyme V1 increased and V3 decreased. ATPase activity increased. In addition, thyroxine altered the apparent structure of the isoenzymes (see figure 7–9). Removal of the thyroid gland had the opposite effect.

It is now accepted that thyroid hormone is responsible for control of myosin isoenzymes in the heart and specifically induces de novo synthesis of the protein. Other hormones — for example, glucocorticoids — had no effect on either the synthesis or the structure of myosin isoenzymes in rat myocardium [113]. It has also been shown that excess thyroid hormone alters the pattern of mitochondrial lipids [114].

The Adrenal Hormones

Adrenal glands are composed of two sections: the medulla and the cortex.

Adrenal Medulla

The catecholamines, epinephrine, and norepinephrine are synthesized by cells in this section of the gland and released into the circulation following the stimulation of the autonomic adrenergic center. The specific effect of catecholamines on the heart is a stimulation of beta receptors of the sarcolemma as was detailed above. The catecholamine-secreting cells, as well as the agents released by the cells, appear to be similar in both developing and mature animals. Exogenous catecholamines have the same effects on developing myocardium as do the endogenous hormones. The lack of effect of adrenal medullary hormones on the time course of contraction and relaxation may therefore encompass one or all of the steps that lie between the stimulation of receptors and the final response.

Adrenal Cortex

The adrenal cortex is not mature in the mammalian fetus. Various zones of this gland do mature during development [115]. The effect of mineralocorticoids, principally aldosterone, is on the control of plasma sodium balance. No effect of this hormone on the myocardium is documented.

Glucocorticoids are best known as anti-inflammatory agents and presumed to be released under stress. They have a specific effect on the metabolism of carbohydrates. In mammals, adrenalectomy leads to death unless replacement therapy is instituted. The animals die of cardiovascular collapse. It has in the past been assumed that the effect of adrenalectomy was on the maintenance of blood pressure [116]. It is now known, however, that treatment of control animals with the synthetic glucocorticoid dexamethasone significantly enhances the mechanical performance of myocardium studied in vitro [117]. Dexamethasone also increased the activity of adenylate cyclase in aged cat myocardium [118] and reversed age-induced alterations in rat myocardial tissue [80]. The pattern of myosin isoenzymes did not change in dexamethasone-treated aged rat heart [113]. Simultaneous treatment with dexamethasone prevented the alterations caused by catecholamine depletion in adult reserpinized cat [117] and in old rat [118].

The specific effect of glucocorticoids on fetal mammalians was reviewed [115] recently. The thrust of many studies has been to elucidate the role of the hormone in the induction of parturition in experimental animals [119, 120, 121, 122]. Three species have been studied: fetal lamb, rhesus monkey, and rat, both in the course of gestation and at term. Among the three species, fetal lamb and rhesus monkey at term appeared to have the same hormone level as did the pregnant mother. At term, a burst of corticosterone appeared in the plasma. This event coincided with an increase in the synthesis of prostaglandin (PgE2) which is presumed to induce parturition in the pregnant mother. In contrast, in fetal rat plasma the level of corticosterone is comparable with that in maternal plasma up to day 19 of gestation. At this time there is a precipitous decline in the circulating level of corticosterone in fetal plasma. At term, the level of corticosterone in the fetal plasma is about 50% of the maternal level. The level remains low at least for the first postnatal weeks and then gradually increases until it reaches the mature level.

Treatment of newborn rats with corticosterone on the second postnatal day leads to enzyme induction in the brain, liver, and heart. For example, the level of aspartate aminotransferase increases in the heart [123]. Glucocorticoids cross the sarcolemmal membrane of the heart and enter the cytosol [124, 125, 126] where corticosterone binds to a receptor protein. The receptor-hormone complex enters the nucleus causing an increase in the level of mRNA and in polymerase activity through a multi-step interaction [127]. There appear to be differences between rat and rabbit glucocorticoid receptors [125]. It has been suggested that the most important effect of glucocorticoid may be on the low-affinity calcium binding sites on the external surface of the sarcolemma [128]. The significance of these apparently abnormal interactions with fetal rat myocardium remains to be explained.

The daily plasma corticosterone level in mice [129] is not comparable with that in the rats [130]. The level of corticosterone is low (1ug/100ml) in newborn mice, increases rapidly on the 12th postnatal day, and by the 20th postnatal day reaches 10 to 15 ug/100ml. The level declines thereafter and adjusts to 2.6 ug/100ml. A description of the mechanical properties of mouse myocardium during this period is not available. It remains to be seen if the behavior of mouse myocardium resembles the abnormal mechanical behavior of rat heart.

The level of glucocorticoids so carefully measured in fetal lamb and rhesus monkey (see above) was not followed postnatally, and plasma levels during development are unknown. Nor has the force-frequency response been examined in these species. It is presumed that, like other mammalian myocardium, newborn lamb and monkey exhibit a positive force-frequency response.

It has become clear that there is a close relationship between circulatory corticosterone levels and the ability of the myocardium to respond to stress. The relationship is particularly dramatic when one compares daily circulating level of corticosterone in neonatal rat [130] with the mechanical performance of the myocardium [131]. Fetal rat myocardium exhibits a positive force-frequency response. The response reverts to negative soon after birth, and a maximum negative force-frequency response is observed about the 14th postnatal day (see figure 7–10). Beyond this point the force-frequency response becomes less negative. By the fourth to sixth postnatal weeks, there is virtually no negative force-frequency response within the experimental range [131]. In newborn piglet myocardium, a slight negative force-frequency is sometimes observed [132]. This negative response reverts to a positive response when the calcium concentration in the bathing solution is doubled. A similar calcium-dependent reversal of response is not seen in the newborn rat. Newborn rat myocardium does, however, respond strongly to doubling calcium concentration in the tissue bath (see figure 7–11). However, when the force-frequency response was studied, it remained negative with a pattern similar to that seen in normal calcium (see figure 7–12). It appears that calcium concentration was rate-limiting in piglet myocardium and that perhaps calcium channels were the rate-limiting step in rat myocardium.

The hypothesis that the calcium channels were rate-limiting in the response of newborn rat myocardium to stress was tested recently in order to determine whether glucocorticoids are involved in the function or control of calcium chan-

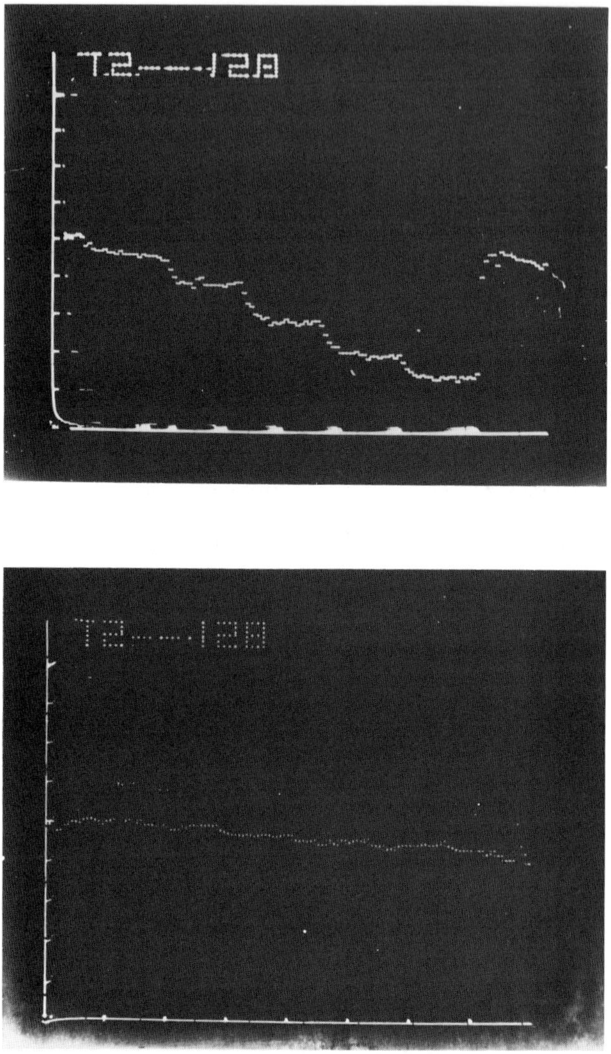

Figure 7–10. Force-frequency response recorded from a developing (15-day-old), and a mature (two-month-old) rat myocardium. Top: A highly negative force-frequency response was present as the rate of stimulation was increased stepwise from 10/min. to 60/min. Bottom: The same procedure was almost without effect on a two-month-old rat heart preparation.

nels [47]. It was found that the negative force-frequency response characteristic of newborn rat heart was reversed to a positive response characteristic of other mammals when rats were injected with cortisol on the second postnatal day. Cortisol and the synthetic glucocorticoid dexamethasone were equally effective in bringing about the reversal in 12-day-old rat myocardium [75].

It should be mentioned that dexamethasone, like other glucocorticoids, crosses the sarcolemmal membrane [127], binds to a receptor protein, and is carried to the

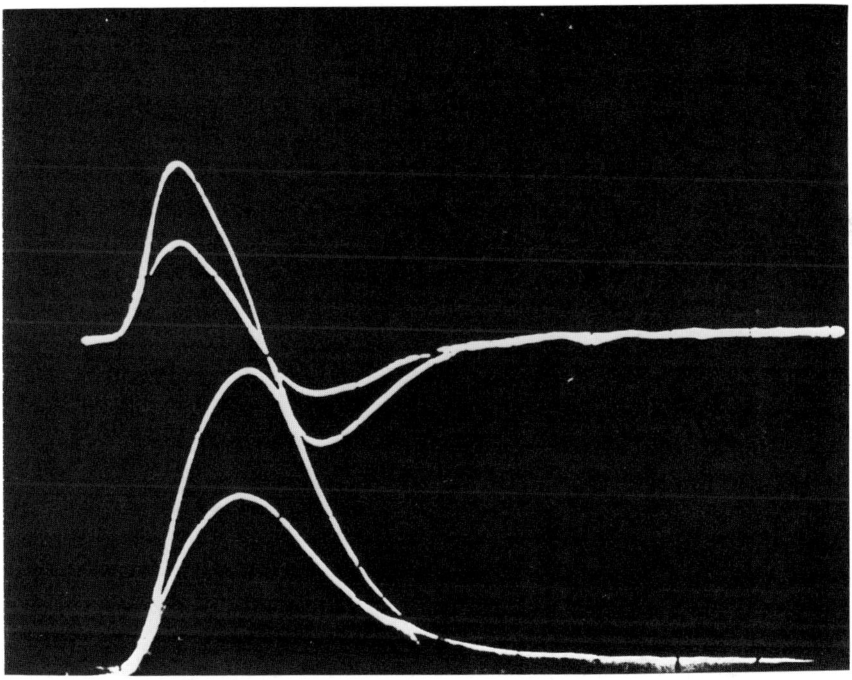

Figure 7–11. Effects of doubling calcium concentration from 2.7 mM to 5.4 mM on steady-state contractile tension of a 15-day-old rat heart.

nucleus where it presumably can induce protein synthesis as it does in other systems.

A systematic study of the role of glucocorticoids in the differentiation of myocardium during fetal and postnatal development is not available. However, these observations on the relationship between myocardial function and circulating glucocorticoids in fetal and newborn rats and on the effects of hormone treatment on mechanical performance emphasize the need for a new approach to the mechanism of regulation of development in mammalian myocardium.

The Hypophyseal Hormones

ACTH

The synthesis of glucocorticoids is under feedback control of ACTH, a hypophyseal hormone. Alterations in hormone level in fetal rat (by combined maternal adrenalectomy and fetal hypophysectomy) retards the development of the zona

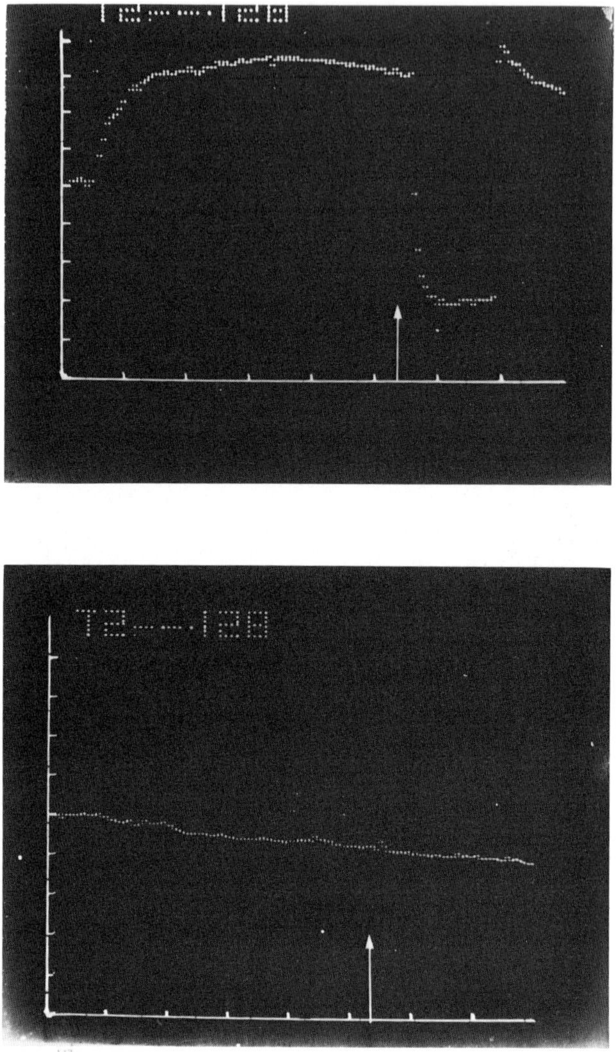

Figure 7–12. Effects of high calcium and stress on developing and mature rat hearts. The preparations were subjected to identical procedures. Calcium concentration was first increased from 2.7 mM to 5.4 mM. After equilibration, the stimulation rate was increased in one step from 10/min. to 60/min. for 20 consecutive stimuli, ana back to 10/min. Top: in the developing heart, high calcium increased peak tension (T2). However, at high rate of stimulation (arrow), the muscle failed to the same extent as it did in normal calcium (see figure 7–10, top). Bottom: in the mature heart, high calcium slightly inhibited contractile tension. However, increasing stimulation rate was without effect (compare figure 7–10, bottom).

glomerulosa in the fetal adrenals [133]. Evidence for a controlling role of ACTH in the maturation of adrenal cortex in fetal rhesus monkey has also been presented [134]. In the newborn rat, it was demonstrated that ACTH can induce the release of corticosterone on the second postnatal day. Control was not fully effective, however, until after the second postnatal week [135]. This observation raises the question whether in the rat the abnormal myocardial force-frequency response in early life results from immaturity of the hypophyseal cells responsible for synthesis and release of ACTH. An age-dependent response to ACTH has been demonstrated [136]. When adrenal glands from 60-day-old and 340-day-old rats were exposed to ACTH and cAMP, the older glands accumulated less cAMP and released less corticosterone.

Growth Hormone

The presence of growth hormone is required for normal function and activity of glucocorticoids [138]. A reducing enzyme in serum inactivates corticosterone. It is thought that interaction of growth hormone with ACTH inhibits enzymatic reduction and maintains the active form of the glucocorticoid [138].

Other Factors

The role of beta-adrenergic receptors on the regulation of contractility was detailed above. The sympathetic and parasympathetic autonomic nervous systems also regulate myocardial contractility via a feedback mechanism. The development of the parasympathetic system and its role in the modulation of sympathetic system is thus of interest. There seems to be species differences in the order of genesis of sympathetic and parasympathetic receptors. It was shown that in fetal mice, between the 12th and 14th gestation days, acetylcholine induced marked bradycardia. The response to epinephrine was minimal during this period [139]. Consequently, cholinergic responses appear before the adrenergic responses in the mouse. The order of development appears to be reversed in the rat since adrenergic receptors were observed on the tenth gestation day and cholinergic receptors appeared one and one-half days later [100].

Effects of Oxygen and PH

Several reports indicate that fetal myocardium is more resistant to hypoxia than either the newborn or adults of the species [5]. Studies reported with fetal lamb indicate that the in utero oxygen concentration is twice as high as the fetus requires for normal function [140]. The Na-K ATPase is equally sensitive to anoxia in both newborn and adult rat myocardium [141]. There are striking differences in the pH-dependence of myosin ATPase extracted from newborn puppies and adult dog myocardia [142]. In the adult preparations a shift in pH from 7.0 to 6.5 shifted the pCa-myosin relationships about 0.4 pCa units (i.e., to a higher calcium concentration). No difference was seen in the fetal myosin preparation under the same conditions. The insensitivity of fetal myosin to pH changes may be related to the tolerance of hypoxia observed in newborn myocardium.

MECHANISM OF CONTRACTION

The commonly accepted theory of myocardial contraction is based upon the "sliding filaments" proposal suggested by Huxley [143] for skeletal muscle. Since electron micrographs show a similar sarcomeric structure in both skeletal and cardiac muscle fibers, it is assumed that the course of contraction and relaxation is similar in both types of muscle. Briefly stated, the wave of activity generated in the pacemaker area and carried by the conducting fibers depolarizes the cardiac sarcolemma and passes along the T tubules. At the site of the junction of the T tubules and the SR system (the diad), the calcium channels open and calcium crosses the membrane. This calcium, known as "trigger calcium," causes the release of stored calcium from the sarcoplasmic reticulum by an as yet unknown mechanism. The phenomenon is referred to as "calcium-induced calcium release," and the latter calcium is called "activator calcium." Activator calcium binds to troponin C (Tn C), removing the inhibition exerted by troponin I (Tn I) on the interaction between actin and myosin. Actin is then free to interact with the myosin heads. The myosin heads (cross-bridges) bind to a second actin on the thin filaments as soon as the previous actin-myosin bond is broken. This action moves the thin filaments towards the center of the sarcomere and shortens the muscle without altering the length of the myofilaments. Bond formation is the source of the force generated by the muscle. Total force generated at any given time is proportional to the number of bonds formed. The velocity of shortening is proportional to the rate of cross-bridge cycling.

According to this theory, relaxation supervenes when calcium is sequestered from the cytosol by the pumping action of SR system. As more calcium is removed, more bonds break and the muscle returns to a resting state when the free calcium concentration in the cytosol is about 10^{-6} M.

The foregoing brief account of the major molecular events in muscle contraction illustrates the central role of calcium in the mechanical performance of muscle. Although pure actin and pure myosin can interact in the absence of calcium if ATP is present, the "contraction" (superprecipitation) is very slow and the system does not relax spontaneously. It is clear that the regulatory proteins (the tropomyosin-troponin system) play an essential role in separating actin and myosin until a wave of activity starts a new cycle. Calcium is therefore a key element in the promotion of contraction.

The physical and chemical properties of calcium relevent to the multipurpose role of the metal in a wide variety of biological systems was recently reviewed [144]. It appears that the size of the calcium ion and its hydrated shell make calcium ideal for interaction with living materials, particularly proteins. Among the various mono- and divalent cations, calcium most readily "fills" the cavities in proteins. Other divalent cations such as strontium and barium substitute only poorly for calcium. It is thus evident that at least two systems compete for free calcium in the cytosol: (1) Tn C, which binds calcium to promote contraction, and (2) the calcium sequestering elements whose function it is to remove calcium from the cytosol. Calcium will be preferentially bound to Tn C as long as the calcium concentration of the cytosol is in the range of the affinity constant of the binding sites on the protein. Calcium dissociates from Tn C when the SR reduces the cytosolic calcium concentration sufficiently. The smooth operation of the contraction-relaxation cycle is hinged upon a single interaction (phosphorylation). It is remarkable that phosphorylation of sarcolemmal elements promotes the entry of calcium into the cytosol. While phosphorylation of the pump proteins enhances the calcium sequestering function of the SR and the sarcolemmal pumps, simultaneous phosphorylation of Tn I reduces the affinity of Tn C for calcium. Thus the same device lowers the calcium affinity of Tn C and raises the efficiency of the pumps. Phosphorylation produces the fine tuning necessary for a smooth contraction and relaxation cycle.

PHOSPHATE BINDING PROTEINS AND THEIR REGULATORY MECHANISM

Incubation of cardiac muscle homogenate with radioactive phosphate and the appropriate enzymes results in phosphorylation of a number of proteins which have been separated and identified. Only a few were shown to have a regulatory function in the contraction and relaxation of myocardium.

1. A 23,000-dalton protein found in the sarcolemmal fraction, calciductin, is closely related to the calcium channels [69]. Phosphorylation of calciductin occurs readily and leads to an increase in the activity of calcium channels.
2. A second phosphorylated protein is the Ca-Mg ATPase, the calcium pump [84, 86, 87, 88]. This protein also is more active in the phosphorylated form and is assumed to remove trigger calcium from the cytosol.
3. The contractile elements actin and myosin as well as the regulatory proteins also may be phosphorylated. However, there is no evidence that phosphorylation of either actin or myosin in the myocardium alters the function of these proteins. Phosphorylation of Tn I (as explained earlier) is an important regulator of contractility.
4. The SR calcium pump is active in the phosphorylated form, sequestering free calcium. This action is closely related to the regulation of contractility.

MECHANISM OF PHOSPHORYLATION

Two known mechanisms are involved in phosphorylation of proteins regulating myocardial contractility: cAMP-dependent protein kinase and calcium-calmodulin-dependent protein kinase.

cAMP-Dependent Protein Kinase

Both phospholamban and calciductin are phosphorylated by such kinases. Activation of the kinases is initiated by stimulation of beta adrenergic receptors on the sarcolemma.

Calcium-Calmodulin-Dependent Protein Kinase

This enzyme is activated by calcium and acts in parallel with the cAMP-dependent system. The mechanical responses resulting from activation of the two kinases are different in the following ways: (1) the catecholamine-initiated response abbreviates time to peak tension (TPT) in adult myocardium, while the calcium-initiated response has no effect on contraction-relaxation time; (2) the catecholamine-initiated response invariably causes spontaneous activity in the adult myocardium while the calcium-initiated response does not. On the other hand, developing myocardium responds identically to both catecholamines and high calcium. The reasons for these differences of response are not known.

SUMMARY

Myocardial contraction entails a complex system of sequential and simultaneous processes that are finely tuned via feedback systems to maintain smooth function. Although it is said that the heart responds in "all or none" fashion, it is known that the contractile state of the heart can change from moment to moment to accommodate the demands of the body. The sarcolemma, contractile elements, and regulatory proteins as well as the metabolic elements of the myocardium are all involved in the maintenance of the contractile state. The development of this organ is not complete at birth, and in fact in mammalian heart at birth a significant number of the myocytes are totally undifferentiated. Relatively little is known, at the molecular level, of the various structural systems in developing heart. Much less is understood about the mechanism of function of the immature tissue. There is still much to be learned about the mode of regulation of contractility in developing mammalian heart.

REFERENCES

1. Friedman, W.F. 1972. The intrinsic properties of the developing heart. *Prog. in Cardiovas. Diseases* 15:87–111.
2. Rowe, M.I., and Arango, A. 1977. A critical appraisal of the common "neonatal" experimental subjects. *J. Ped. Surg.* 12:409–412.
3. van Petten, G.R.; Harris, Meard, G.J. 1976. Development of fetal cardiovascular responses to alpha-adrenergic agonists. In *Fetal and newborn cardiovascular physiology,* ed. L.D. Longo and D.D. Reneau, pp. 153–166. Garland Publishers.
4. Clark, J.B., and Clark, C.M. 1982. The growth and metabolism of developing heart. In *Biochemical development of the fetus and neonate,* ed. C.T. Jones, pp. 167–212. Elsever Biomed. Press.
5. Penefsky, Z.J. 1983. Perinatal development of cardiac mechanisms. In *Perinatal cardiovascular function,* ed. N. Gootman and P.M. Gootman, pp. 109–200.
6. Baker, J.B.E. 1953. Some observations upon perfused human fetal hearts. *J. Physiol.* 120:122–128.
7. Coltart, D.J., and Spilker, B.A. 1972. Development of human foetal inotropic responses to catecholamines. *Experientia* 285:525–526.
8. Andersson, K.E.; Gennser, G.; and Nilsson, E. 1970. Influence of contraction rate, coronary perfusion and diastolic pressure on the mechanical performance of isolated fetal heart. *Acta Physiol. Scand.* Suppl. 353, 80:5–19.
9. Gennser, G., and Nilsson, E. 1968. The relation between the action potential and the active state in human fetal myocardium and its dependence on muscle length and contraction frequency. *Acta Physiol. Scand.* 73:42–53.
10. Myklebust, R.; Soestersdal, T.S.; Engedal, H.; and Ulstein, M. 1978. Ultrastructural studies on the human embryonic and adult hypertrophied heart. *Anat. Embryol.* 152:127–140.
11. Kranias, E.G., and Solaro, R.J. 1982. Phosphorylation of troponin I and phospholamban during catecholamine stimulation of rabbit heart. *Nature* 298:182–184.
12. Johnson, J.D.; Collins, J.H.; Robinson, S.P.; and Potter, J.D. 1980. A fluorescent probe study of calcium binding to the calcium-specific sites of cardiac troponin and troponin C. *J. Biol. Chem.* 255:9635–9640.
13. Holroyde, M.J.; Robertson, S.P.; Johnson, Solaro, R.J.; and Potter, J.D. 1980. The calcium and magnesium binding sites on cardiac troponin and their role in the regulation of myofibrillar adenosin triphosphatase. *J. Biochem.* 255:11688–11693.

14. Potter, J.D., and Gergley, J. 1975. The calcium and magnesium binding sites on troponin and their role in regulation of myofibrillar adenosin triphosphatase. *J. Biol. Chem.* 250:4628–4633.
15. Robertson, R.J.; Johnson, J.D.; Holroyde, M.J.; Kranias, E.G.; Potter, J.D.; and Solaro, R.J. 1982. The effect of troponin I phosphorylation on the calcium binding proteins of the calcium regulatory sites of bovine cardiac troponin. *J. Biol. Chem.* 257:260–263.
16. Solaro, R.J.; Robertson, S.P.; Johnson, J.D.; Holroyde, M.J.; and Potter, J.D. 1981. Tropônin I phosphorylation: a unique regulator of the amount of calcium required to activate cardiac myofibers. *Cold Spring Harbor Conference on Cell Proliferation: Protein Phosphorylation.* 8:901–911.
17. Demaille, J.G., and Pechere, J.F., Group. 1983. The control of contractility by protein phosphorylation. *Adv. in Cyclic Nucleotides Res.*, ed. P. Greengard and G.A. Robison, pp. 337–371.
18. Potter, J.D.; Robertson, S.P.; and Johnson, J.D. 1981. Magnesium and the regulation of muscle contraction. *Fed. Proc.* 40:2653–2656.
19. Robertson, S.P.; Johnson, J.D.; and Potter, J.D. 1981. The time course of calcium exchange with calmodulin, troponin, parvalbumin, and myosin in response to transient increases in calcium. *Biophys. J.* 34:559–569.
20. Solaro, R.J. The regulatory proteins of developing heart. Manuscript submitted for publication.
21. Rockstein, M.; Chesky, J.A.; and Lopez, T. 1978. Calcium sensitivity of myocardial actomyosin ATPase in young and mature male fischer rats. A brief note. In *Mechanism of ageing and development,* ed. R. Goldman and M. Rockstein, pp. 413–416.
22. Perry, S.V. 1979. The regulation of contractile activity of muscle. *Biochem. Soc. Trans.* 7:593–617.
23. Johnson, D.J.; Robinson, D.E.; Robertson, S.P.; Schwartz, A.; and Potter, J.D. 1981. Calcium exchange with troponin and the regulation of muscle contraction. In *The regulation of muscle contraction: excitation-contraction coupling,* ed. A.D. Grinnel, pp. 241–259. Acad. Press.
24. Martin, F.; Gabrion, J.; and Cavadore, J.C. 1981. Myosin filaments assembly-disassembly is controlled by myosin light chain phosphorylation-dephosphorylation. *FEBS Lett.* 131:235–238.
25. Waters, J. 1981. Changes in rat cardiac myosin during development and in culture. *J. Molec. Cell. Cardiol.* 13:1011–1013.
26. Wolf, H., and Hoffman, F. 1980. Purification of myosin light chain kinase from bovine cardiac muscle. *Proc. Natl. Acad. Sci., USA* 77:5852–5855.
27. Walsh, H.P.; Vallet, B.; Autrie, F.; and Demaille, J.G. 1979. Purification and characterization of bovine cardiac calmodulin-dependent myosin light chain kinase. *J. Biol. Chem.* 254:12135–12149.
28. Walsh, M.P.; Vallet, B.; Cavadore, J.C.; and Demaille, J.G. 1980. Homologous calcium binding proteins in the activation of skeletal, cardiac and smooth muscle myosin light chain kinases. *J. Biol. Chem.* 255:335–337.
29. Kopp, S.J., and Barany, M. 1979. Phosphorylation of the 19,000 dalton light chain of myosin in perfused rat heart under the influence of negative and positive inotropic agents. *J. Biol. Chem.* 254:12007–12012.
30. Westwood, S.A., and Perry, S.V. 1981. The effect of adrenaline on phosphorylation of the p-light chain of myosin and troponin I in the perfused rabbit heart. *Biochem. J.* 197:185–193.
31. Zak, R.; Chizzonite, R.A.; Everett, A.W.; and Clark, W.A. 1982. Study of ventricular isomyosins during normal and thyroid hormone induced growth. *J. Molec. Cell. Cardiol.* Suppl. 3, 14:111–117.
32. Chizzonite, R.A., and Zak, R. Synthesis of isomyosin heavy chains in developing rat ventricle is dependent on endogenous thyroid hormone. submitted for publication.
33. Schwartz, K.; Lecarpentier, Y.; Martin, J.L.; Lompre, M.; Mercadier, J.J.; and Swynghedauw, B. 1981. Myosin isoenzymic distribution correlation with speed of myocardial contraction. *J. Molec. Cell. Cardiol.* 13:1071–1075.
34. Kranias, E.G.; Bilezikian, L.M.; Potter, J.D.; Piascik, M.T.; and Schwartz, A. 1980. The role of calmodulin in regulation of cardiac sarcoplasmic reticulum phosphorylation. *Ann. N.Y. Acad. Sci.* 356:279–291.
35. Louis, C.F., and Moffitt, M. 1982. Characterization of calmodulin-mediated phosphorylation of cardiac muscle sarcoplasmic reticulum. *Arch. Biochem. Biophys.* 218:109–118.

36. Caroni, P., and Carafoli, E. 1981. The calcium pumping ATPase of the heart sarcolemma: characterization, calmodulin dependence and partial purification. *J. Biol. Chem.* 256:3263–3270.
37. Kirchberger, M.A., and Antonetz, T. 1982. Calmodulin-mediated regulation of calcium transport, and calcium-magnesium-activated activity in isolated cardiac sarcoplasmic reticulum. *J. Biol. Chem.* 257:5685–5691.
38. Rowland, E., and Scordilis, S.P. Calmodulin levels during the development of normal and disgenic Mice. Submitted for publication.
39. Chamberlain, B.K.; Levitsky, D.O.; and Fleischer, S. 1983. Characterization of canine cardiac sarcoplasmic reticulum with improved calcium transport proteins. *J. Biol. Chem.* 258:6602–6609.
40. Brandt, N.R.; Brunschwig, J.P.; and Lallanzio, F.A. Identification of T tubules/junctional sarcolemma in cardiac muscle. Submitted for publication.
41. Page, E., and Buecker, J.L. 1981. Development of diadic junctional complexes between sarcoplasmic reticulum and plasmalemma in rabbit left ventricular myocardial cells: morphometric analysis. *Circ. Res.* 48:519–522.
42. Hoerter, J.; Mazet, F.; and Vassort, G. 1981. Perinatal growth of the rabbit cardiac cell: possible implications for the mechanism of relaxation. *J. Molec. Cell. Cardiol.* 13:725–740.
43. Mitchell, R.D.; Saito, A.; Palade, P.; and Fleischer, S. 1983. Morphology of the isolated triads. *J. Cell. Biol.* 96:1017–1029.
44. Boland, R., and Martonosi, A. 1974. Development changes in the composition and function of sarcoplasmic reticulum. *J. Biol. Chem.* 240:612–623.
45. Nayler, W.G., and Fassold, E. 1977. Calcium accumulating and ATPase activity of cardiac sarcoplasmic reticulum before and after birth. *Cardiovas. Res.* 11:231–237.
46. Fabaito, A., and Fabiato, F. 1978. Calcium induced release of calcium from the sarcoplasmic reticulum of skinned cells from adult human, dog, cat, rabbit, rat and frog hearts and from fetal and newborn rat ventricles. In *Calcium transport and cell function. Aa. NY Acad. Sci.* 307:491–522.
47. Penefsky, Z.J.; Wolfe, J.M., and Greengard, O. Effects of cortisol on developing rat myocardium. Manuscript in preparation.
48. George, B.L.; Shimizu, T.; and Jarmakani, J.M. 1980. Age-related differences in the effects of caffein on cardiac e-c coupling. *Circ.* Suppl. 3, 62:11–25.
49. Lepeuch, C.J.; Guilleux, J.C.; and Demaille, J.G. 1980. Phospholamban phosphorylation in the perfused rat heart is not solely dependent on beta adrenergic stimulation. *FEBS Lett.* 114:165–168.
50. De Meis, L., and Inesi, G. 1982. ATP synthesis by sarcoplasmic reticulum ATPase following calcium, pH, temperature and water activity jump. *J. Biol. Chem.* 257:1289–1294.
51. Kirchberger, M.A.; Tada, M.; Repke, D.; and Katz, A.M. 1972. Cyclic adenosine monophosphate-dependent protein kinase stimulation of calcium uptake by canine microsomes. *J. Molec. Cell. Cardiol.* 4:673–680.
52. Kirchberger, M.A.; Tada, M.; and Katz, A.M. 1974. Adenosine 3-5monophosphate dependent protein kinase catalyzed phosphorylation reaction and its relationship to calcium transport in cardiac sarcoplasmic reticulum. *J. Biol. Chem.* 249:6166–6173.
53. Tada, M.; Kirchberger, M.A.; Repke, D.; and Katz, A.M. 1974. The stimulation of calcium transport in cardiac sarcoplasmic reticulum by adenosine 3-5monophosphate-dependent protein kinase. *J. Biol. Chem.* 249:6174–6180.
54. Kirchberger, M.A.; Tada, M.; and Katz, A.M. 1975. Phospholamban: a regulatory protein of the cardiac sarcoplasmic reticulum. In *Recent Advances in Studies on Cardiac Structure and Metabolism,* ed. A. Fleckenstein and N.S. Dhalla, pp. 103–115.
55. Tada, M.; Kirchberger, M.A.; and Katz, A.M. 1975. Phosphorylation of a 22,000 dalton component of the cardiac sarcoplasmic reticulum by adenosine 3-5monophosphate-dependent protein kinase. *J. Biol. Chem.* 250:2640–2647.
56. Kirchberger, M.A., and Raffe, A. 1977. Decrease in calcium transport associated with phosphoprotein phorphorylase-catalyzed dephosphorylation of cardiac sarcoplasmic reticulum. *J. Cyclic. Nucleotide Res.* 3:45–53.
57. Bilezikian, L.M.; Kranias, E.G.; Potter, J.D.; and Schwartz, A. 1980. Calmodulin stimulated phosphorylation of cardiac sarcoplasmic reticulum. *Fed. Proc.* 39:1663.

58. Bilezikian, L.M.; Kranias, E.G.; Potter, J.D.; Schwartz, A. 1981. Studies on phosphorylation of canine cardiac sarcoplasmic reticulum by calmodulin-dependent protein kinase. *Circ. Res.* 49:1356–1362.
59. Kirchberger, M.A., and Antonetz, T. 1982. Phospholamban: dissociation of 22,000 molecular weight protein of cardiac sarcoplasmic reticulum into 11,000 and 5,500 molecular weight forms. *Biochem. Biophys. Res. Comm.* 105:152–156.
60. Tada, M.; Inui, M.; Yamada, M.; Kadoma, M.; Kuzuya, T.; Abe, H.; and Kakiuchi, S. 1983. Effects of phospholamban phosphorylation catalyzed by adenosine 3-5monophosphate and calmodulin-dependent protein kinase on calcium transport ATPase of cardiac sarcoplasmic reticulum. *J. Molec. Cell. Cardiol.* 15:335–346.
61. Novak, E.; Drumond, G.I.; Skala, J.; and Hahn, P. 1977. Developmental changes in cyclic AMP, protein kinase, phosphorylase kinase and phosphorylase in liver, heart and skeletal muscle of rat. *Arch. Biochem. Biophys.* 150:511–518.
62. Penefsky, Z.J.; Barry, C.R.; Buckley, N.M.; and Sorrenson, A.L. 1980. Mechanical and electrical properties of developing swine myocardium. *The Physiologist,* 23:117.
63. Carafoli, E. 1972. Mitochondria and sarcoplasmic reticulum in the calcium shifts in cardiac cells. *Colloq. INSERM Surcharges Card.* 193–204.
64. Carafoli, E., and Crompton, M. 1976. Calcium ions and mitochondria. *Symp. Soc. Exp. Biol.* 30:89–115.
65. Grinwald, P.M., and Nayler, W.G. 1981. Calcium entry in the calcium paradox. *J. Molec. Cell. Cardiol.* 13:867–880.
66. Fran, J.S., and Rich, T.L. 1983. Calcium depletion and repletion in rat heart: age-dependent changes in sarcolemma. *Am. J. Physiol.* 245:H343–H353.
67. Shigenobu, K., and Sperelakis, N. 1974. Failure of development of fast sodium channels during organ culture of young embryonic heart. *Developmental Biol.* 38:326–330.
68. Walsh, D.A.; Clippinger, M.S.; Sivaramakrisnan, S.; and McCullough, T.E. 1979. Cyclic adenosine monophosphate dependent and independent phosphorylation of sarcolemmal membrane proteins in perfused rat heart. *Biochem.* 18:871–877.
69. Rinaldi, M.L.; Capony, J.P.; and Demaille, J.G. 1982. The cyclic AMP-dependent modulation of cardiac sarcolemmal slow calcium channels. *J. Molec. Cell. Cardiol.* 14:279–289.
70. Cachelin, A.B.; de Peyer, J.E.; Kokubun, S.; and Reuter, H. 1983. Calcium channel modulation by 8-bromocyclic AMP in cultured heart cells. *Nature* 304:862–864.
71. Gargouil, Y.-M.; Coraboeuf, E.; Tricoche, R.; and Wallon, G. 1959. Evaluation des electrogrammes cardiac chez le rat au cours de La croissance. *c.R. Seances Soc. Biol.* CLIII:450–453.
72. Couch, J.R.; West, T.; and Hoff, H. 1969. Development of the action potential of the prenatal rat heart. *Circ. Res.* 24:19–31.
73. Robinson, R.B. 1982. Action potential characteristics of rat cardiac cells do not change with time in culture. *J. Molec. Cell. Cardiol.* 14:367–370.
74. van Ginneken, A.C.G., and Jongsma, H.J. 1983. Slow inward current in aggregates of normal rat heart cells, and its contribution to steady state current-voltage relationship. *Pflugers Arch.* 397:265–271.
75. Penefsky, Z.J., and McCann, F.V. Electrical and mechanical changes in newborn S-D rat myocardium. Effects of glucocorticoids. Manuscript in preparation.
76. Binah,; Legato, M.J.; Danilo, P.; and Rosen, M. 1983. Developmental changes in the cardiac effects of amrinone in the dog. *Circ. Res.* 52:747–752.
77. Penefsky, Z.J.; Sorenson, A.L.; Barry, C.R.; and Buckley, N.M. 1980. Postnatal development of mechanical and electrical response in swine myocardium. *J. Molec. Cell. Cardiol.* Suppl. 1, 12.
78. Penefsky, Z.J., and McCann, F.V. Electrical and mechanical responses of aging S-D rat myocardium. Manuscript in preparation.
79. Forester, G.V., and Mainwood, G.W. 1974. Ouabain sensitivity in the rat myocardium: correlation with a model of subcellular calcium movement. In *Myocardial biology,* ed. N.S. Dhalla, pp. 273–279. University Park Press.
80. Penefsky, Z.J., and Scott, W.N. 1982. Alterations in mechanical responses of aged rat myocardium: effects of dexamethasone. *Comp. Biochem. Physiol.* 72A:559–569.
81. Allen, J.C.; Entman, M.L.; and Schwartz, A. 1975. The nature of the transport adenosine triphosphatase-digitalis complex. VIII. The relationship between in vivo formed (H-ouabain

sodium potassium adenosine triphosphatase) complex and ouabain-induced positive inotropism. *J. Pharmacol. Exp. Ther.* 192:105–112.
82. Schwartz, A.; Lindenmeyer, G.E.; and Allen, J.C. 1972. The Na-K ATPase membrane transport system: importance in cellular function. In *Current topics in membrane and transport,* ed. F. Bronner and A. Kleinzeller, pp. 1–82. New York: Academy Press.
83. Langer, G.A.; Brady, A.J.; Tan, S.T.; and Serena, S.D. 1975. Correlation of the glycoside response, the force staircase, and the action potential configuration in the neonatal rat. *Circ. Res.* 36:744–752.
84. Penefsky, Z.J. 1983. Effects of inotropic agents on adrenalectomized rat heart muscle. *J. Molec. Cell. Cardiol.* Suppl. 4, 15:36.
85. Caroni, P., and Carafoli, E. 1980. An ATP-dependent ca pumping system in dog heart sarcolemma. *Nature* 283:765–767.
86. Caroni, P., and Carafoli, E. 1981. The ca-pumping ATPase of the heart sarcolemma by a phosphorylation-dephosphorylation process. *J. Biol. Chem.* 256:9371–9373.
87. Lamers, J.M., and Steinis, J.T. 1980. Phosphorylation of low molecular weight proteins in purified preparations of rat heart sarcolemma and sarcoplasmic reticulum. *Biochem. Biophys. Acta* 624:443–459.
88. Jones, L.R.; Maddock, S.W.; and Hattaway, D.R. 1981. Membrane localization of myocardial type II cAMP-dependent protein kinase activity. *Biochem. Biophys. Acta* 641:212–253.
89. Dzurba, A.; Ganguly, P.K.; Beamish, R.E.; and Dhalla, N.S. 1983. Stimulation of calcium pump activity in heart sarcolemma by timolol. *Can. J. Physiol. Pharmacol.* 61:240–244.
90. Caroni, P., and Carafoli, E. 1981. Regulation of ca-pumping ATPase of heart sarcolemma by a phosphorylation-Dephosphorylation process. *J. Biol. Chem.* 256:9371–9373.
91. Caroni, P.; Reinaldi, L.; and Carafoli, E. 1980. Charge movements during Na-Ca exchange in heart sarcolemmal vesicles. *Proc. Natl. Acad. Sci. USA* 77:6354–6358.
92. Reuter, H., and Seitz, N. 1968. The dependence of calcium efflux from cardiac muscle on temperature and ion composition. *J. Physiol.* 195:451–470.
93. Philipson, K.D., and Nishimoto, A.Y. 1981. Efflux of calcium from cardiac sarcolemmal vesicles: influence of external Ca and Na. *J. Biol. Chem.* 256:3698–3702.
94. Jundt, H.; Porzig, H.; Reuter, H.; and Stucki, J.W. 1975. The effects of substances releasing intracellular calcium ions on sodium dependent calcium efflux from guinea pig auricles. *J. Physiol.* 246:229–253.
95. Katz, A.M. 1972. Increased Ca entry during the plateau of the action potential: possible mechanism of cardiac glycoside action. *J. Molec. Cell. Cardiol.* 4:87–89.
96. Langer, G.A. 1982. Calcium exchange in the heart. *Ann. Rev. Physiol.* 44:435–449.
97. Bridge, J.H.B., and Bassingthwaighte, J.B. 1983. Uphill sodium transport driven by an inward calcium gradient in heart muscle. *Science:* 219:178–180.
98. Chapman, R.A. 1983. Control of cardiac contractility at the cellular level. *Am. J. Physiol.* 245:H535–H552.
99. Martin, S.; Levey, B.A.; Levey, G.S. 1973. Development of the cardiac beta adrenergic receptors in fetal rat heart. *Biochem. Biophys. Res. Comm.* 54:949–954.
100. Rebkin, M.A.; Shepard, T.H.; and Dyer, D.C. 1976. Autonomic receptors of the early rat embryo heart: growth and development. *Proc. Soc. Exp. Biol. Med.* 15:799–803.
101. Buckley, N.M.; Gootman, P.M.; Yellin, E.L.; and Braseau, P. 1979. Age-related cardiovascular effects of catecholamines in anesthetized piglets. *Circ. Res.* 45:282–292.
102. Walker, M.T.A. 1978. Initial investigations into adrenaline accumulation and adrenergic response in cultured neonatal rat heart cells. *Br. J. Pharmacol.* 62:185–193.
103. Gennser, G., and Studnitz, W.V. 1975. Noradrenaline synthesis in human fetal heart. *Experientia* 31:1422–1424.
104. Fleisch, J. 1981. Age-related decrease in beta adrenoceptors activity of the cardiovascular system. *Trends in Pharmacological Science* 2:337–340.
105. Au, T.L.S.; Collins, G.R.; Walker, M.J.A. 1980. Rate, force, and cyclic adenosine 3-5 monophosphate responses to (-)adrenaline in neonatal rat heart tissue. *Br. J. Pharmacol.* 69:601–608.
106. Whitsett, J., and Darovec-Beckerman, C. 1981. Developmental aspects of beta adrenergic receptors and catecholamine-sensitive adenylate cyclase in rat myocardium. *Ped. Res.* 15:1363–1369.
107. Dail, W.G., Jr., and Almer, G.C. 1973. Localization and correlation of catecholamine containing cells with adenylate cyclase and phosphodiesterase activitiets in the human fetal heart. *Anat. Rec.* 177:265–288.

108. Gennser, G., and Nilsson, E. 1970. Response to adrenaline and changes of contraction frequency in early human fetal hearts. *Experientia* 26:1105–1107.
109. Kohrman, A.F. 1973. Pattern of development of adenyl cyclase activity and norepinephrine responsiveness in the rat. *Ped. Res.* 7:575–581.
110. Geis, W.P.; Takooler, G.J.; Priola, D.V.; and Friedman, W.F. 1975. Factors influencing neurohormonal control of the heart in the newborn dog. *Am. J. Physiol.* 228:1685–1689.
111. Rinaldi, M.L.; Lepeuch, C.J.; and Demaille, J.G. 1981. The epinephrine-induced activation of the cardiac slow calcium channel is mediated by the cAMP dependent phosphorylation of calciductin, a 23,000 mr sarcolemmal protein. *FEBS Lett.* 129:277–281.
112. Brun, G.; Flockerzi, V.; Hoffman, F.; Osterrieden, W.; and Trautwein, W. 1983. Injection of catalytic subunit of cAMP dependent protein kinase into isolated cardiac myocytes. *Pflugers Arch.* 398:147–154.
113. Penefsky, Z.J.; Chizzonite, R.; and Zak, R. Lack of response to glucocorticoids by myosin isoenzymes. *Manuscript in preparation.*
114. Hoch, F. 1982. Thyroid control over biomembranes. VIII. Heart muscle mitochondria from L-Triiodothyronine injected rats. *J. Molec. Cell. Cardiol.* 14:81–90.
115. Jost, A. 1966. Problems of fetal endocrinology: the adrenal gland. *Recent. Progress in Hormone Res.* 22:541–569.
116. Lefer, A.M. 1968. Influence of corticosteroids on mechanical performance of isolated rat papillary muscle. *Mm. J. Physiol.* 214:518–524.
117. Penefsky, Z.J., and Kahn, M. 1971. Inotropic effects of dexamethasone in mammalian heart muscle. *Eur. J. Pharmacol.* 15:259–266.
118. Penefsky, Z.J., and Scott, W.N. 1980. Age-related defects in mammalian myocardium; effect of dexamethasone. *Mt. Sinai J. Med.* 47:117–125.
119. Cohen, A. 1973. Plasma corticosterone in the fetal rat. *Horm. Metab. Res.* 5:66.
120. Cohen, A. 1976. Adrenal and plasma corticosterone levels in the pregnant fetal and neonatal rat, in perinatal period. *Horm. Metab. Res.* 8:474–478.
121. Jaffe, R.B.; Seron-Ferre, M.; and Huhtaniemi, I. 1977. Regulation of the primate fetal adrenal gland and testis in vitro and in vivo. *J. Steroid Biochem.* 8:479–490.
122. Setalo, G., and Nakane, P.K. 1976. Functional differentiation of the fetal anterior pituitary cells in the rat. *Endocr. Exp.* 10:155–166.
123. Greengard, O. 1975. Cortisol treatment of neonatal rats: effects on enzymes in kidney, liver, and heart. *Biol. of Neonate.* 27:352–360.
124. Lan, N.C.; Nguyen, T.; Cathala, G.; Wolf, M.E.; Kallman, P.A.; Wegnez, M.; Martial J.A.; and Baxter, J.D. 1982. Molecular mechanisms of glucocorticoid hormones action. *J. Molec. Cell. Cardiol.* Suppl. 3, 14:43–48.
125. Giannopoulos, G. 1975. Early events in the action of glucocorticoids in developing tissues. *J. Steroid Biochem.* 6:623–631.
126. Daniels, V.G.; Hardy, R.N.; Malinowska, K.W.; and Nathanielsz, P.W. 1972. Adrenocortical hormones and absorption of macromolecules by the small intestines of the young rats. *J. Endocr.* 52:405–406.
127. Seleznev, Y.M.; Danilov, S.M.; Probradjensky, S.N.; Volkova, N.G.; Kuznestova, L.A.; Smirnov, V.N.; and Volcheck, A.G. 1978. Glucocorticoid binding proteins of rat heart cytosol and their possible role in the transfer of glucocorticoids into nuclei. *J. Molec. Cell. Cardiol.* 10:877–891.
128. Seleznev, I.M., and Martynov, A.V. 1982. Permissive effect of glucocorticoids in catecholamine action in the heart: possible mechanism. *J. Molec. Cell. Cardiol.* Suppl. 3, 14:49–58.
129. Diez, J.A.; Sze, P.Y.; and Ginsburg, B.E. 1976. Postnatal development of mouse plasma and brain corticosterone levels: new findings contingent upon the use of a competitive protein-binding assay. *Endocrinology* 98:1434–1442.
130. Allen, C., and Kendall, J.W. 1967. Maturation of the circadean rhythm of plasma corticosterone in the rat. *Endocrinology* 80:926–930.
131. Penefsky, Z.J. 1983. Perinatal changes in contractility of rat myocardium. *Biophys. J.* 41:256a.
132. Penefsky, Z.J.; Buckley, N.M.; Robinson, T.; and Sorrenson, A.L. Mechanical properties of developing swine myocardium. Submitted for publication.
133. Jack, P.M., and Milner, D.G. 1975. Effects of decapitation and ACTH on somatic development of rabbit fetus. *Biol. Neonate.* 26:195–204.
134. Kittinger, G.W.; Beamer, N.B.; Hagemenas, E.; Hill, J.D.; Baughman, W.L.; and Ochsner, A.J. 1972. Evidence for autonomous pituitary-adrenal function in the near term fetal rhesus (macaca mulata). *Endocrinol.* 91:1037–1044.

135. Allen, C., and Kendall, J.W. 1967. Maturation of the circadean rhythm of plasma corticosterone in the rat. *Endocrinol.* 80:926–930.
136. Prichett, J.F.; Sartrin J.L.; Marple, D.N.; Harper, W.L.; and Till, M.I. 1979. Interaction of aging with in vitro adrenocortical responsiveness to ACTH and cyclic AMP. *Horm. Res.* 10:96–103.
137. Challis, J.R.G.; Jones, C.T.; Robinson, J.S.; and Thorburn, G.D. 1977. Development of fetal pituitary-adrenal function. *J. Steoid Biochem.* 8:471–478.
138. Colby, H.D.; Caffrey, J.L.; and Kitay, J.I. 1973. Interaction of G.H. and ACTH in the regulation of adrenocortical secretion in rats. *Endocrinol.* 93:189–192.
139. Widenthal, K. 1973. Maturation of responsiveness to cardio-active drugs. *J. Clin. Invest.* 62:2250–2258.
140. Wilkening, R.B., and Meschia, G. 1983. Fetal oxygen uptake, oxygenation and acid-base balance as a function of uterine blood flow. *Am. J. Physiol.* 244:H749–H755.
141. Schimizu, T.; Nakanishi, T.; Uemura, S.; and Jarmakani, J. 1983. Effects of hypoxia on Ouabain inhibition of sodium pump in newborn rat myocardium. *Am. J. Physiol.* 244:H756–H762.
142. Park, I.S.; Michael, L.H.; and Driscoll, D.J. 1981. Comparative response of developing canine myocardium to inotropic agents. *Am. J. Physiol.* 242:H13–H18.
143. Huxley, A.F. 1957. Muscle strücture and theories of contraction. *Prog. Biolphys.* 17:255–318.
144. Carafoli, E., and Crompton, M. 1978. The regulation of intracellular calcium. In *Current topics in membranes and transport,* ed. A. Kleinzeller and R. Bronner, pp. 151–215.

8. ISOMYOSIN SHIFTS IN NORMAL AND INDUCED CARDIAC GROWTH

KETTY SCHWARTZ AND JEAN-JACQUES MERCADIER

Myosin is the most abundant protein of the contractile apparatus, representing approximately 60% of the total myofibrillar proteins. Each molecule of myosin is composed of two heavy chains (MHC, 200,000 daltons), two regulatory or phosphorylatable light chains (LC2,17,000 to 20,000 daltons), and two alkali light chains, non-phosphorylatable (LC1,29,000 to 22,000 daltons and, in skeletal muscles, LC3 15,000 to 16,000 daltons). The whole molecule can be cleaved in several regions or domains: sub-fragments 1 and 2 (S1 and S2), heavy meromyosin (HMM), and light meromyosin (LMM).

The existence of several myosin isoforms or isomyosins that differ in heavy and light sub-unit composition and in ATPase activity was first demonstrated in adult fast and slow skeletal and in cardiac muscles. Barany [1] showed, for various adult skeletal muscles, that there was a correlation between maximal shortening speed of the muscle and myosin ATPase activity, whether this activity was measured in the presence of actin or was the calcium-ATPase of the myosin alone. This idea, i.e., that a multiplicity of isomyosins reflects the difference in intrinsic contractility of skeletal muscles, was important for a better understanding of the different physiological behavior of different fiber types. Experiments with cross-reinnervation, stimulation at various frequencies (c.f. [2] for review), and developing skeletal muscles [3] later provided striking evidence that a given skeletal muscle may show different patterns of isomyosin expression, depending on the functional needs.

Much less was known concerning isomyosin transitions within the heart muscle, probably because myosins from hearts in different physiological and patho-

M.J. Legato (ed.) THE DEVELOPING HEART, pp. 149–171.

logical states are very similar in overall structure and size. The past five years, however, has been a period of rapid development in this field. The use of new and sensitive biochemical methods together with the advent of recombinant DNA technology, has now made it possible to recognize the existence of several cardiac isomyosins. The aim of this review is to summarize the data on the changes in cardiac isomyosin expression in different stages of normal and pathological cardiac growth. Attention will be focused (1) on the number of isomyosins (and thus genes) and on their distribution within the cardiac tissue, (2) on the regulation of their expression, and (3) on the functional significance of isomyosin redistributions.

ANALYTICAL APPROACHES TO DISTINGUISH CARDIAC ISOMYOSINS

Isomyosins comprise a family of closely related proteins. The characterization of new myosin isoenzymes must be unequivocal, and it is only through a combination of different analytical approaches that it is possible to achieve a satisfactory analysis. The origin of the multiplicity of isomyosins may reside in the small sub-units, the heavy chain sub-unit, or both. It is thus necessary to use sensitive procedures which can distinguish the different sub-units of the myosin molecule.

Electrophoresis

The simplest way to identify the type of myosin is polyacrylamide gel electrophoresis in the presence of SDS (sodium dodecyl sulfate), which detects differences in the apparent molecular weights of polypeptides. Due to the large size of the heavy chain (200,000 daltons), differences at their level are difficult to assess precisely. In contrast, at the level of the light chains (around 25,000 daltons), small differences can be detected. For example, myosins of the two major classes of skeletal muscles, fast and slow, contain light chains of different apparent molecular weights, and this technique has been largely used to define a muscle type biochemically [2]. However, the resolution is rather limited. Improved resolution was obtained by Whalen et al. [4] who used the O'Farrell technique of two-dimensional gel electrophoresis.

A major step in electrophoretic analysis of isomyosins came about with the description of conditions permitting the electrophoresis of myosins in their native state. This method was set up independently by Hoh et al. [5] and by d'Albis et al. [6]. Analyses are carried out in pyrophosphate buffer, and only minute amounts of cardiac tissue are necessary (3 to 4 mg). It is possible to measure directly, in the gels, the calcium-activated ATPase activity of each isomyosin, which makes this technique the only one allowing separate enzymatic measurements in a mixture of several isomyosins. The inconvenience is that some isoforms are more sensitive to alkaline denaturation than others, and thus enzymatic differences might be artificially increased. Electrophoresis of native myosins has led to the separation of up to five isomyosins in the same heart. However, caution must be exercised in the interpretation of results, since charge differences can reside in the small sub-units, the heavy-chain sub-units, or both.

Immunological Methods

Classical polyclonal immunoglobulins were first prepared, and recently monoclonal antibodies have been obtained by several teams.

Concerning the polyclonal antibodies, the antisera were prepared against native or denatured myosins or heavy meromyosins, using proteins from different sources as immunogens [7, 8, 9. 10]. Immunoglobulins highly specific of one myosin type were isolated from the antisera by affinity chromatography on columns of heterologous insolubilized myosins [10, 11]. With this type of antibody, direct and indirect immunofluorescence on cryostat sections was widely used by Schiaffino and colleagues. It has the advantage of localizing each isoform at the level of individual myocytes, and of visualizing minor quantities of stained fibers. However, it is not possible to quantify the degree of structural relatedness between the various isomyosins nor to localize differences within the molecules. We have investigated such parameters using microcomplement fixation, a technique known for its sensitivity to small structural differences. This technique enabled us to show for the first time the presence of a new myosin species appearing in cardiac hypertrophy due to chronic mechanical overload [8, 12] and to localize constant and variable regions of the heavy sub-units [9, 13]. Enzyme-linked immunosorbent assays (ELISA) in solid phase on polysterene plates or combined with gel electrophoresis (immunoblots) are now being developed in several laboratories. The main advantage is their sensitivity, which is of the same order of magnitude as that of radioimmunoassays but which does not require the use of radioactive material. There have been numerous improvements in the ELISA techniques [14], and in the future their applications in the field of isomyosins will almost certainly increase. Thus far, in non-competitive or competitive assays, they have been used to distinguish atrial and ventricular myosins [15], and to quantify the proportion of one myosin type among a mixture [16].

Monoclonal antibodies are also used currently as immunological probes to discriminate between several isomyosins. The first set of monoclonal antibodies was developed by Clark et al. [17], who immunized BALB/c mice against cardiac myosins from chicken (18d embryos) and rabbit. These antibodies were used mainly in immunoprecipitation tests and radioimmunoassays. Monoclonal antibodies prepared against human cardiac myosins are also now available [18, 19].

Polypeptide Mapping

Partial enzymatic or chemical degradation of native myosin followed by one-dimensional analysis of the products was used very early on to identify myosin isoenzymes in different muscle types. However, the technique does not have sufficient resolution, and cleavage may be influenced by the sub-unit composition of the molecule. Denaturation of the myosins with SDS before enzymatic digestion and analysis of the resulting polypeptides by two-dimensional gel analysis have now made this technique a valid and sensitive probe of heavy chain structure [3]. The comparison of peptide fragments was made even more useful when antibodies were also added in these experiments. The principle is to prepare a replica of the polypeptide pattern by diffusion or electroblotting to nitrocellulose sheets, and to react the replica with selected antibodies. The antigen-antibody complexes are then visualized by ELISA assays. By selection of appropriate antibodies, it is possible to visualize specific and discrete domains of the myosin heavy chain, to compare their structure and to study their arrangement in one or the other isoform [13].

Genetic Analysis

Recently developed recombinant DNA technology has made possible isolation of individual myosin genes from higher organisms, including man, in quantities large enough for their study and manipulation. Two laboratories have cloned mammalian cardiac myosin heavy-chain DNA sequences. Briefly (figure 8–1), an in vitro synthesized double-stranded DNA molecule, called cDNA, is copied from a population of messenger RNA (mRNA) which carries the information for the synthesis of myosin heavy chain. This cDNA is introduced into a small circular piece of bacterial DNA, a plasmid. The resulting plasmid containing the cDNA is then used to transform bacteria. The next step is to identify the clones which contain the sequence of the myosin gene that is under investigation. The most common technique for this is screening by colony filter hybridization, followed by nucleotide sequence analysis, hybridization to RNA and DNA extracted from different striated muscles, restriction mapping analysis, and nuclease S1 mapping. Mahdavi et al. [20] have identified four cDNA clones as cardiac-specific. These clones comprise the sequences of the 430 carboxyl terminal amino acids, the complete untranslated 3'end, and a portion of the poly(A) track of two myosin heavy-chain mRNA. Sinha et al. [21] have also isolated two clones containing sequences for cardiac myosin heavy chains, the initial poly(A)RNA being obtained from adult rabbit hearts. The two clones specify a segment of the myosin heavy chain corresponding to the COOH-terminal portions of sub-fragment 1 (S1) and to a portion of sub-fragment 2 (S2). These clones represent an excellent tool for the study of the number of different cardiac myosin genes and possible mechanisms involved in gene switching.

NUMBER OF CARDIAC ISOMYOSINS

Attempts to identify the number of myosin isoenzymes in a given muscle require a precise knowledge of the number of each individual sub-units: heavy chains, akali light chains, and regulatory light chains.

Heavy Chains

Two different types of heavy chain were identified in the cardiac tissue of each animal species. These two types are referred to as α and β, (or as V1 and V3; see below), according to the terminology of Hoh et al. [5]. This set of cardiac myosin heavy chain has been demonstrated in the rat, rabbit, pig, and beef, and also in man [5, 10, 15, 22, 23, 24]. Although highly homologous, α and β differ in their

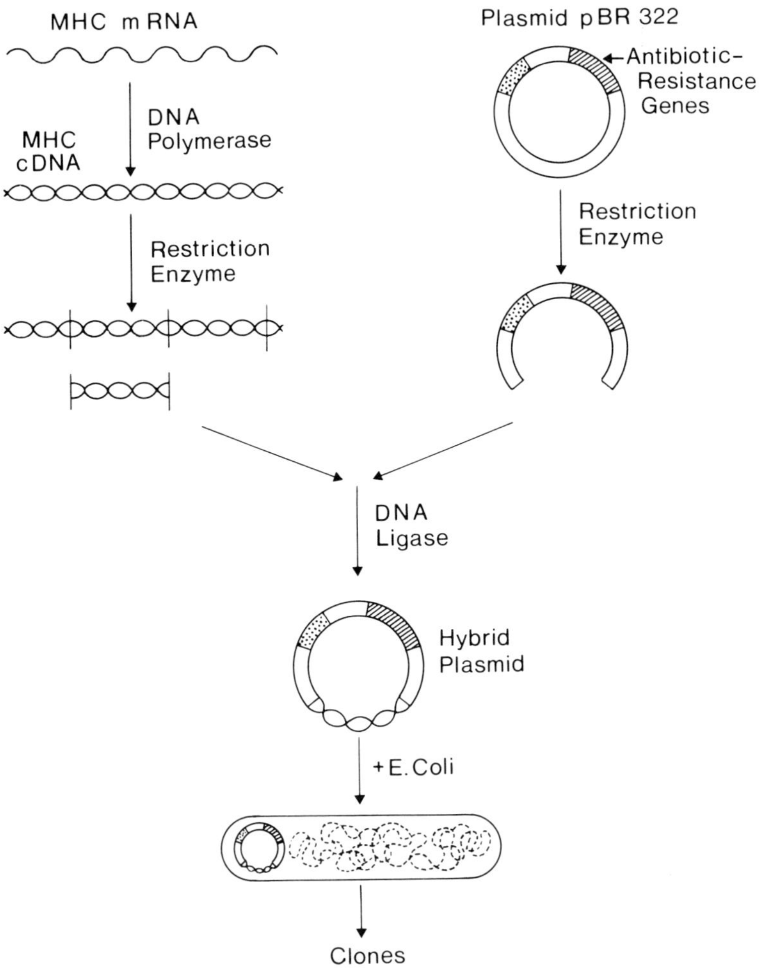

Figure 8–1. A schematic diagram of myosin heavy-chain gene cloning in plasmid pBR 322. Myosin heavy-chain messenger RNA (MHC mRNA) is purified from poly(A)RNA. The clones containing inserts complementary to MHC mRNA are selected by several genetic techniques (see text).

peptide maps, their immunological properties, and in the nucleotide sequence of their respective genes. Comparisons of the peptide maps, with or without specific antibodies, showed that the differences were distributed all along the molecules, at the levels of S1 and S2, heavy meromyosin (HMM), and light meromyosin (LMM) [11, 13, 23, 25]. This has been confirmed by gene analysis, although only selected regions of each gene have been analyzed. In the rabbit, the known sequences cover roughly the middle portion of the heavy chain (S1 and S2) [21]. They exhibit 90% homology, with respect to both the nucleotide and the amino acid sequences. The 10% divergence appears as three tetra-, one tri-, and several single nucleotide mismatches. Comparison with direct amino acid sequence analysis of rabbit fast skeletal muscle myosin reveals 79% homology for the β clone and 83% for the α clone. In the rat, it is the LMM portion of the heavy chain that has been analyzed [20]. The nucleotide sequences of the two genes show very close similarities in the coding region (95% homology) and clear-cut differences in the untranslated 3'ends. One of the clones contains an extra sequence in this untranslated 3'end, indicating heterogeneity in the polyadenylation site. The 5% difference in the coding region resides in a small fragment where there is a 30% base mismatch. The derived amino acid sequences (corresponding to the LMM fragment) have 97% homology, and there are also striking similarities at this level with rabbit skeletal myosin. This sequence conservation in LMM is also found among myosin extracted from animal species as distant as nematode, sea urchin, rat, and man, and is probably relevant to the formation and maintenance of the coiled-coil secondary structure of the LMM and to the organization of the myosin molecules in the thick filament.

Taken together, the above data clearly show that a minimum of two myosin heavy chains exist in mammalian cardiac tissue, and that they are more homologous in their LMM region than in their HMM region. It is, of course, possible that other undetected forms exist and that the cardiac heavy chains may be even more diverse. At the present time, however, there is no conclusive experimental evidence for an additional heterogeneity.

Light Chains

A characteristic feature of the mammalian adult heart is the difference in apparent molecular weights of atrial and ventricular light chains. For example, canine atrial light chains exhibit molecular weights of 26,000 and 21,000 daltons as compared to 28,000 and 18,500 for their ventricular counterparts [26, 27]. The same differences were consistently found in more than seven animal species [27]. Whalen et al. [3, 28] demonstrated that embryonic skeletal muscles contain a distinct form of alkali light chain, LCemb, and that an identical or at least very similar protein existed in embryonic rat and beef ventricular myosin. This fetal light chain was thoroughly compared with the adult atrial form on two-dimensional electrophoresis [29]. It cannot be distinguished from the adult atrial LC1, suggesting that, in fact, two different types of alkali LC exist in the heart, an atrial or embryonic type and a ventricular type (ALC1 and VLC1, respectively, according to the terminology of Price et al. [29]). At the level of the phosphorylatable light chains, atrial and ventricular myosins also differ, containing ALC2 and VLC2 sub-units, respectively. In humans each of these light chains can be resolved under particular electrophoretic conditions into more than one component [29]. The precise nature of all the electrophoretic bands — proteolysis or altered charges due to experimental conditions or new isoforms or genetic differences — is not clear. Three se-

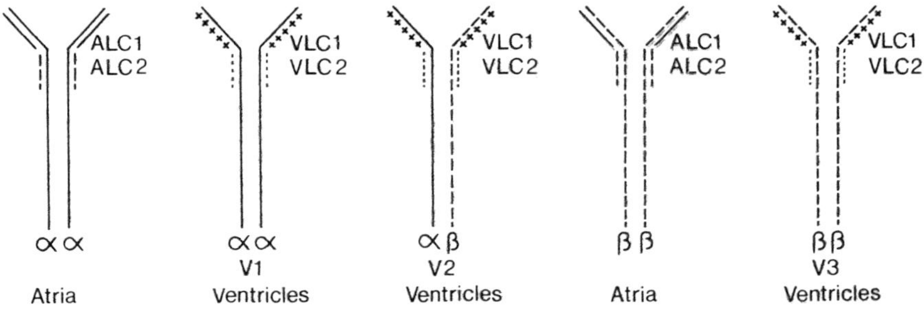

Figure 8–2. Sub-unit composition of cardiac isomyosins. ALC1: atrial light-chain 1 (or embryonic light chain); ALC2: atrial light-chain 2. VLC1: ventricular light-chain 1; VLC2: ventricular light-chain 2; α and β refer to the two heavy-chain types.

quence micro-heterogeneities were also found at the level of human ventricular LC2, perhaps also due to genetic differences [30].

The situation with regard to cardiac light chains thus indicates a minimum of four different proteins, two alkali light chains, ALC1 and VLC1, and two regulatory light chains, ALC2 and VLC2.

Isomyosins

The possible simultaneous presence of two heavy-chain and four light-chain forms might allow the formation of a number of different myosin isozymes if all heavy and light chains could independently associate. So far (figure 8–2), three different heavy-chain isozymes have been characterized in rat and rabbit ventricles, V1, V2, and V3 [5, 22, 31, 32, 33, 34]. This terminology, which was provided by Hoh et al. [5], refers to mobility under electrophoresis in the native state, V1 having the highest mobility. V1 and V3 are composed of αα and ββ homodimers, respectively, while V2 has been postulated to be an αβ heterodimer [25, 31]. V1 and V3 exhibit a three- to five-fold difference in their rate of ATP hydrolysis: both the calcium-activated and actin-activated ATPase activities of V3 are lower than those of V1 [5, 12, 22, 32, 33, 34, 35]. In the atria (figure 8–2), two different forms were also found, A1 and A2, probably corresponding to an association of α type heavy chain with the atrial or ventricular light chains [5, 23]. In the ventricles of larger animal species, only one band of myosin can be resolved under the same electrophoretic conditions [22, 23, 24, 36, 37]. Evidence has, however, also been presented for the existence of α and β chains [23, 24]. By analogy with the terminology of Hoh et al. [5], the α and β chains were also called V1 and V3, respectively. In humans, the available results also imply a higher enzymatic activity for V1.

These results suggest that the general scheme for all mammals is three heavy-chain isozymes and two light-chain isozymes, making a total of five isomyosins.

Functional differences in terms of ATPase activity reside in the heavy-chain isozymes, with the rate of turnover of ATP being higher with V1 than with V3.

NORMAL CARDIAC GROWTH

The ratio of cardiac isomyosins is developmentally regulated at the level of both the light and heavy sub-units.

Heavy Chains

In all species studied (rat, mouse, rabbit, guinea pig, pig, dog, beef and man), V3 is the most abundant isomyosin in the fetus [22, 31, 38] (figure 8–3). It represents 75% of total myosin at 18 days in utero for the mouse, 75% and 70% at 16 and 20 days for the rat, 64% and 67% at 20 and 28 days for the rabbit. In man, values ranging from 92% to 99% have been reported at around 30 weeks of gestation [24]. Birth is a transition period during which V1 appears [22, 31]. Postnatal development depends on the species. In the rat, and probably also in the mouse, V1 completely replaces V3, so that three-week-old animals present V1 exclusively. V1 remains predominant throughout life, and V2 and V3 reappear with aging [5, 39]. In the rabbit, the peak of V1 comes approximately one week after birth; and at that time, V1 represents about 50% of the total myosin. V1 then disappears rapidly and adult animals contain only V3 [22, 31]. In larger mammalian species, ventricular myosin is essentially V3, and there are only very small amounts of V1 [22, 23, 36, 37]. Precise quantification of V1 has been performed in only a few human hearts: four hearts contained less than 2% V1, one contained 6%, and another 15%. No obvious relation was found with age [24]. This low level in humans was confirmed by immunofluorescence studies which indicated small amounts of stained cells with specific antibodies [15, 40]. In bovines as well, only a few myocytes are stained with anti-V1 immunoglobulins [10].

The situation is different in the atria. Evidence is accumulating that the left atria of adults of all animal species are composed mainly of α heavy chain and thus contain V1 [23, 24, 41]. In the right atria, this V1 type is also predominant, but in some specific localizations — the crista terminalis, for example — a mixed pattern of isoforms is found [42]. Nothing is known so far concerning the embryonic atria.

All these results show that the pattern of myosin heavy-chain isoenzymes during ontogenic development varies with the species. In rats, the fetal ventricular form completely disappears after birth, and adult animals contain the same heavy chains in the atria and the ventricles. In other species, the fetal ventricular form persists in the ventricles throughout life, and the atria of adult animals, which contain V1, are different from the ventricles.

Very little is known about the level at which myosin heavy-chain gene expression is regulated. Recently, the level of each of the two cardiac myosin heavy-chain mRNA was quantified in rats during normal cardiac development. It was found that the pattern of expression of myosin heavy-chain genes at the mRNA level was in agreement with the pattern of accumulation of the corresponding isomyo-

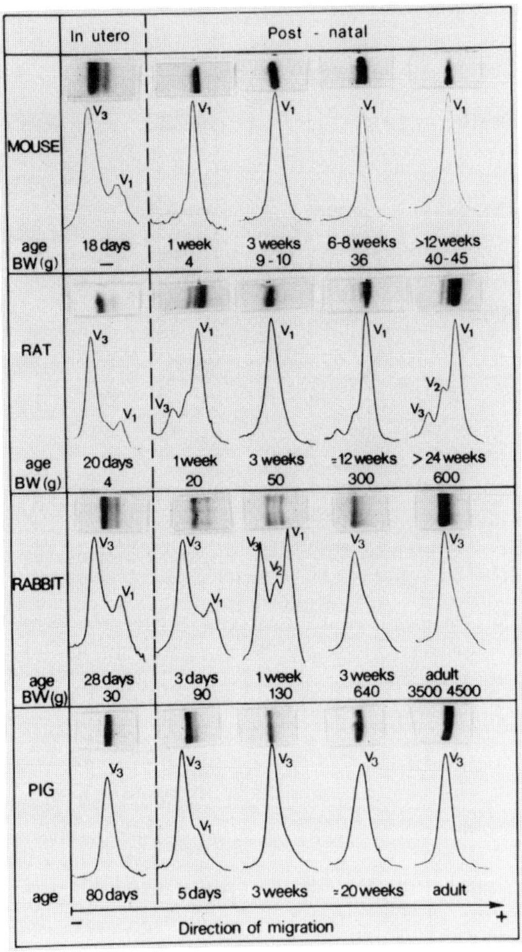

Figure 8–3. Age-dependent evolution of cardiac isomyosins of mouse, rat, rabbit, and pig. Typical electrophoretic patterns and densitometric profiles obtained from both ventricles of fetuses and from left ventricles of neonates and adults. Reprinted from Lompré et al. [22] with permission.

sins [43]. This strongly suggests that the regulation of isozymic transitions of the ventricular myosin heavy chains during development is due to factors which control the amount of the corresponding mRNA in the ventricular cell.

Light Chains

The profile of myosin light chain also changes during normal development of the heart (figure 8–4). The fetal ventricles of rat, beef, and humans contain LC1, LCemb, and LC2 [24, 28, 29, 38]. LCemb represents about 25% of the total LC1 type in 20-day embryonic rat ventricles, and around 50% in fetal human hearts at 20-week gestation. It is the only component in human fetal atrial myosin. This

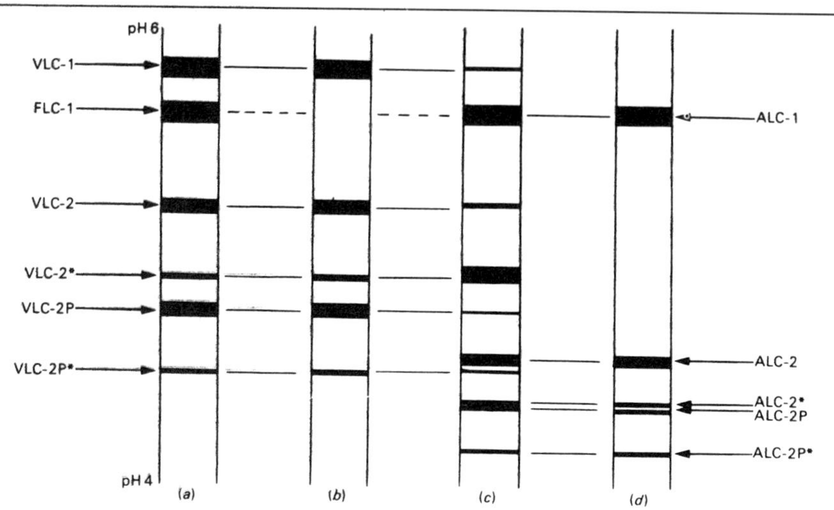

Figure 8–4. Schematic representation of isoelectric-focusing gels of myosin light chains from human hearts. (*a*): normal fetal ventricular myosin; (*b*): normal adult ventricular myosin; (*c*): myosin from severe cardiac-overload adult right or left atria (*d*): normal adult right or left atria. Reprinted from Cummins [44] with permission.

fetal light chain disappears completely from the ventricles in the postnatal period [29, 44]. Conversely, the other LC1 component of fetal ventricles progressively increases in this tissue, and it is the only form found in adult ventricles.

A striking feature is that the two embryonic subunit forms, LCemb and V3 heavy chain, are not necessarily linked: LCemb disappears after birth in large animals, whereas V3 remains. The level of specific light chain mRNA has not yet been quantified. If regulation for the light chain is also at the transcriptional level, it could mean that the elements responsible for heavy- and light-chain mRNA activation are not the same. It is also possible that the genome organization is not the same for the two sub-units.

CARDIAC GROWTH INDUCED BY HEMODYNAMIC OVERLOADS

Increase in hemodynamic load on the heart, due either to pathologic changes in the cardiovascular system or to increased physiological activity, rapidly induces cardiac growth. If the animals are young (three-week-old rats, for example) at the time of onset of work overload, accelerated nuclear divisions followed by cytokinesis accompany heart enlargement. In adults, cardiac hypertrophy can be explained by an increase in the size of existing myocytes. Intrinsic contractile state and myosin ATPase activities of such hearts has been the subject of considerable investigation for more than 15 years. In the last five years, considerable progress has been made in the description of isomyosin changes which can account for the alterations in both myosin enzymatic activity and muscle-shortening velocity.

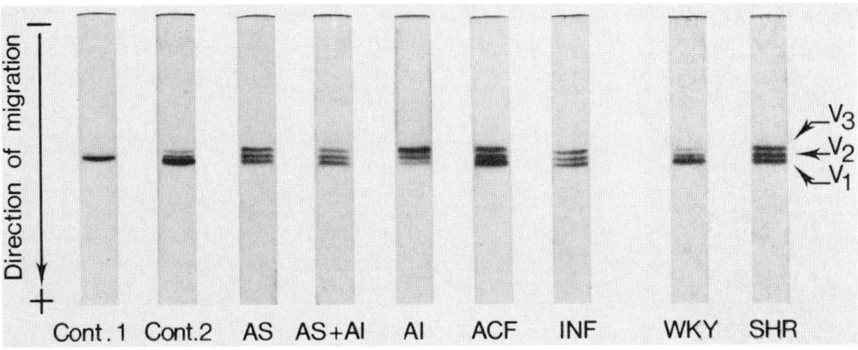

Figure 8–5. Gel electrophoresis in non-dissociating conditions of native myosin from left ventricles of two controls and of several types of overloaded rat hearts. Body weights of control 1 and control 2 were, respectively, 375g and 590g. AS: aortic stenosis; AI: aortic incompetence; ACF: aorto caval fistula; INF: overload of the non-infarcted area after left coronary artery ligation; WKY: Wistar Kyoto rats; SHR: spontaneously hypertensive rats. Reprinted from Mercadier et al. [39] by permission of the American Heart Association Inc.

The change in heavy-chain cardiac isomyosins in response to work overload depends on (1) the potential of the tissue to regulate myosin isozymic composition and (2) the nature of the inciting stimulus. In rats (figure 8–5), almost all models of experimental systolic overload or of combined systolic and diastolic overload (including the spontaneously hypertensive strains) induce an isomyosin redistribution from V1 to V3 [12, 39, 45, 46, 47]. V3 synthesis occurs in young animals as well as in adults, and it is a very early response which is detected as soon as three days after the onset of work overload [48, 49]. Due to the initial low level of V3, the potential for increase is large: V3 can represent as much as 80% of myosin from hypertrophied hearts, and its amount is correlated with the degree of hypertrophy. In rabbits, the potential for increase is much smaller, about 30% in young animals, but the same type of response also occurs after pressure overloads [32]. Very recent experiments showed that, in human hypertrophied atrial tissue (which normally contains high levels of V1), there is also a marked decrease in V1 and a significant increase in V3 [Mercadier et al., in preparation]. In contrast, in pig and human hypertrophied ventricles, no marked differences were found compared to controls, which contain mainly V3 [24, 36, 37, 50]. However, humans seem to lose the small amount of V1 that they normally have, since this form was detected neither in autopsy material of patients suffering from hypertensive disease nor in per-operative biopsies of patients with valvular disease [40, 51]. Thus, adaptation to pressure overload in cardiac tissue with initial high levels of V1 (or low levels of V3) is mediated through increased synthesis of V3. Pure volume overloads, on the contrary, do not have the same effect: a normal isomyosin pattern was found in a certain number of hypertrophies due to aorto-caval fistula in rats [39]. Indeed, physical training, which also results in increased volume load, has the opposite effect since it increases the synthesis of V1 and depresses that of V3 [46]. This shift was shown to arise predominantly from the exercise component and not from the stress [52]. In these hypertrophy models, isomyosin redistribution and hypertrophy are completely dissociated adaptational responses.

Isomyosin redistributions also occur at the level of the light chains: hypertrophied human atria there is a transition from atrial to ventricular myosin light chains (loss of ALC1 and appearance of VLC1), which is correlated with the degree of hypertrophy [44] (figure 8–4). Contradictory observations were reported in the ventricles: some authors found reappearance of the embryonic LC1 (or ALC1) in hypertrophied ventricles [50], but this form was also found by others in normal hearts [53].

ALTERATIONS IN CARDIAC GROWTH DUE TO HORMONAL IMBALANCES

Cardiac protein synthesis may be affected by many hormonal factors. Isomyosin changes have been documented during thyrotoxicosis, chemically induced diabetes, and gonadectomy.

The effects of administration of thyroid hormones on cardiac isozymes are to stimulate synthesis of V1 and to decrease that of V3 [5, 15, 21, 31, 32, 33, 41, 54, 55]. Conversely, hypothyroidism is associated with rapid replacement of V1 isomyosin by V3. The rapid accumulation of V1 is due to both a rapid increase in the fractional synthesis rate of α heavy chain and a marked decrease in that of β heavy chain [56]. Measurements of the amounts of each myosin heavy-chain mRNA showed that in the normal young rabbit and in hypothyroid rats and rabbits [21, 43], β transcripts predominate whereas thyrotoxic ventricles contain α transcripts. This strongly suggests that the relative concentrations of each isomyosin is regulated directly by the transcriptional activity of the cardiac genes (figure 8–6) [21, 43]. The combined effects of thyroid hormones on the synthesis of the two myosin heavy chains has been found consistently in all animal species studied (rat, rabbit, pig, guinea pig). As would be expected from the initial isomyosin pattern, the extent of the changes varies with the species. In the rat, administration of thyroid hormones has very little influence. After thyroidectomy or hypophysectomy, however, there is a switch in isomyosin composition to a predominance of the V3 form, an effect which is reversed by thyroxine administration [5]. By contrast, administration of thyroid hormones in rabbits stimulates the synthesis of V1. The complete change which occurs in those various experimental conditions and which was first recognized from differences in one- and two-dimensional peptide maps [54, 55] greatly facilitated structural comparisons between α and β heavy

---

Figure 8–6. Nuclease S1 mapping of cardiac RNA from normal, hypothyroid, and thyrotoxic rabbits, with myosin heavy-chain cDNA. In (*A*) the cloned sequences are specific of β-type heavy chain (plasmid pMHCB174) and in (*B*), they are specific of the α-type (pMHC252). The hybridizations were performed with 30 ug of total RNA from hypothyroid rabbit ventricles (lanes 1–4) or 5 μg of poly (*A*) RNA from normal (langes 5–8) or hypothyroid (9–12) rabbit ventricles. Lanes 1, 5, and 9: undigested probe; 2, 6, and 10, 1,000 units of nuclease S1; 3, 7, and 11, 2,500 units of nuclease S1. The hybrids were subjected, after digestion, to gel electrophoresis and autoradiography. Reprinted from Sinha et al. [21] with permission.

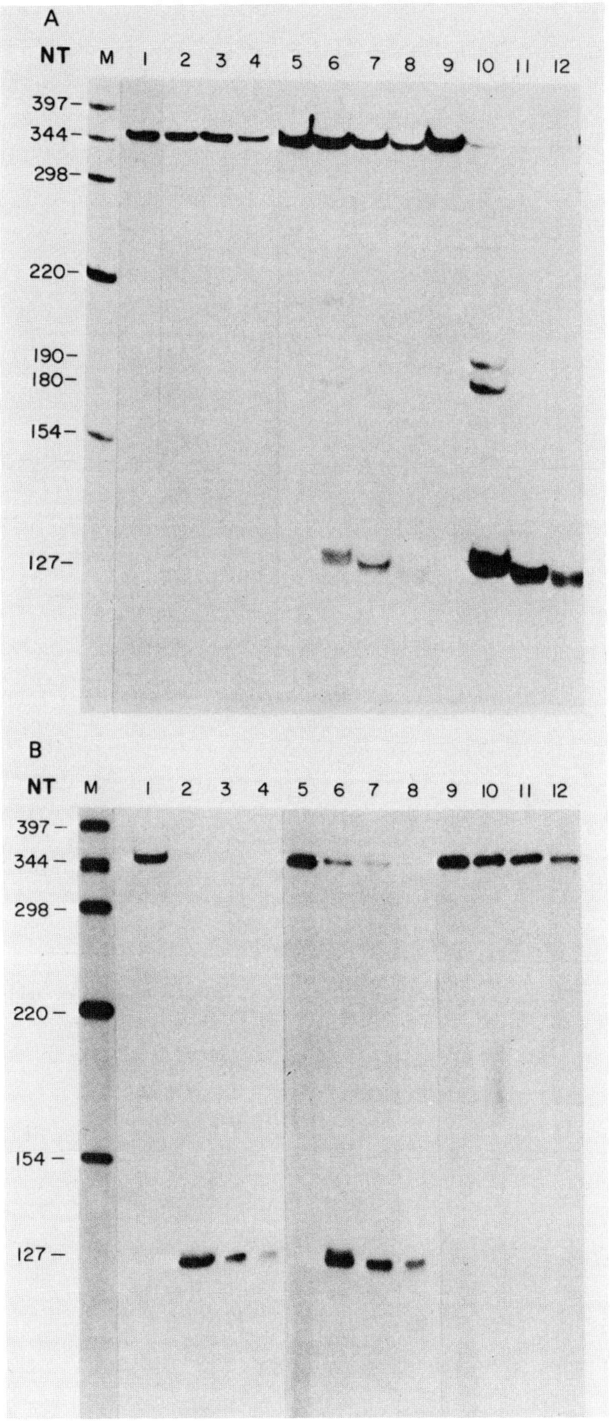
A
NT M 1 2 3 4 5 6 7 8 9 10 11 12
397–
344–
298–
220–
190–
180–
154–
127–
B
NT M 1 2 3 4 5 6 7 8 9 10 11 12
397 –
344 –
298 –
220 –
154 –
127 –

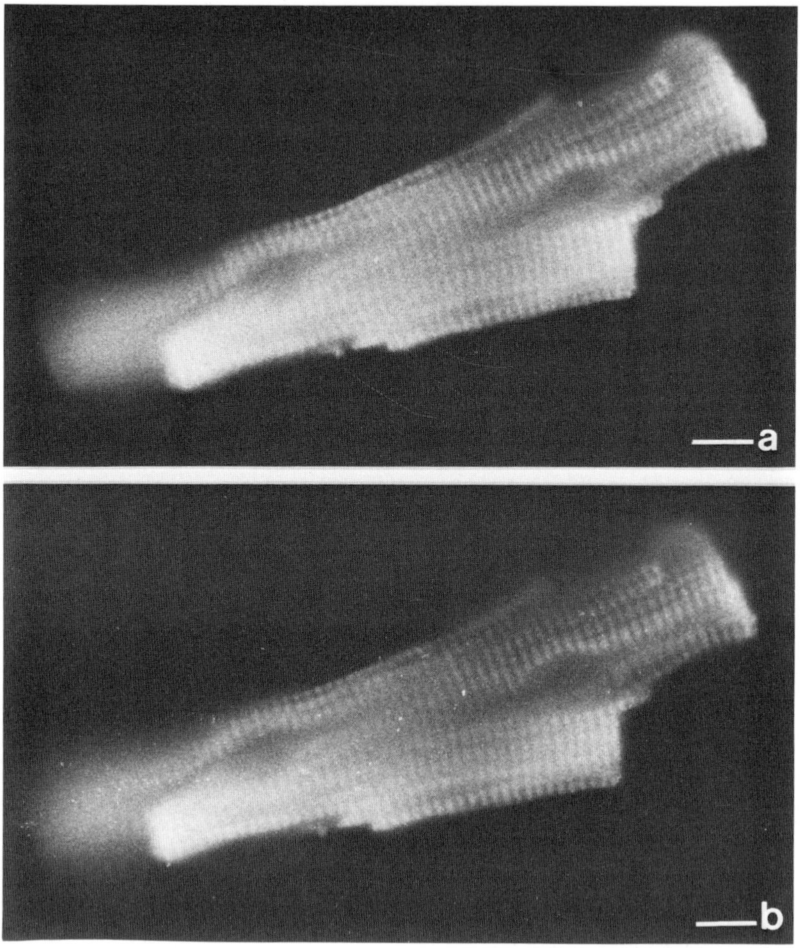

Figure 8–7. Double indirect immunofluorescence micrograph of a myocyte isolated from a four-week-old rat heart, one week after aortic stenosis was performed. Incubations were performed as in [63] with anti-V3 myosin immunoglobulins revealed by fluorescein (*a*) and anti-V1 myosin immunoglobulins revealed by rhodamine (*b*). The anti-V3 myosin were a gift from S. Schiaffino and S. Sartore and the anti-V1 myosin was prepared as in [11]. Bar = 10 um.

chains. In rats and rabbits, V3 myosin from fetuses and hypophysectomized adults are identical, both at the protein [11, 31] and gene [21, 43] levels. These observations promoted the notion of the principle of two basic types of cardiac myosin heavy chains. The role of thyroid hormones during the isomyosin transformations that occur in normal heart during development is not clear. At birth, neonate rats from hypothyroid mothers do not synthesize the V1 form [Samuel et al., in preparation]. The burst of α heavy-chain synthesis is correlated, in rabbits, with a rapid increase in thyroid hormones. Later on, however, while the levels of T3 and

T4 remain constant, the proportion of V1 declines with age [56]. And yet, adult rabbits respond to thyroxine administration by synthesizing V1.

Changes in cardiac isomyosins in other hormonal imbalances are less documented. Streptozocin-diabetes in the rat has the same effect as hypothyroidism: it stimulates the synthesis of V3 [34, 57]. This finding appears, in fact, to be related to the diabetogenic effect of streptozocin, since it is prevented by prior injection of 3-0-methylglucose [34]. Moreover, isomyosin redistribution can be reverted to control levels by insulin administration [57]. In vitro translation of specific mRNA from hearts of hypophysectomized and diabetic rats produces the same myosin heavy chain, as judged by one-dimensional peptide map analysis [58]. This again is in favor of only one V3 form, whatever the physiologic or pathologic situation. Though the diabetic animals have depressed values of circulating thyroid hormones, the increase in V3 cannot be explained solely in terms of hypothyroidism.

Abnormalities in cardiac isomyosins have also been reported in the alterations of cardiac growth induced by sex hormones: gonadectomy decreases V1, and the effects are greater in male than in female rats [59].

DISTRIBUTION OF ISOMYOSINS WITHIN THE HEART

The expression of myosin heavy-chain genes varies among cardiac myocytes, depending on the functional demand. This heterogeneous distribution was first demonstrated by immunocytochemical studies with antibodies specific of α heavy chain in the rabbit: in normal animals, this heavy chain appears to be more abundant in muscle fibers from the right ventricles and in sub-epicardial layers [15]. In the bovine heart also, some reactive fibers are present in the right ventricle [10]. A reciprocal pattern of distribution was observed in the rat with antibodies specific of rat V3: the partially stained fibers are more numerous in the left ventricle and in subendocardial layers [15]. Work overload is characterized by a gradient of isomyosin redistribution. For example, in response to aortic stenosis, the isomyosin redistribution takes place first in the papillary muscles, then in the left ventricular wall, and only after in the right ventricle [60, 61]. In contrast, thyroid-modulated shifts are homogeneous within the ventricles: the isomyosin compositionn of the papillary muscle is equal to that of its associated ventricle [62].

The heterogeneous distribution of V1 and V3 isomyosins in different myocytes was also found by double immunolabelling of isolated myocytes with anti-V1 and anti-V3 immunoglobulins [63]. Three types of myocytes were found in adult rat ventricular myocardium: one type contained only V1, the second contained only V3, and the third contained both V1 and V3. However, within single myocytes, V1 and V3 heavy chains are identically distributed, with the two heavy-chain forms being closely intermingled throughout the entire length of the cell. Histochemical detection of V1 and V3 (involving pre-incubation at different pH before ATPase assays) also showed that both heavy chains coexist in the same cell [64]. Neither the immunochemical nor the histochemical method allow determination of whether or not the double-labeled cells contain a natural V2 heterodimer or V1 and V3 homodimers closely mixed within one myofibril. The interesting point is that the structural linkage between the two heavy chains seems to be a general

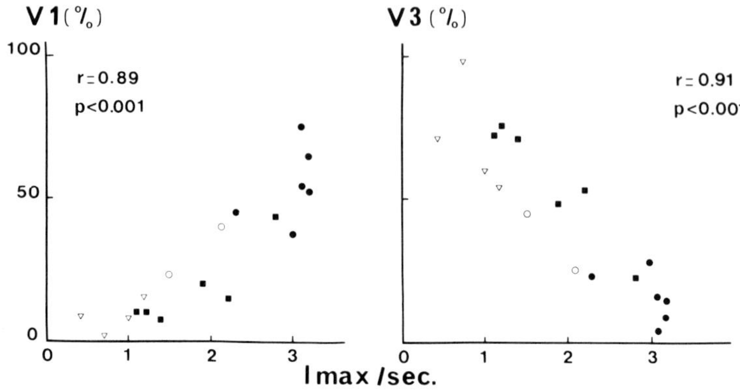

Figure 8–8. Relationship between the maximum velocity of shortening (Vmax expressed as Lmax/sec) and the relative amount of isomyosins V1 and V3. Both types of measurements were performed on the same papillary muscles of rat hearts. (●): four-to six-month-old controls; (○): one-to-three-year-old controls; (□) cardiac hypertrophy induced by aortic constriction; (▽) hypophysectomized rats. Statistical significance of the regression lines: (*a*): $r = 0.89$, $p < 0.001$; (*b*): $r = 0.91$, $p < 0.001$. Reprinted from Schwartz et al. [73] with permission.

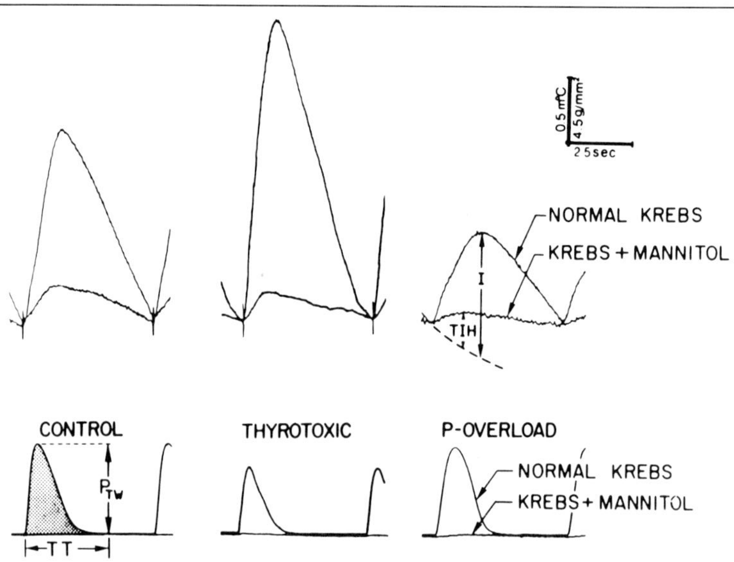

Figure 8–9. Records of right ventricular papillary muscle heat production (above) and isometric force (below) for control, thyrotoxic, and pressure-overloaded rabbit hearts. For both heat production and force records, the upper trace was obtained in normal Krebs-Ringer solution and the lower trace in 2.5 × hyperosmotic mannitol Krebs-Ringer solution. TT: twitch-time; Ptw: isometric force; TIH: tension-independent heat; I: initial heat. Reprinted from Alpert and Mulieri [78] with permission.

feature of the cardiac tissue, since it was found both in whole hearts and in cultured cells [63]. This means that in spite of their enzymatic differences, V1 and V3 must operate within the constraints of a single cell. Winegrad et al. [65] have proposed that selection of one or the other form is under cAMP regulation: β-adrenergic stimulation would select V1 and thus force generators would be primarily fast. On the contrary, V3 would be dependent on the α-adrenergic system. Although this hypothesis is an attractive one for selection of one myosin type in the rat myocardium, additional work is necessary to determine whether it can be applied in another form in hearts which contain only V3.

FUNCTIONAL SIGNIFICANCE OF ISOMYOSINS SHIFTS

The correlation found by Barany [1] in skeletal muscle between shortening speed of muscle fiber and myosin enzymatic activity can be explained by the inverse relationship between force and velocity of muscle contraction described by the Hill equation [66] and by the interpretation given for this relation by Huxley [67]. In brief, Hill's relation links the active tension which a muscle fiber is able to develop to the initial speed of fiber shortening. The maximum active tension corresponds to the isometric twitch where no muscle shortening is observed. When total load is decreased, the muscle shortens with an initial speed that increases as the load decreases. Extrapolation to zero load gives the initial speed of fiber shortening at zero load (Vmax), which is independent of initial muscle length and accounts for intrinsic muscle fiber contractility. This tension-speed relation also has been shown to apply to heart muscle [68]. According to Huxley's model of contraction, tension-speed relation can be explained by the movements of actin-myosin cross-bridges. For a given load, active tension depends on initial muscle length; i.e., on the degree of thick and thin filament overlap and therefore on the number of attached cross-bridges. When the load decreases, initial fiber-shortening velocity increases corresponding to the smaller number of attached cross-bridges. The maximum velocity of the unloaded muscle is an intrinsic property of the population of cross-bridges determined by the detachment rate for resisting cross-bridges, which itself depends on myosin ATPase activity.

In cardiac muscle, a correlation has been found between myosin ATPase activity and Vmax [69, 70, 71]. Moreover, it was shown that the rate constant of the transient initial phase of ATP hydrolysis by myosin, which proceeds in the same speed range as that of the contractile process itself, also correlated with Vmax. This is further support for the close connection between mechanical Vmax and the rate of ATP hydrolysis by myosin [72].

The discovery of three rat cardiac isomyosins together with the demonstrated difference in their intrinsic Ca^{2+}-stimulated ATPase activity (V1> V3) (see above) have permitted the direct linking of myosin phenotype and fiber contractility under various conditions of cardiac growth. A correlation has been demonstrated in the same papillary muscle between Vmax and myosin isoenzymic pattern for normal hearts of young rats, old rats, rat hearts submitted to chronic mechanical overload, and hearts from hypophysectomized animals [73]. A similar correlation was demonstrated between (1) the isoenzymic pattern and myofibrillar ATPase activity of the left ventricle and (2) the Vmax of trabeculae dissected from the left ventricular wall of pressure overloaded rat hearts and the hearts of rats submitted to physical training [74].

The functional significance of myosin isoenzyme redistribution in terms of contraction economy has been especially well documented by the work of Alpert et al. [75, 76, 77, 78]. Using a new technology based on the development of a highly sensitive model of thermopile, these authors have been able to separate, among the total heat resulting from muscle contraction, the initial heat from recovery heat, and to distinguish, in initial heat, the part specifically related to tension development. Tension-dependent heat (THD) is a measure of the number of moles of ATP split by the cycling cross-bridge heads during a contraction-relaxation cycle, with the force developed being a function of the unitary strength of the cross-bridge, the rate of cross-bridge cycling, and the duration of the "on" time in the force-generating configuration. These authors have clearly established that TDH is decreased with pressure overloaded rabbit hearts or hypothyroid rat hearts (i.e., with the V3 myosin isotype), demonstrating a decreased cross-bridge cycling rate, an increase in the "on" time, and an increase in the "off" time resulting in a greater economy of force development, i.e., a better efficiency of isometric twitch [75, 76, 77, 78]. Conversely, hyperthyroid rabbit hearts (i.e., with the V1 myosin isotype) are associated with increased TDH, indicating an increase in the cross-bridge cycling rate, and a decrease in the "on" and "off" time, resulting in less economy than normal in force development [75, 78]. These conclusions are in agreement with Kissling et al., who reported a reduced oxygen consumption of the pressure-overloaded rat heart [79]. We can summarize as follows: normal rat or hyperthyroid rabbit hearts seem to sacrifice economy for the ability to develop tension rapidly. This enables the myocardium to beat at rapid rates to fulfill peripheral oxygen demands. Conversely, hypothyroid or pressure-overloaded rat and rabbit heart develop tension more economically than normal myocardium but at the expense of contraction speed.

Concerning factors which trigger the expression of a precise myosin phenotype when the myocardium faces new functional requirements, a possible explanation in the case of pressure overload (80) is based upon Goldspink's [81] and Nwoye's and Goldspink's [82] observations on skeletal muscle. In brief, pressure overload decreases the muscle's instantaneous shortening speed and thus the efficiency of muscle contraction, with optimal efficiency corresponding to that which accompanied the normal shortening velocity before overload. This decrease in efficiency (by mechanisms still unknown) would induce the expression of a more efficient myosin phenotype. As for the mechanisms which trigger the V1 isotype expression with hyperthyroidism, a direct action on the myocyte has been suggested [83]. However, an indirect action cannot be excluded, since hyperthyroid animals have a high heart rate and high cardiac output which constitute a volume overload. The action of other hormonal factors in the course of normal postnatal development cannot be excluded either [22]. Postnatal myocardial tailoring occurs in response to normally occurring changes in the endocrine and hemodynamic environment [84]. In addition, we must point out that the expression of a specific myosin isotype may not be the only gene expression alteration which accounts for the alteration in muscle contractility and muscle economy observed with new functional requirements. Furthermore, the myosin isotype present within the heart cell is probably not the only determinant of muscle-fiber alterations in contractility observed in various physiological and pathological conditions; indeed, intracellular processes such as free calcium level or extracellular events such as collagen production may also play a role [74].

CONCLUDING REMARKS

New insights into the question of regulation of cardiac contractility during normal and induced cardiac growth have been provided by recognition of the plasticity of cardiac myosin. The shift from high ATPase activity form V1 to low activity form V3 is a true molecular adaptation which allows a slow and economical tension development. This adaptation presumably occurs in hypertrophy due to pressure overload in all conditions where V1 is initially predominant, i.e., rat ventricles and the atria of other animal species. A similar but less pronounced adaptation occurs in normal postnatal development. Regulation seems to be at the transcriptional level, but further investigation will be necessary to identify the elements responsible for such control. The finding that at the level of myosin, adaptation of the adult rat heart to work overload is mediated through synthesis of the fetal form can now be used as an example for studying phenotypic switches of other proteins, and thus for defining other adaptational mechanisms of cardiac tissue to normal development and/or compensatory challenges. A salient feature is the dissociation between activation of isomyosin transitions and activation of total protein synthesis in myocyte enlargement. There are not yet sufficient data to determine the exact time for the onset of both mechanisms. It is clear, however, that they do not depend on the same signals. Moreover, the switches from fetal to adult program (or vice versa) do not seem to be coordinated in the different cardiac chambers or even in adjacent myocytes. These features now make the heart an excellent model for studying the more basic aspects of regulation of gene expression. It can also be expected that rapidly expanding information from other multigene families in other eukaryotic systems will result in a better understanding of normal and pathologic cardiac development.

REFERENCES

1. Barany, M. 1967. ATPase activity of myosin correlated with speed of muscle shortening. *J. Gen. Physiol.* 50:197–216.
2. Pette, D., ed. 1980. *Plasticity of muscle.* Berlin, New York: Walter de Gruyter.
3. Whalen, R.G.; Sell, S.M.; Butler-Browne, G.S.; Schwartz, K.; Bouveret, P.; and Pinset-Härström, I. 1981. Three myosin heavy-chain isozymes appear sequentially in rat muscle development. *Nature* 292:805–809.
4. Whalen, R.G.; Butler-Browne, G.S.; and Gros, F. 1978. Identification of a novel form of myosin light chain present in embryonic muscle tissue and cultured muscle cells. *J. Mol. Biol.* 126:415–431.

5. Hoh, J.H.; McGrath, P.A.; and Hale, P.T. 1978. Electrophoretic analysis of multiple forms of rat cardiac myosin: effect of hypophysectomy and thyroxine replacement. *J. Mol. Cell. Cardiol.* 10:1053–1076.
6. D'Albis, A.; Pantaloni, C.; and Bechet, J.J. 1979. An electrophoretic study of native myosin isoenzymes and their subunit content. *Eur. J. Biochem.* 99:261–272.
7. Pierobon-Bormioli, S.; Sartore, S.; Vitadello, M.; and Schiaffino, S. 1980. Slow myosins in vertebrate skeletal muscle: an immunofluorescence study. *J. Cell Biol.* 85:672–681.
8. Schwartz, K.; Bouveret, P.; Bercovici, J.; and Swynghedauw, B. 1978. An immunochemical difference between myosins from normal and hypertrophied rat hearts. *FEBS Lett.* 93:137–140.
9. Schwartz, K.; Lompre, A.M.; Bouveret, P.; Wisnewsky, C.; and Swynghedauw, B. 1980. Use of antibodies against dodecylsulfate — denatured heavy meromyosins to probe structural differences between muscular myosin isoenzymes. *Eur. J. Biochem.* 104:341–346.
10. Sartore, S.; Gorza, L.; Pierobon-Bormioli, S.; Dalla Libera, L.; and Schiaffino, S. 1981. Myosin types and fiber types in cardiac muscle. I. Ventricular myocardium. *J. Cell Biol.* 88:226–233.
11. Schwartz, K.; Lompré, A.M.; Bouveret, P.; Wisnewsky, C.; and Whalen, R.G. 1982. Comparisons of rat cardiac myosins at fetal stages, in young animals and in hypothyroid adults. *J. Biol. Chem.* 257:14412–14418.
12. Lompré, A.M.; Schwartz, K.; d'Albis, A.; Lacombe, G.; Thiem, N.Y.; and Swynghedauw, B. 1979. Myosin isoenzyme redistribution in chronic heart overload. *Nature* 282:105–107.
13. Lompré, A.M.; Han, K.; Bouveret, P.; Richard, C.; and Schwartz, K. 1984. Comparison of the tryptic digestion pattern of Subfragments-1 from V1 and V3 rat cardiac isomyosins. *Eur. J. Biochem.* 139:459–765.
14. Avrameas, S.; Druet, P.; Masseyeff, R.; and Feldmann, G., eds. 1983. *Immunoenzymatic techniques.* Amsterdam, New York, Oxford: Elsevier.
15. Schiaffino, S.; Gorza, L.; and Sartore, S. 1983. Distribution of myosin types in normal and hypertrophic hearts. In *Perspectives in cardiovascular research,* ed. N. Alpert Vol. 7, pp. 149–166. New York: Raven Press.
16. Mercadier, J.J.; Bouveret, P.; Wisnewsky, C.; de la Bastie, D.; and Schwartz, K. 1983. Simplified methodology using non-competitive ELISA assay for the quantitation of myosin isoforms. In *Immunoenzymatic techniques,* ed. S. Avrameas, P. Druet, R. Masseyeff, and G. Feldmann, pp. 329–332. Amsterdam, New York, Oxford: Elsevier.
17. Clark, W.A.; Everett, A.W.; Ficht, F.W.; Frogner, K.S.; Jakovcic, S.; Rabinowitz, M.; Warner, A.M; and Zak, R. 1980. Characterization of monoclonal antibodies directed against determinants of cardiac myosin heavy chain. *Biochem. Biophys. Res. Commun.* 95:1680–1686.
18. Léger, J.; Dechesne, C.; Bouvagnet, P.; and Léger, J.J. 1983. Distribution and characterization of myosin variants in normal and hypertrophied human hearts. *J. Mol. Cell. Cardiol.* Suppl. 2, 15:37 (abstr.).
19. Yazaki, Y.; Tsuchimochi, H.; Kuro-o, M.; Kurabayashi, M.; Isobe, M.; Ueda, S.; Nagai, T.; and Takaku, F. 1983. Distribution of myosin isozymes in human atrial and ventricular myocardium. *J. Mol. Cell. Cardiol.* Suppl. 2 15:32 (abstr.).
20. Mahdavi, V.; Periasamy, M.; and Nadal-Ginard, B. 1982. Molecular characterization of two myosin heavy chain genes expressed in the adult heart. *Nature* 297:659–664.
21. Sinha, A.M.; Umeda, P.K.; Kavinsky, C.J.; Rajamanickam, C.; Hsu, H.J.; Jakovcic, S.; and Rabinowitz, M. 1982. Molecular cloning of mRNA sequences for cardiac α and β form myosin heavy chain: expression in ventricles of normal, hypothyroid and thyrotoxic ventricles. *Proc. Natl. Acad. Sci. USA* 79:5847–5851.
22. Lompré, A.M.; Mercadier, J.J.; Wisnewsky, C.; Bouveret, P.; Pantaloni, C.; d'Albis, A.; and Schwartz, K. 1981. Species and age-dependent changes in the relative amounts of cardiac myosin isoenzymes in mammals. *Develop. Biol.* 84:286–290.
23. Clark, W.A.; Chizzonite, R.A.; Everett, A.W.; Rabinowitz, M.; and Zak, R. 1982. Species correlations between cardiac isomyosins. *J. Biol. Chem* 257:5449–5454.
24. Mercadier, J.J.; Bouveret, P.; Gorza, L.; Schiaffino, S.; Clark, W.A.; Zak, R.; Swynghedauw, B.; and Schwartz, K. 1983. Myosin isoenzymes in normal and hypertrophied human ventricular myocardium. *Circ. Res.* 53:52–62.
25. Hoh, J.F.Y.; Yeoh, G.P.S.; Thomas, M.A.W.; and Higginbottom, L. 1979. Structural differences in the heavy chains of rat ventricular myosin isozymes. *FEBS Lett.* 97:330–334.
26. Long, L.; Fabian, F.; Mason, D.T.; and Wikman-Coffelt, J. 1977. A new cardiac myosin characterized from the canine atria. *Biochem. Biophys. Res. Comm.* 76:626–635.

27. Syrovy, I.; Delcayre, C.; and Swynghedauw, B. 1979. Comparison of ATPase activity and light subunits in myosins from left and right ventricles and atria in seven mammalian species. *J. Mol. Cell. Cardiol.* 11:1129–1135.
28. Whalen, R.G., and Sell, S.M. 1980. Myosin from fetal hearts contains the skeletal muscle embryonic light chain. *Nature* 286:731–733.
29. Price, K.M.; Littler, W.A.; and Cummins, P. 1980. Human atrial and ventricular myosin light-chain subunits in the adult and during development. *Biochem. J.* 191:571–580.
30. Klotz, C.; Léger, J.J.; and Elzinga, M. 1982. Comparative sequence of myosin light chains from normal and hypertrophied human hearts. *Circ. Res.* 50:201–209.
31. Chizzonite, R.A.; Everett, A.W.; Clark, W.A.; Jakovcic, S.; Rabinowitz, M.; and Zak, R. 1982. Isolation and characterization of two molecular variants of myosin heavy-chain from rabbit ventricle: change in their content during normal growth and after treatment with thyroid hormone. *J. Biol. Chem.* 257:2056–2065.
32. Litten, R.Z.; Martin, B.J.; Low, R.B.; and Alpert, N.R. 1982. Altered myosin isozyme patterns from pressure-overloaded and thyrotoxic hypertrophied rabbit hearts. *Circ. Res.* 50:846–856.
33. Martin, A.F.; Pagani, E.D.; and Solaro, R.J. 1982. Thyroxine-induced redistribution of isoenzymes of rabbit ventricular myosin. *Circ. Res.* 50:117–124.
34. Malhotra, A.; Penpargkul, S.; Fein, F.S.; Sonnenblick, E.H.; and Scheuer, J. 1981. The effect of streptozotocin-induced diabetes in rats on cardiac contractile proteins. *Circ. Res.* 49:1243–1250.
35. Pope, B.; Hoh, J.F.Y.; and Weeds, A. 1980. The ATPase activity of rat cardiac myosin isoenzymes. *FEBS Lett.* 118:205–208.
36. Schier, J.J., and Adelstein, R.S. 1982. Structural and enzymatic comparison of human cardiac muscle myosins isolated from infants, adults and patients with hypertrophic cardiomyopathy. *J. Clin. Invest.* 69:816–825.
37. Wisenbaugh, T.; Allen, P.; Cooper, IV, G.; Holzgrefe, H.; Beller, G.; and Carabello, B. 1983. Contractile function, myosin ATPase activity and isozymes in the hypertrophied pig left ventricle after chronic progressive pressure overload. *Circ. Res.* 53:332–341.
38. Whalen, R.G.; Sell, S.M.; Eriksson, A.; and Thornell, L.E. 1982. Myosin subunit types in skeletal and cardiac tissues and their developmental distribution. *Develop. Biol.* 91:478–484.
39. Mercadier, J.J.; Lompré, A.M.; Wisnewsky, C.; Samuel, J.L.; Bercovici, J.; Swynghedauw, B.; and Schwartz, K. 1981. Myosin isoenzymic changes in several models of rat cardiac hypertophy. *Circ. Res.* 49:525–532.
40. Schiaffino, S.; Gorza, L.; Sartore, S.; Valfré, C.; and Pauletto, P. 1983. Adaptive changes in cardiac isomyosins as visualized by immunofluorescence. In *Cardiac adaptation to hemodynamic overload, training and stress,* ed. R. Jacob, R.W. Gülch, and G. Kissling, pp. 101–103. Darmstadt: Steinkopff Verlag.
41. Dalla Libera, L., and Sartore, S. 1981. Immunological and biochemical evidence for atrial-like isomyosin in thyrotoxic rabbit ventricle. *Biochim. Biophys. Acta* 669:84–92.
42. Gorza, L.; Sartore, S.; and Schiaffino, S. 1982. Myosin types and fiber types in cardiac muscle. II. Atrial myocardium. *J. Cell. Biol.* 95:838–845.
43. Lompré, A.M.; Nadal-Ginard, B.; and Malidairi, V. 1984. Expression of the cardiac ventricular α and β myosin heavy-chain genes is developmentally and hormonally regulated. *J. Biol. Chem.* 255:6437–6446.
44. Cummins, P. 1982. Transitions in human atrial and ventricular myosin light-chain isoenzymes in response to cardiac-pressure-overload-induced hypertrophy. *Biochem. J.* 205:195–204.
45. Gorza, L.; Pauletto, P.; Pessina, A.L.; Sartore, S.; and Schiaffino, S. 1981. Isomyosin distribution in normal and pressure overloaded rat ventricular myocardium: an immunohistochemical study. *Cir. Res.* 49:1003–1009.
46. Rupp, H. 1981. The adaptative changes in the isoenzyme pattern of myosin from hypertrophied rat myocardium as a result of pressure overload and physical training. *Basic Res. Cardiol.* 76:79–88.
47. Martin, A.F.; Haithcoat, J.L.; and Dowell, R.T. 1983. Redistribution of ventricular myosin isoenzymes in neonatal and adult rat heart in response to a chronic pressure overload. In *Perspectives in cardiovascular research,* ed. N. Alpert, Vol. 7, pp. 359–371. New York: Raven Press.
48. Bugaisky, L.B.; Siegel, E.L.; and Whalen, R.G. In press. Myosin isozyme changes in the heart following constriction of the aorta of 25 day old rat. *FEBS Lett.*
49. Mercadier, J.J.; Lompré, A.M.; Bouveret, P.; Samuel, J.L.; Rappaport, L.; Swynghedauw, B.;

and Schwartz, K. 1983. Myosin isoenzymic distribution in hypertrophied rat and human hearts. In *Cardiac adaptation to hemodynamic overload, training and stress,* ed. R. Jacob, R.W. Gülch, and G. Kissling, pp. 104–112. Darmstadt: Steinkopff Verlag.

50. Tuchschmid, C.R.; Srihari, T.; Hirzel, H.O.; and Schaub, M.C. 1983. Structural variants of heavy chains of atrial and ventricular myosins in hypertrophied human hearts. In *Cardiac adaptation to hemodynamic overload, training and stress,* ed. R. Jacob, R.W. Gülch, and G. Kissling, pp. 123–128. Darmstadt: Steinkopff Verlag.
51. Schwartz, K.; Mercadier, J.J.; de la Bastie, D.; Bouveret, P.; Wisnewsky, C.; Schneider, J.M.; Younes, A.; and Swynghedauw, B. 1983. Left ventricular isomyosins and isoCPK in normal and hypertrophied rat and human hearts. *J. Mol. Cell. Cardiol.* Suppl. 2, 15:34 (abst.).
52. Rupp, H.; Felbier, H.R.; and Jacob, R. 1983. Blood pressure and cardiac myosin heterogeneity in the rat as influenced by swimming and emotional excitation. In *Cardiac adaptation to hemodynamic overload, training and stress,* ed. R. Jacob, R.W. Gülch, and G. Kissling, pp. 274–281. Darmstadt: Steinkopff Verlag.
53. Mercadier, J.J.; Whalen, R.G.; and Schwartz, K. 1983. Myosin light-chain patterns of normal and hypertrophied human left ventricles. *J. Mol. Cell. Cardiol.* Suppl. 2, 15:66 (abstr.).
54. Flink, I.L., and Morkin, E. 1977. Evidence for a new cardiac myosin species in the thyrotoxic rabbit. *FEBS Lett.* 81:391–394.
55. Flink, I.L.; Rader, J.H.; and Morkin, E. 1979. Thyroid hormone stimulates synthesis of a cardiac myosin isozyme. *J. Biol. Chem.* 254:3105–3110.
56. Everett, A.W.; Chizzonite, R.A.; Clark, W.A.; and Zak, R. 1983. Relationship of changes in molecular forms of myosin heavy chains to endogenous level of thyroid hormone during postnatal growth. In *Perspectives in cardiovascular research,* ed. R.C. Tarazi and J.B. Dunbar, Vol. 8, pp. 83–93. New York: Raven Press.
57. Dillmann, W.H. 1980. Diabetes mellitus induces changes in cardiac myosin of the rat. *Diabetes* 29:579–582.
58. Dillmann, W.H.; Barrieux, A.; and Reese, G. 1984. Effect of diabetes and hypothyroidism on cardiac myosin heavy chain synthesized in-vivo and in-vitro. *J. Biol. Chem.* 259:2035–2038.
59. Malhotra, A.; Schaible, T.; and Scheuer, J. 1982. Abnormalities in cardiac myosin after gonadectomy (abstr.). *Circulation* Suppl. II, 66:259.
60. Mercadier, J.J.; Lecarpentier, Y.; Delcayre, C.; Lompré, A.M.; Swynghedauw, B.; and Schwartz, K. 1982. Mécanismes biochimiques de l'adaptation myocardique dans l'hypertrophie cardiaque. *Arch. Mal. Coeur.* 75:1179–1186.
61. Schwartz, K.; Lompré, A.M.; d'Albis, A.; Lacombe, G.; Thiem, N.V.; and Swynghedauw, B. 1980. Myosin chronic cardiac overload. In *Plasticity of muscle,* ed. D. Pette, pp. 569–580. Berlin, New York: Walter de Gruyter.
62. Loiselle, D.S.; Wendt, I.R.; and Hoh, J.F.Y. 1982. Energetic consequences of thyroid-modulated shifts in ventricular isomyosin distribution in the rat. *J. Musc. Res. Cell. Motil.* 3:5–23.
63. Samuel, J.L.; Rappaport, L.; Mercadier, J.J.; Lompré, A.M.; Sartore, S.; Triban, C.; Schiaffino, S.; Schwartz, K. 1983. Distribution of myosin isozymes within single cardiac cells. An immunohistochemical study. *Circ. Res.* 52:200–209.
64. Weisberg, A.; Winegrad, S.; Tucker, M.; and McClellan, G. 1982. Histochemical detection of specific isozymes of myosin in rat ventricular cells. *Circ. Res.* 51:802–809.
65. Winegrad, S.; McClellan, G.; Tucker, M.; and Lin, L. 1983. Cyclic AMP regulation of myosin isozymes in mammalian cardiac muscle. *J. Gen. Physiol.* 81:749–765.
66. Hill, A.V. 1938. The heat of shortening and the dynamic constants of muscle. *Proc. R. Soc. Lond. (Biol.) Series B* 126:136–195.
67. Huxley, A.F. 1957. Muscle structure and theories of contraction. *Prog., Biophys. Chem.* 7:225–318.
68. Sonnenblick, E.H. 1962. Force-velocity relations in mammalian heart muscle. *Am. J. Physiol.* 202:931–939.
69. Delcayre, C., and Swynghedauw, B. 1975. A comparative study of heart myosin. ATPase and light subunits from different species. *Pflugers Arch.* 355:39–47.
70. Hamrell, B.B., and Low, R.B. 1978. The relationship of mechanical Vmax to myosin ATPase activity in rabbit and marmot ventricular muscle. *Pflugers Arch.* 377:119–124.
71. Carey, R.A.; Bove, A.A.; Coulson, R.L.; and Spann, J.F. 1979. Correlation between cardiac muscle myosin ATPase activity and velocity of muscle shortening. *Biochem. Med.* 21:235–245.
72. Van Thiem, N.; Lacombe, G.; and Swynghedauw, B. 1978. Early phosphate burst of heart myosins: phylogenic variations. *Eur. J. Biochem.* 91:243–248.

73. Schwartz, K.; Lecarpentier, Y.; Martin, J.L.; Lompré, A.M.; Mercadier, J.J.; and Swynghedauw, B. 1981. Myosin isoenzymic distribution correlates with speed of myocardial contraction. *J. Mol. Cell. Cardiol.* 13:1071–1075.
74. Ebrecht, G.; Rupp, H.; and Jacob, R. 1982. Alterations of mechanical parameters in chemically skinned preparations of rat myocardium as a function of isoenzyme pattern of myosin. *Basic Res. Cardiol.* 77:220–234.
75. Alpert, N.R., and Mulieri, L.A. 1981. Heat, mechanics, and myosin ATPase in normal and hypertrophied heart muscle. *Fed. Proc.* 41:192–198.
76. Alpert, N.R., and Mulieri, L.A. 1982. Increased myothermal economy of isometric force generation in compensated cardiac hypertrophy induced by pulmonary artery constriction in the rabbit. *Circ. Res.* 50:491–500.
77. Holubarsch, C.; Goulette, R.P.; Litten, R.Z.; Mulieri, L.A.; and Alpert, N.R. 1983. Myokardiale Energetik und Myosin-Isoenzym-muster der Ratte. *Z Kardiol. Suppl. 1, 72:56/198.*
78. Alpert, N.R., and Mulieri, L.A. 1983. Myocardial myosin isoenzymes and thermomechanical economy. In *Perspectives in cardiovascular research,* ed. R.C. Tarazi and J.B. Dunbar, Vol. 8, pp. 157–166. New York: Raven Press.
79. Kissling, G.; Rupp, H.; Malloy, L.; and Jacob, R. 1982. Alterations in cardiac oxygen consumption under chronic pressure overload: significance of the isoenzyme pattern of myosin. *Basic Res. Cardiol.* 77:255–269.
80. Swynghedauw, B., and Delcayre, 1982. Biology of cardiac overload. In *Pathobiology annual,* ed. H.L. Ioachim, Vol. 12, pp. 137–183. New York: Raven Press.
81. Goldspink, G. 1975. Biochemical energetics for fast and slow muscles. In *Comparative physiology —functional aspects of structural materials,* ed. L. Bolis, H.P. Maddrell, and K. Schmidt-Nielsen, pp. 173–185. Amsterdam: North Holland.
82. Nwoye, L.O., and Goldspink, G. 1981. Biochemical efficiency and intrinsic shortening speed in selected vertebrate fast and slow muscles. *Experientia* 37:856–857.
83. Hoh, J.F.Y., and Egerton, L.J. 1979. Action of triiodothyronine on the synthesis of rat ventricular myosin isoenzymes. *FEBS Lett.* 101:143–148.
84. Rupp, H.; Kissling, G.; and Jacob, R. 1983. Hormonal and hemodynamic determinants of polymorphic myosin. In *Perspectives in Cardiovascular Research,* ed. N.R. Alpert, Vol. 7, pp. 373–383. New York: Raven Press.

9. DEVELOPMENTAL AND ADAPTIVE CHANGES OF ATRIAL ISOMYOSINS

STEFANO SCHIAFFINO, LUISA GORZA, SAVERIO SARTORE, AND LARS-ERIC THORNELL

During embryonic development, the complex events that lead from the primitive tubular heart to the definitive configuration of the adult organ are accompanied by differentiation at the cellular level in which an homogeneous population of mesodermic cells gives rise to several varieties of cardiac muscle cells. These have different and characteristic functional and structural properties. Pacemaker cells, conduction tissue cells, and ordinary working myocardial cells are the traditional designations which stress the functional specialization of the main types of cardiac myocytes. It is now clear that each of these cell populations is itself heterogeneous, though the characterization of the various sub-populations is still incomplete, and the functional significance of this heterogeneity is not yet fully appreciated.

The diversification of cardiac muscle cells begins early in development. It has long been known that tissue fragments or cells isolated from different regions of the primitive tubular heart vary in their intrinsic pulsation rate and in the shape of the action potential [1]. Evidence for muscle cell heterogeneity in the developing and mature heart has come mainly from electrophysiological and morphological studies. Both approaches have intrinsic limitations, however, and do not provide a unambiguous definition of specific cell types. This is perhaps best illustrated by the recurrent controversy about the existence of specialized internodal pathways [2, 3, 4].

Identification of cell-type-specific molecular markers by immunological methods is being widely used as a means of identifying cell populations and tracing their developmental histories in different tissues. Antibodies against specific car-

M.J. Legato (ed.) THE DEVELOPING HEART, pp. 173–189.

diac muscle cell components can provide more reliable criteria for cell typing in the heart and contribute to a better understanding of the molecular basis of cardiac development and function. Myosin appears to be a good candidate for a cell-type-specific marker in cardiac muscle. There are different cardiac isomyosins which are closely related to and in part responsible for the contractile properties of the muscle cells. The distribution of ventricular isomyosins in the normal and hypertrophied heart has been the object of intensive investigation during the last several years. In contrast, the myosin composition of the atrial myocardium has been comparatively less studied. In this chapter we will briefly review previous studies on atrial isomyosins and describe work from our laboratory illustrating the use of antimyosin antibodies to identify muscle cell populations in the developing and mature atrial myocardium. We will also review adaptive changes in response to hemodynamic overload.

MYOCYTE HETEROGENEITY IN ATRIAL MYOCARDIUM

Atria play two basic roles in heart physiology: a mechanical role in augmenting ventricular filling, and an electrophysiological role, in pacing the ventricles. The two functions pertain to distinct but anatomically interconnected tissues: the ordinary or working atrial muscle and the specialized conducting tissues which are responsible for the generation and conduction of electrical impulses. While the unique structural and electrophysiological properties of nodal muscle cells have long been known, only recently have important functional differences between the working atrial and ventricular myocardium been recognized.

In the mammalian heart, atrial muscle contracts more rapidly than ventricular muscle. Force-velocity studies have shown that this is due to a higher maximal velocity of shortening, suggesting differences in the contractile machinery itself [5, 6, 7]. There are also indications that the membrane systems responsible for excitation and for excitation-contraction coupling also differ between atrial and ventricular muscle cells. Atrial myocytes have action potentials with a shorter plateau [8] and show a more developed Ca^{2+}-induced release of Ca^{2+} from the sarcoplasmic reticulum in skinned fiber preparations [9]. However, the molecular basis for these differences between atrial and ventricular muscle cells have not been explored. In contrast, precise information has been gained during the last few years concerning the distinct properties of atrial myosin.

Myosin isolated from mammalian atrial myocardium has been shown to differ from ventricular myosin with respect to ATPase activity, electrophoretic mobility in non-dissociating gels, and in the structure of myosin light and heavy chains [10, 11, 12, 13, 14, 15, 16, 17]. The higher Ca^{2+}- and actin-activated ATPase activity of atrial myosin can account for the higher speed of shortening of atrial compared to ventricular muscle [7]. We have used an immunological approach to investigate the distribution of different cardiac myosin isoforms. In the avian and mammalian heart, atrial and ventricular muscle cells differ in their reactivity toward specific antimyosin antibodies [18, 19, 20, 21]. As shown in figure 9–1, this difference also holds true for the frog heart. Immunofluorescence studies have further shown that atrial and ventricular myocardium contain different myosin isoforms heteroge-

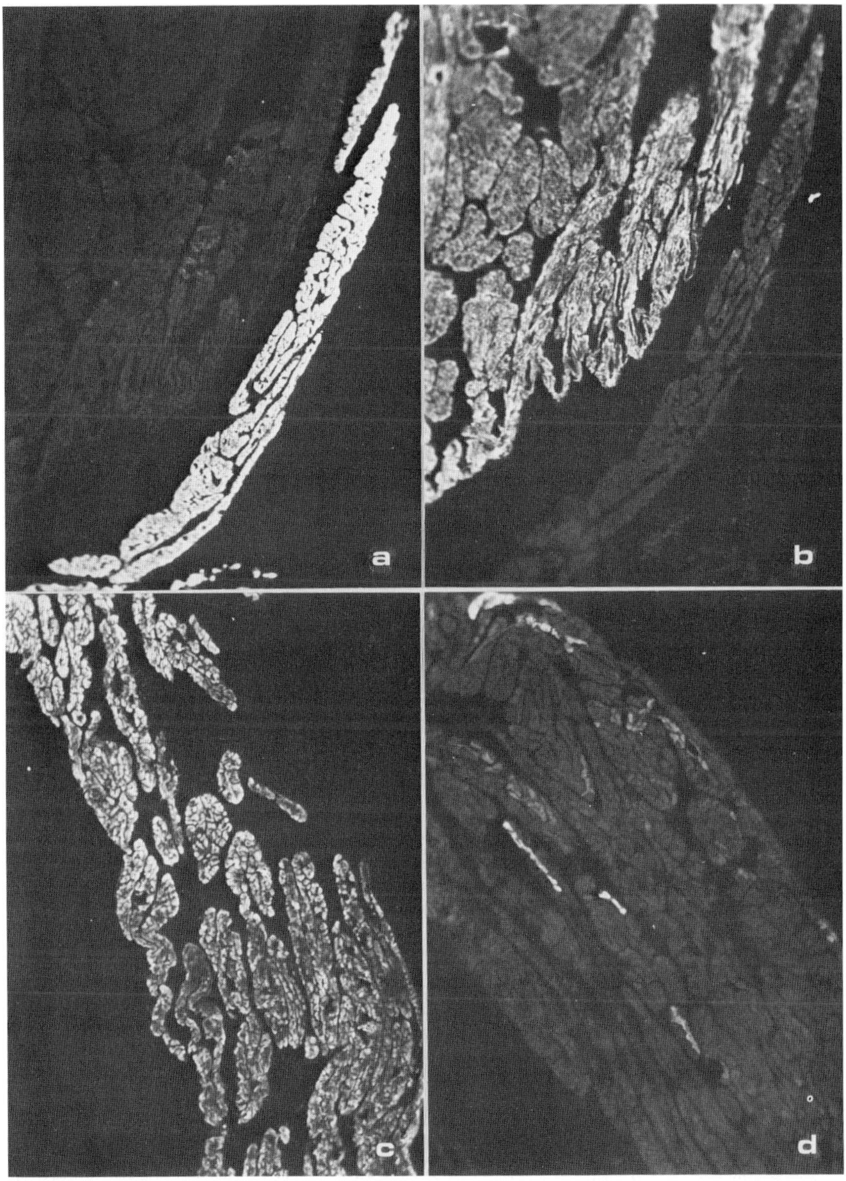

Figure 9–1. Frog heart *a, b*: serial cryosections of atrial (lower right) and ventricular (upper left) myocardium processed for indirect immunofluorescence with anti-bovine atrial myosin (*a*) and anti-chicken ventricular myosin antibodies (*b*). Note the reciprocal staining pattern of atria and ventricles. *c, d*: higher magnification of atrial myocardium stained with the same antibodies showing muscle cell heterogeneity.

neously distributed among cardiac muscle cells. Two distinct types of myosin heavy chain (HC), designated HC_{α} and HC_{β}, have been identified in the ventricular myocardium of the mammalian heart [11, 22]. The two HC's can be variably

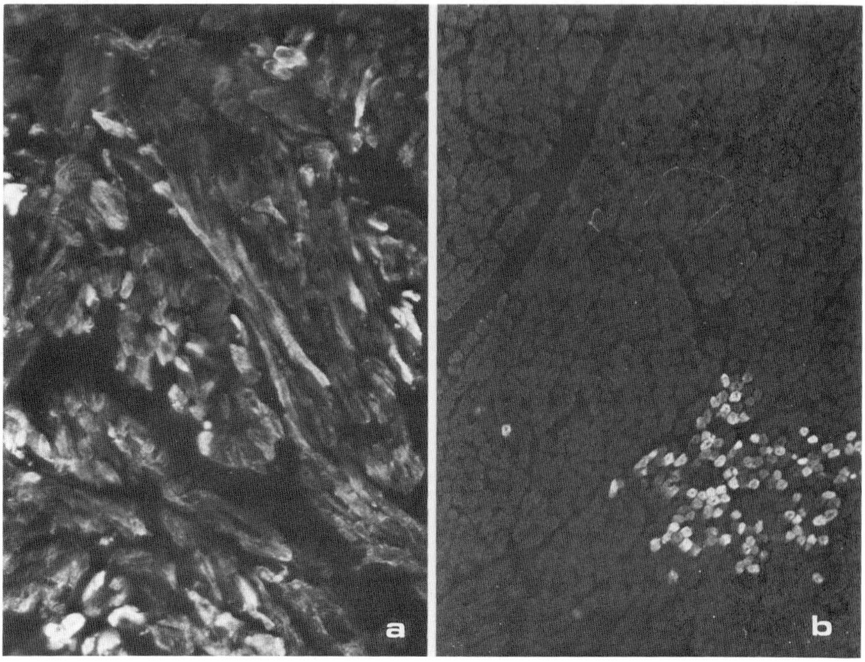

Figure 9–2. *a*: section of rat right auricle stained with anti-guinea pig soleus myosin, an antibody specific for myosin HC_β in the rat heart. *b*: section of bovine left auricle stained with anti-bV, an antibody specific for HC_β in the bovine heart. This particular field shows a group of labeled fibers, which are rare in the bovine left atrium.

associated so to give rise to three distinct isomyosins, distinguishable on the basis of their mobility in non-dissociating gel electrophoresis: the high ATPase V_1 isomyosin, corresponding to the HC_α homodimer; the low ATPase V_3 isomyosin, corresponding to the HC_β homodimer; and the V_2 heterodimer. Several lines of evidence indicate that the major component of atrial myosin HC is similar if not identical to ventricular HC_α [16, 23, 24]. In addition to this α-type myosin HC, a number of atrial myocytes were found to contain a distinct type of myosin HC, immunologically similar to ventricular HC_β. Figure 9–2 shows a section of rat atrial myocardium stained with an antibody, anti-guinea pig soleus myosin, which reacts selectively with ventricular myosin HC_β in the rat heart [20, 25]: a small number of atrial myocytes are labeled; these contain a β-type myosin HC. The existence of β-type HC has also been demonstrated in the bovine heart [26] and in the human heart [27]. It has not been established whether the two atrial myosin HC's recognized by immunological methods correspond to the two atrial isomyosins identified by pyrophosphate gel electrophoresis [11, 17]. Since atrial myosins with different HC composition have not yet been separated, it is not known whether they differ in ATPase activity. Enzyme histochemical methods that demonstrate myosin ATPase activity show that atrial fibers containing myosin HC_α have an alkali stable and acid labile ATPase, while fibers containing HC_β have an alkali labile and acid stable ATPase [26, 28].

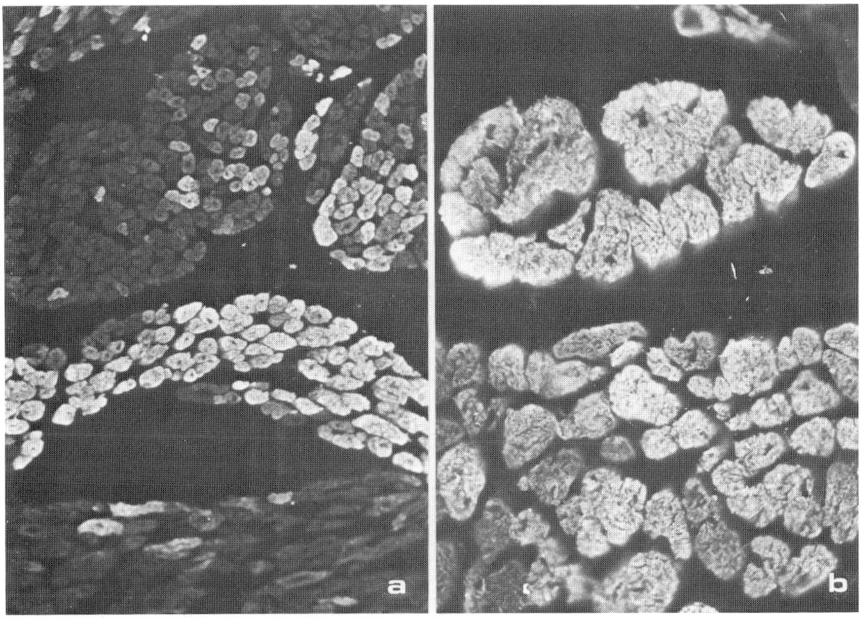

Figure 9–3. Bovine right atrium. Sections from a pectinate muscle (*a*) and the coronary sinus region (*b*) showing numerous fibers labeled by anti-bV.

Immunocytochemical studies indicate that β-type HC's are more abundant in the right compared to the left atrium in different mammalian species. In the bovine and human heart which have been more extensively investigated, fibers staining for β-type HC were frequently grouped in bundles and were especially numerous in the crista terminalis and in the interatrial septum (figures 9–2b and 9–3) [26, 27]. The distribution of β-type myosin HC thus appears to correlate with the distribution of fibers specialized for faster conduction along the internodal tracts [29].

The idea that atrial muscle cells specialized for impulse conduction may have a distinct myosin composition is supported by previous studies on the avian atrial and ventricular Purkinje system [18] and the bovine ventricular Purkinje system [19]. Large fibers with high glycogen content and subendocardial distribution, similar to the Purkinje fibers of the avian atria, are occasionally observed also in bovine atria: they usually stain for β-type myosin HC (figure 9–4). Myosin heterogeneity in the atrial myocardium is further complicated by the recent finding that muscle cells present in the sino-atrial (SA) and atrio-ventricular (AV) nodes of the bovine heart contain a novel type of myosin HC, distinct from the atrial α and β type HC's and immunologically related to myosin HC's present in skeletal muscle during fetal development (Gorza et al., in preparation). Figure 9–5 illustrates the reaction of bovine nodal cells with antibody against bovine fetal skeletal myosin (anti-bF) and reactive with developing skeletal muscle fibers in different mammalian species [30].

In conclusion, as summarized in table 9–1, the bovine atrial myocardium contain at least three antigenically distinct types of myosin HC which are variably

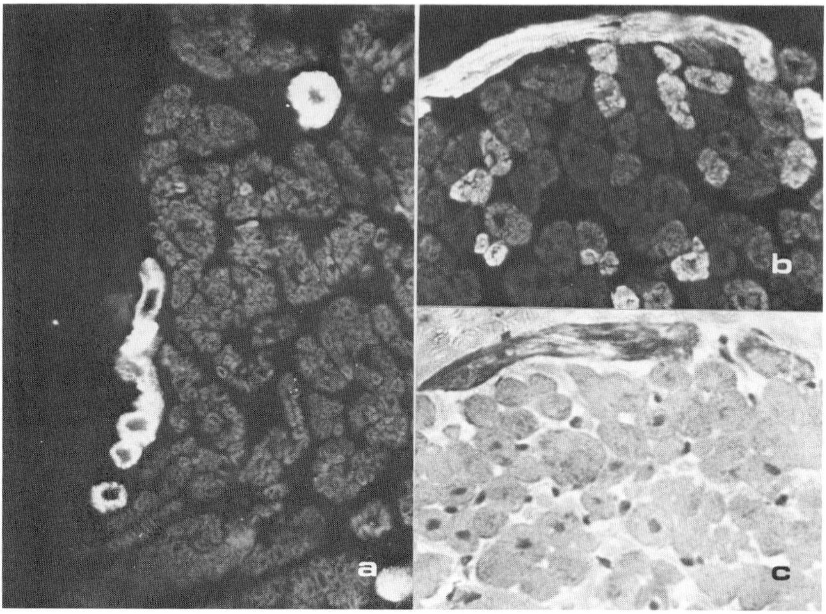

Figure 9–4. Purkinje-like fibers in the atrial myocardium. Sections from duck (*a*) and bovine (*b*, *c*) right atrial myocardium, endocardial surface. *a*: large fibers labeled by anti-anterior latissinus dorsi muscle myosin, an antibody that stains Purkinje fibers specifically in avian heart (see [18]). *b*, *c*: serial sections stained with anti-bV (*b*) and PAS (*c*).

distributed in different atrial regions. The pattern of distribution of the myosin HC isoforms suggests that the different myosin HC's may be related to the functional role of the various cell types, α-type HC being associated with ordinary atrial myocardium, β-type HC with specialized conducting tissue and nodal-type HC with nodal (or pacemaking) tissues. However, this interpretation is probably too simplistic; the correlation between myosin type and cell type is complicated by the frequent coexistence of different HC types within the same cell. Furthermore, it should be stressed that the ordinary left atrial muscle cells, which in the normal heart contain only α-type HC, can express β-type HC as a result of adaptive processes under the effect of hemodynamic overload (see below).

DIFFERENTIATION OF ATRIAL MUSCLE CELLS

Developmental changes of cardiac myosins have recently been the object of several studies, most of which are concerned with ventricular myosins. Fetal ventricular myosin is characterized by the presence of a light chain (LC) isoform undistinguishable on two-dimensional gels from the LC 1_{emb} present in developing skeletal muscle and from the LC 1 present in fetal and adult atria and in adult Purkinje fibers [15, 31, 32, 33]. On the other hand, there is no evidence for the presence of unique forms of cardiac myosin HC's during fetal development. Fetal and adult ventricular myosin HC's are apparently identical in the bovine and human heart [32, 33]. Ventricular myosin HC's present in the rat fetal heart are un-

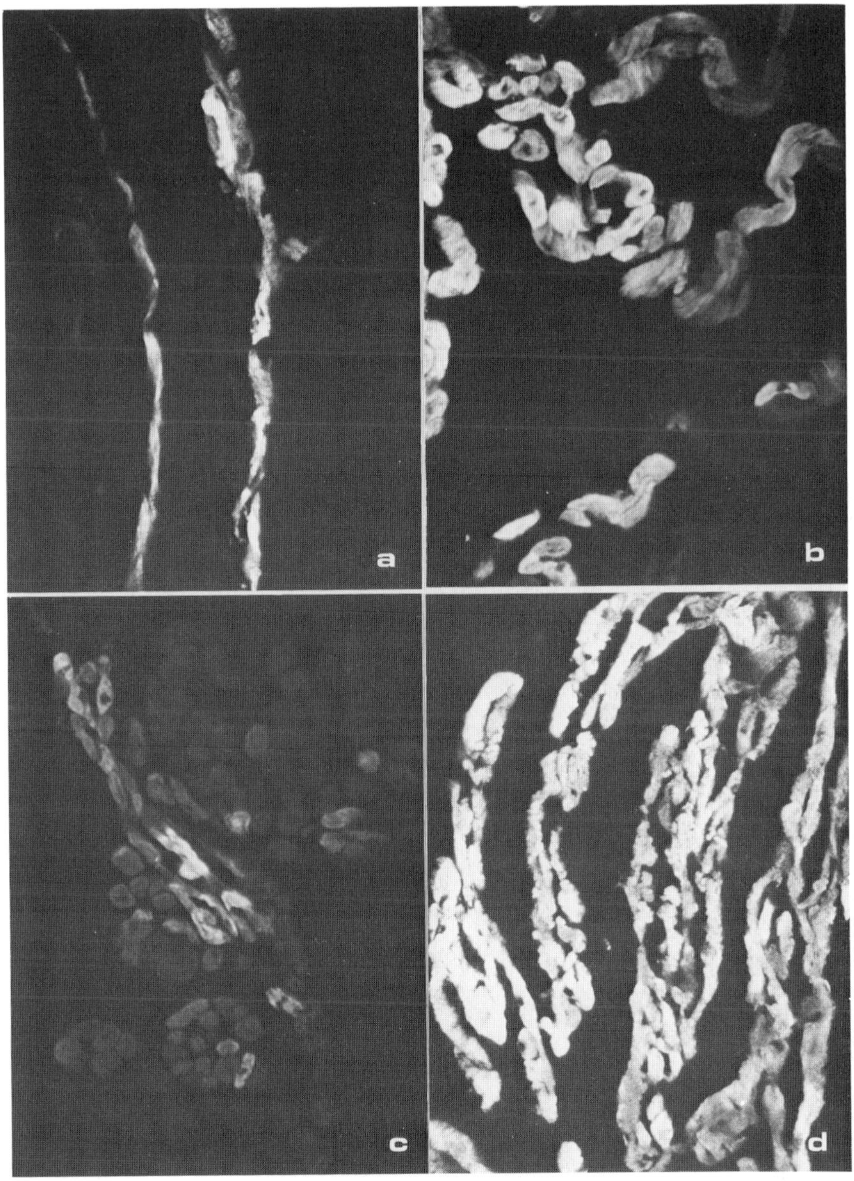

Figure 9–5. Nodal tissue, bovine heart. Sections from the central (*a*) and peripheral (*b*) regions of the sino-atrial node, the origin of the crista terminalis (*c*) and the central region of the atrio-ventricular node (*d*), stained with anti-bF. In the bovine heart this antibody stains exclusively nodal cells and occasional cells in the internodal tracts.

distinguishable by a variety of techniques from the HC_β present in the adult V3 isomyosin [25]. Postnatal development in several mammalian species is characterized by an early and rapid accumulation of HC_α (V1 isomyosin) around birth:

Table 9–1. Myosin HC Types in the Bovine Atrial Myocardium

HC Type	Antibody[a]	Distribution of HC Types
α-type	anti-bA	Almost all atrial fibers (rare nonreactive fibers in transitional regions between SA and AV nodes and in atrial myocardium)
β-type	anti-bV	Right atrium, especially crista terminalis, and interatrial septum; rare reactive fibers in left atrium
Nodal-type	anti-bF	SA and AV nodes

[a]Antibody used to identify the various HC types. The antibodies are described in [19] (anti-bA), [26] (anti-bV), and [30] (anti-bF).

HC_α becomes the predominant HC isoform throughout life in the mouse and in the rat, while it progressively decreases in other species like the rabbit [11, 20, 34, 35]. The early postnatal changes in the relative synthesis of HC_α and HC_β in the ventricular myocardium correlate with and probably result from parallel changes in endogenous levels of thyroid hormone [36]. Hemodynamic changes that accompany the transition from fetal to adult circulation may also modulate the regional distribution of myosin HC isoforms in the developing ventricles. To summarize, developmental changes of ventricular myosin consist in the progressive disappearance of LC 1_{emb} and in characteristic species-dependent variations in the expression of the two types of HC's, HC_α and HC_β.

Little information is available about developmental changes in atrial isomyosins. The LC pattern of atrial myosin is essentially similar in the fetal and adult heart. A minor ventricular-like LC2 component is occasionally seen in the fetal or adult heart [15, 31, 37]. To our knowledge there is no biochemical study on atrial myosin HC's during development.

Using antibodies specifically reactive with different types of cardiac myosin HC's and immunofluorescence techniques, we have examined the myosin composition of the developing atria in various species. In the mature chicken heart, atrial myocardium is selectively stained by anti-fast skeletal myosin antibodies and is unreactive with anti-ventricular myosin antibodies, whereas ventricular myocardium shows the reversed staining pattern [18]. As shown in figure 9–6 in the four-day-old chick embryo, atrial myocardium reacts with both antibodies, the definitive pattern of antimyosin immunoreactivity being established in the nine-day embryo. At this stage only a minor population of atrial myocytes is labeled by anti-ventricular myosin. The heterogeneous distribution of different types of myosin HC in the embryonic chicken atria appears to be an early and permanent differentiation process since embryonic atrial myocytes grown in vitro also display a variable pattern of reactivity toward specific antimyosin antibodies (figure 9–7). We have previously reported the existence of muscle cell heterogeneity in cultures from chick embryonic ventricles and rat neonatal ventricles [38, 39]. The tissue culture system may offer unique possibilities to investigate the factors responsible for regulating isomyosin expression in cardiac muscle cells and the stability of their differentiated state.

In the mature bovine heart most atrial myocytes contain α-type myosin HC; they display alkali stable and acid labile myosin ATPase activity and react with

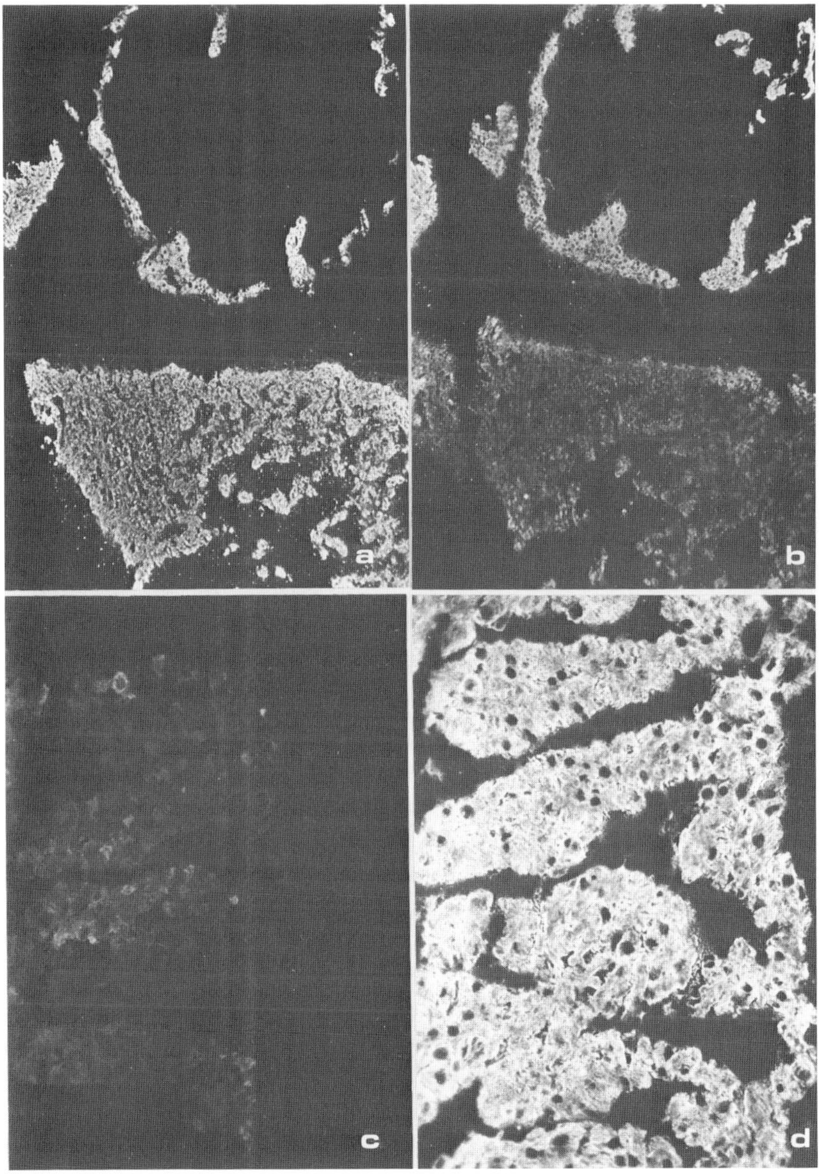

Figure 9–6. Embryonic chicken heart. Serial section from 4.5-day (*a, b*) and 9-day (*c, d*) chick embryos. Indirect immunofluorescence with anti-chicken ventricular myosin (*a, c*) and anti-chicken pectoralis muscle myosin (*b, d*). Note that atrial muscle cells are all stained by anti-chV in the 4.5-day embryo (*a*) whereas most of them are unstained in the 9-day embryo (*c*).

anti-bovine atrial but not with anti-ventricular myosin antibodies [26]. The difference in myosin composition between atrial and ventricular myocardium is already established in the heart of four-week-old bovine embryos (figure 9–8; see also [40]). Even at these stages of embryonic development, however, a number of

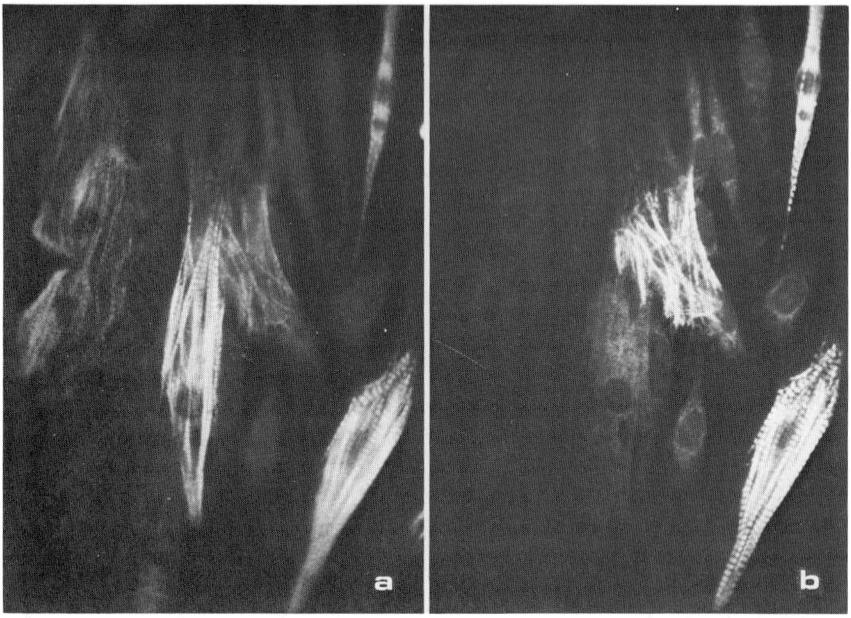

Figure 9–7. Cultured atrial muscle cells from embryonic chicken heart. Muscle cells dissociated from 14-day chick embryo and grown in vitro for three days (see [38]). Double immunofluorescence with fluorescein-labeled anti-chicken ventricular myosin (*a*) and rhodamine-labeled anti-chicken pectoralis myosin (*b*). Note the variable response of different atrial muscle cells to the two antimyosin antibodies.

muscle cells in the right atrium react also with anti-ventricular myosin antibodies, and therefore contain a β-type myosin HC (figure 9–9). Clusters of muscle cells reactive with anti-fetal skeletal myosin, possibly the precursors of sino-atrial nodal cell, can be identified in the sinus venosus region of the developing bovine heart (figure 9–9). The basic events leading to the segregation of different isomyosins in different regions of the bovine heart must therefore occur at very early stages, namely during the first month of embryonic development. Minor variations in the myosin HC composition of the bovine atrial myocardium take place during the late stages of fetal development. In the rat heart there are significant isomyosin shifts in the atrial myocardium during postnatal development. The left atrium of newborn and one-month-old rat contain an occasional fiber reactive for β-type myosin (figure 9–10). Labeled fibers are more numerous in the adult rat atria, and most atrial fibers are reactive with variable intensity in old rats (figure 9–10). This progressive accumulation of β-type HC in the rat atrial myocardium is analogous to the parallel shift toward HC_{β} (V3 isomyosin) that occurs with age in the ventricles [20, 21, 41].

PLASTICITY OF THE ATRIAL MYOCARDIUM

The idea that long-term adaptive changes in cardiac muscle contractility may entail isomyosin shifts has become a widely accepted notion during the last five

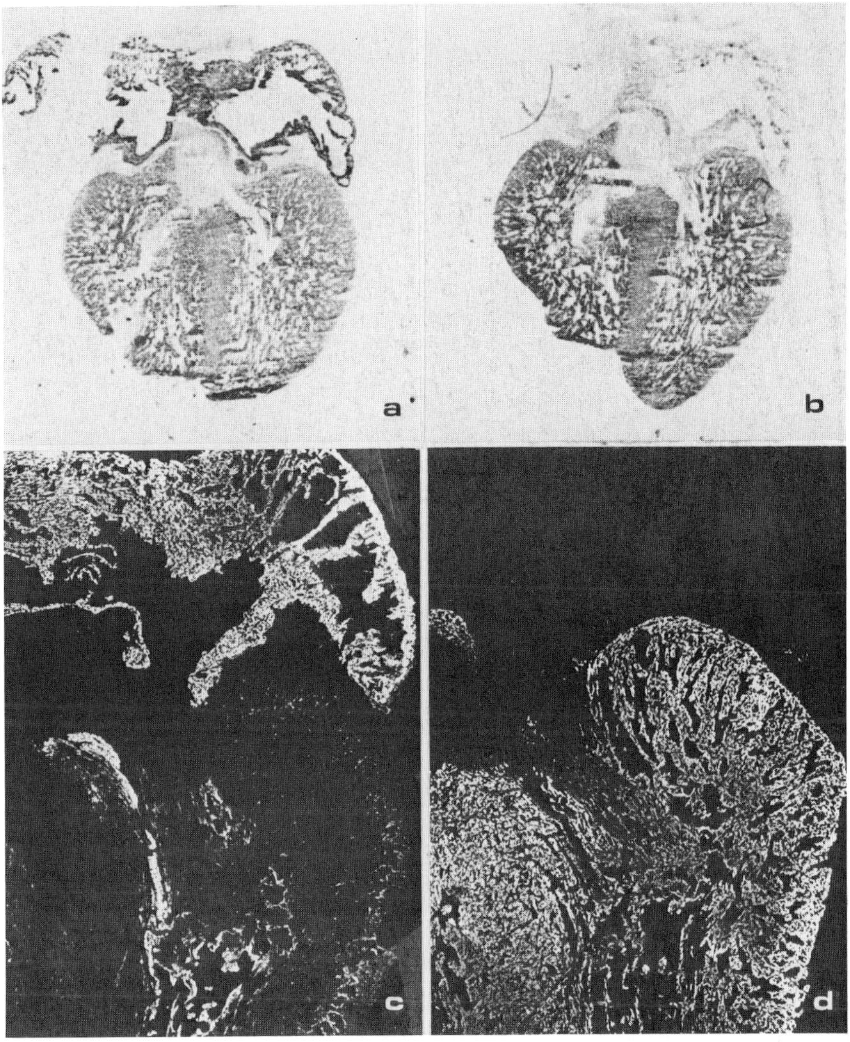

Figure 9–8. Bovine fetal heart. Serial sections of the heart from a six-week fetus processed for the histochemical demonstration of myosin ATPase after preincubation at pH 10.4 (*a*) and 4.6 (*b*), and stained with anti-bA (*c*) and anti-bV (*d*). The whole heart is shows in *a* and *b*, while only part of left atrium (upper field) and left ventricle (lower field) in shown in *c* and *d*. Note that atrial muscle cells stain intensely for myosin ATPase after alkaline preincubation and very weakly after acid preincubation, and are labeled by anti-bA whereas they are completely unreactive with anti-bV.

years. The application of new techniques has led to the demonstration that different isomyosins are present in the ventricular myocardium of the mammalian heart and that their relative concentration changes during development and under the effect of various types of stress. In the ventricular myocardium the transition from HC_α to HC_β (V1-to-V3 isomyosin shift) induced by pressure overload [41] and the β-to-α transition (V3-to-V1 isomyosin shift) induced by thyroid hormone

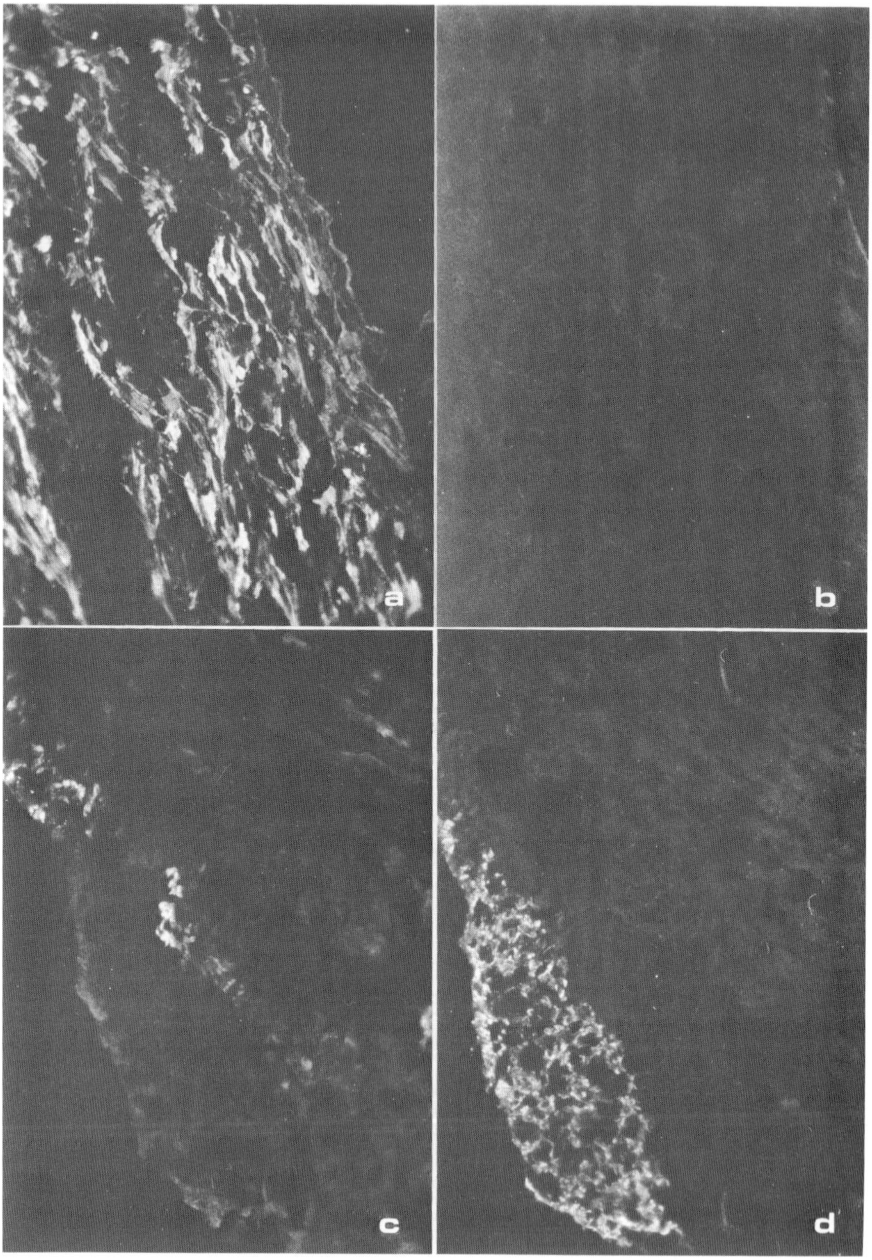

Figure 9–9. Bovine fetal heart. Serial sections from the right atrium (*a*, *b*) and from the sinus venosus region (*c*, *d*) of an eight-week fetus. Indirect immunofluorescence with anti-bV (*a*, *c*) and anti-bF (*b*, *d*). A cluster of muscle cells reactive with anti-bF and unreactive with anti-bV are present in the sinus venosus region. Fibers reactive with anti-bV and unreactive with anti-bF are distributed in different regions of the right atrium.

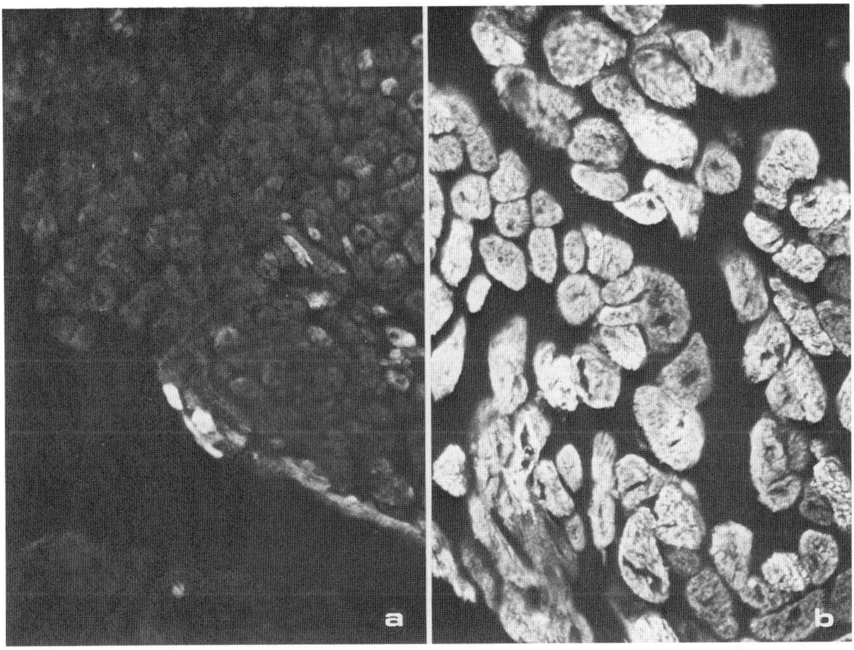

Figure 9–10. Rat atrial myocardium. Sections of left atrium from one-month-old (*a*) and one-year-old (*b*) rats stained with anti-guinea pig soleus myosin. Fibers reactive for myosin HC_β are more numerous in the old rat.

[11] can account for the changes in myosin ATPase activity and in cardiac muscle contractility observed under these conditions (for review see [42]). Compared to the large amount of information obtained on adaptive ventricular isomyosin shifts, there are very few experimental studies on atrial myosins. In a comparative study of atrial and ventricular myosins from normal, thyrotoxic and thyroidectomized rabbits, no change was observed in atrial myosin with respect to ATPase activity or in electrophoretic mobility in pyrophosphate gels [43]. In contrast to the response of the ventricular myocardium, the increased rate of synthesis of rabbit atrial myosin HC induced by thyroid hormone was not accompanied by any change in atrial myosin HC type as determined by radioimmunoassay with specific monoclonal antibodies [35]. Moreover, no difference in the light chain pattern or in the electrophoretic mobility of atrial myosin was observed in hypertrophied atria after aortic banding in young rats [37]. Perhaps surprisingly, the only evidence of atrial myosin changes comes from human studies.

In patients with mitral valve stenosis, the left atrial myocardium undergoes hypertrophy, followed by progressive chamber dilatation with fibrosis. Cummins [44] observed a change in the myosin light-chain pattern of the hypertrophied atria, with the appearance of ventricular-like light-chain isoforms. The degree of the transformation was related to the level of mean left atrial pressure and involved

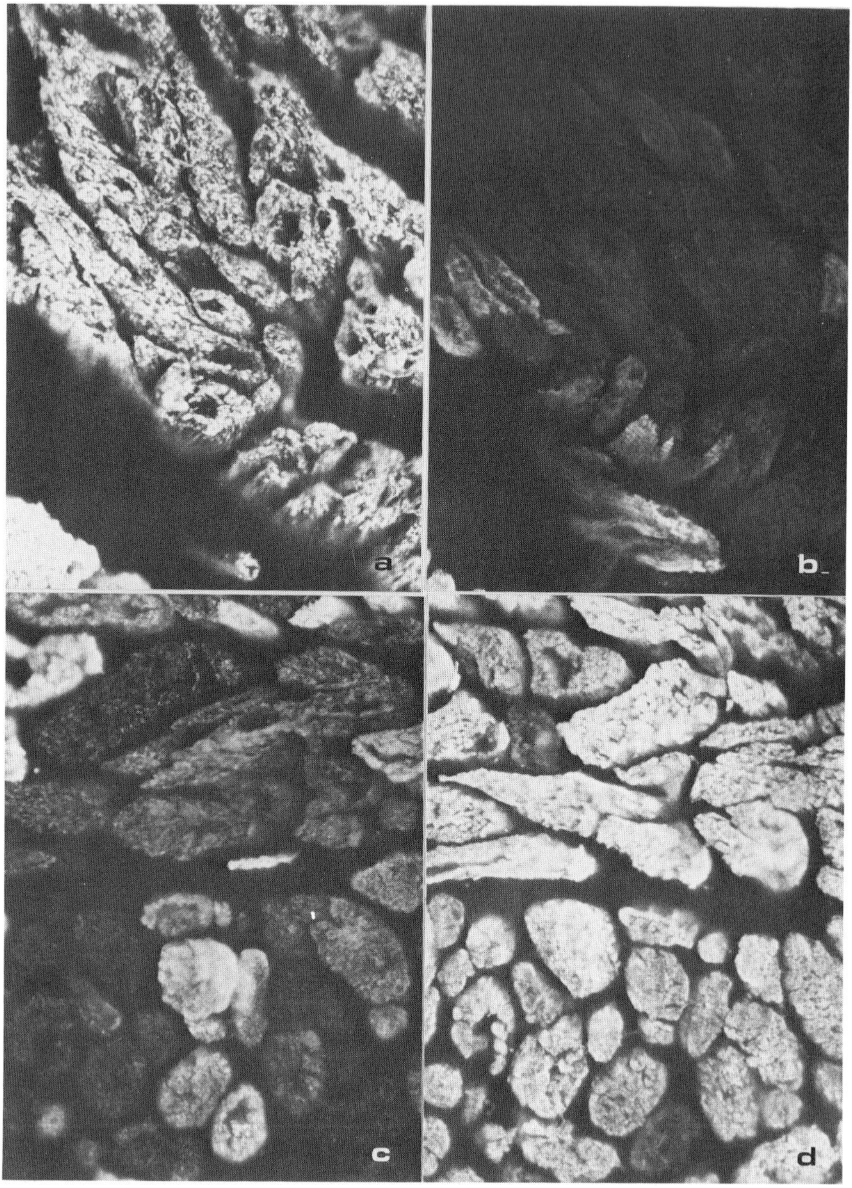

Figure 9–11. Human atrial myocardium. Serial sections of left atrium from a normal heart (*a, b*) and from the heart of a patient with mitral stenosis (*c, d*). Indirect immunofluorescence with anti-bA (*a, c*) and with anti-human pectoralis myosin (*b, d*), an antibody specific for β-type myosin HC in the human heart (see [27] and [45]). Note the striking reversal of the immunoreactivity pattern in the hypertrophic atria.

especially one of the LC 2 isoforms (the phosphorylatable light chains). Using antibodies specific for human HC_{α} and HC_{β}, we have been able to demonstrate that the HC's are also transformed in hypertrophied human atria, with the normally prevailing HC_{α} being substituted by HC_{β} [27, 45]. Immunofluorescence studies show that the degree of α-to-β transition is quite variable in different fibers of the same sample (figure 9–11) and in different samples from the same chamber. In order to obtain a quantitative evaluation of the isomyosin shift in hypertrophied atria, we developed a micromethod to extract myosin from single cryosections which were serial to those processed for immunofluorescence. Quantitative enzyme immunoassays performed on these myosins showed that the HC_{β} concentration, which is negligible in normal left atria, may rise to 90% in certain hypertrophied specimens [45].

Further experimental studies on the effect of different types of mechanical overload on atrial myosins are obviously necessary to explain the apparently contradictory results from human and animal studies. On the basis of the available information it appears that the potential of adult atrial myocytes for adaptive changes in myosin composition may vary according to species, age, and type of inciting stress, and may not necessarily be identical to that of the ventricular myocardium to the same stress. That atrial and ventricular myocytes may differ in their adaptive reactions to stress or injury is illustrated by the much greater capacity of atrial myocytes to resume DNA synthesis and increase number of nuclei and ploidy in the adult rat [46, 47]. The myosin changes observed in human atrial muscle during hypertrophy secondary to mitral valve stenosis bear some analogy to myosin shifts induced by pressure overload in the rat or rabbit heart. Although an α-to-β myosin HC transition takes place in both situations, myosin light-chain changes also occur in the overloaded human atria but not in the hypertrophied rat and rabbit ventricles.

ACKNOWLEDGMENTS

This work was supported by Ministero della Pubblica Istruzione and CNR, Gruppo Cardiorespiratorio (grants no. 82.02169.04 and CT82.00204.04). We gratefully acknowledge the contribution of Dr. S. Pierobon Bormioli to comparative studies on atrial myosins. We thank Maurizio Moretto and Massimo Fabbri for excellent technical assistance and Fiorella Burgio for secretarial work.

REFERENCES

1. De Haan, R.L. 1965. Morphogenesis of the vertebrate heart. In *Organogenesis,* ed. R.L. De Haan and H. Ursprung, pp. 377–419. New York:
2. Hoffman, B.F. 1979. Fine structure of internodal pathways. *Am. J. Cardiol.* 44:385–386.
3. Sherf, L., and James, T.N. 1979. Fine structure of cells and their histologic organization within internodal pathways of the heart: clinical and electrocardiographic implications. *Am. J. Cardiol.* 44:345–369.
4. Anderson, R.H.; Becker, A.E.; Tranum-Jensen, J.; and Janse, M.J. 1981. Anatomicoelectrophysiological correlations in the conduction system — a review. *Br. Heart J.* 45:67–82.

5. Korecky, B., and Michael, L.H. 1974. Regional differences in contractions of mammalian hearts. In *Recent advances in studies on cardiac structure and metabolism,* Vol. 4, ed. N.S. Dhalla, pp. 77–87. Baltimore: University Park Press.
6. Urthaler, F.; Walker, A.A.; Hefner, L.L.; and James, T.N. 1975. Comparison of contractile performance of canine atrial and ventricular muscles. *Circ. Res.* 37:762–671.
7. Wikman-Coffelt, J.; Refsum, H.; Hollosi, G.; Rouleau, L.; Chuck, L., and Parmley, W.W. 1982. Comparative force-velocity relation and analyses of myosin of dog atria and ventricles. *Am. J. Physiol.* 243:H391–H397.
8. Katz, A.M. 1977. *Physiology of the heart.* New York: Raven Press.
9. Fabiato, A. 1982. Calcium release in skinned cardiac cells: variations with species, tissue, and development. *Federation Proc.* 41:2238–2244.
10. Long, L.; Fabian, F.; Mason, D.T.; and Wikman-Coffelt, J. 1977. A new cardiac myosin characterized from the canine atria. *Biochem. Biophys. Res. Commun.* 76:626–635.
11. Hoh, F.Y.; McGrath, P.A.; and Hale, P.T. 1978. Electrophoretic analysis of multiple forms of rat cardiac myosin: effects of hypophysectomy and thyroxine replacement. *J. Mol. Cell. Cardiol.* 10:1053–1076.
12. Flink, J.L.; Rader, J.H.; Banerjee, S.K.; and Morkin, E. 1978. Atrial and ventricular cardiac myosins contain different heavy chain species. *FEBS Lett.* 94:125–130.
13. Syrovy, I.; Delcayre, C.; and Swynghedauw, B. 1979. Comparison of ATPase activity and light subunits in myosins from left and right ventricles and atria in seven mammalian species. *J. Mol. Cell. Cardiol.* 11:1129–1135.
14. Yazaki, Y.; Ueda, S.; Nagai, R.; and Shimada, K. 1979. Cardiac atrial myosin adenosine triphosphatase of animals and humans: distinctive enzymatic properties compared with cardiac ventricular myosin. *Circ. Res.* 45:522–527.
15. Price, K.M.; Littler, W.A.; and Cummins, P. 1980. Human atrial and ventricular myosin light-chain subunits in the adult and during development. *Biochem. J.* 191:571–580.
16. Dalla Libera, L., and Sartore, S. 1981. Immunological and biochemical evidence for atrial-like isomyosin in thyrotoxic rabbit ventricle. *Biochem. Biophys. Acta* 669:84–92.
17. Clark, W.Q.; Chizzonite, R.A.; Everett, A.W.; Rabinowitz, M.; and Zak, R. 1982. Species correlations between cardiac isomyosins. *J. Biol. Chem.* 257:5449–5454.
18. Sartore, S.; Pierobon-Bormioli, S.; and Schiaffino, S. 1978. Immunohistochemical evidence for myosin polymorphism in the chicken heart. *Nature* 274:82–83.
19. Sartore, S.; Gorza, L.; Pierobon-Bormioli, S.; Dalla Libera, L.; and Schiaffino, S. 1981. Myosin types and fiber types in cardiac muscle. 1. Ventricular myocardium. *J. Cell. Biol.* 88:226–233.
20. Gorza, L.; Pauletto, P.; Pessina, A.C.; Sartore, S.; and Schiaffino, S. 1981. Isomyosin distribution in normal and pressure overloaded rat ventricular myocardium: an immunohistochemical study. *Circ. Res.* 49:1003–1009.
21. Schiaffino, S.; Gorza, L.; and Sartore, S. 1983. Distribution of myosin types in normal and hypertrophic heart: an immunocytochemical approach. In *Perspectives in cardiovascular research,* Vol 7, *Myocardial hypertrophy and failure,* ed. N.R. Alpert, pp. 149–166. New York: Raven Press.
22. Hoh, J.F.Y.; Yeoh, G.P.S.; Thomas, M.A.V.; and Higginbottom, L. 1979. Structural differences in the heavy chains of rat ventricular myosin isoenzymes. *FEBS Lett.* 97:330–334.
23. Sartore, S.; Dalla Libera, L.; and Schiaffino, S. 1979. Fractionation of rabbit ventricular myosin by affinity chromatography with insolubilized antimyosin antibodies. *FEBS Lett.* 106:197–201.
24. Clark, W.A.; Everett, W.A.; Chizzonite, R.A.; and Zak, R.A. 1983. Classification and characterization of cardiac isomyosins. *J. Mol. Cell. Cardiol.* Suppl. 2, 15:25.
25. Schwartz, K.; Lompre, A.M.; Wisnewsky, C.; and Whalen, R.G. 1983. Comparisons of rat cardiac myosins at fetal stages in young animal and in hyperthyroid adults. *J. Biol. Chem.* 257:14412–14418.
26. Gorza, L.; Sartore, S.; and Schiaffino, S. 1982. Myosin types and fiber types in cardiac muscle. 2. Atrial myocardium. *J. Cell. Biol.* 95:838–845.
27. Gorza, L.; Mercadier, J.J.; Schwartz, K.; Thornell, L.E.; Sartore, S.; and Schiaffino, S. In press. Myosin types in the human heart: an immunofluorescence study of normal and hypertrophied atrial and ventricular myocardium. *Circ. Res.*

28. Thornell, L.E., and Forsgren, S. 1982. Myocardial cell heterogeneity in the human heart with respect to myosin ATPase activity. *Histochem. J.* 14:479–490.
29. James, T.N., and Sherf, L. 1971. Specialized tissues and preferential conduction in the atria of the heart. *Am. J. Cardiol.* 28:414–426.
30. Sartore, S.; Gorza, L.; and Schiaffino, S. 1982. Fetal myosin heavy chains in regenerating muscle. *Nature* 298:294–296.
31. Whalen, R.G., and Sell, S.M. 1980. Myosin from fetal hearts contains the skeletal muscle embryonic light chain. *Nature* 286:731–733.
32. Whalen, R.G.; Sell, S.M.; Eriksson, A.; and Thornell, L.E. 1982. Myosin subunit types in skeletal and cardiac tissues and their developmental distribution. *Develop. Biol.* 91:485–490.
33. Schier, J.J., and Adelstein, R.S. 1982. Structural and enzymatic comparison of human cardiac muscle myosin isolated from infants, adults and patients with hypertrophic cardiomyopathy. *J. Clin. Invest.* 69:816–825.
34. Lompré, A.M.; Mercadier, J.J.; Wisnewsky, C.; Bouveret, P.; Pantaloni, C.; D'Albis, A.; and Schwartz, K. 1981. Species- and age-dependent changes in the relative amounts of cardiac myosin isoenzymes in mammals. *Devel. Biol.* 84:286–290.
35. Everett, A.W.; Clark, A.; Chizzonite, R.A.; and Zak, R. 1983. Change in synthesis rates of a- and b-myosin heavy chains in rabbit heart after treatment with thyroid hormone. *J. Biol. Chem.* 258:24121–2425.
36. Everett, A.W.; Chizzonite, R.A.; Clark, W.A.; and Zak, R. 1983. Relationship of changes in molecular forms of myosin heavy chains to endogenous level of thyroid hormone during postnatal growth. In *Perspectives in cardiovascular research,* Vol. 8, ed. R.C. Tarazi and J.B. Dunbar. New York: Raven Press.
37. Bugaisky, L.B.; Siegel, E.; and Whalen, R.G. 1983. Myosin isozyme changes in the heart following constriction of the ascending aorta of 25-day old rat. *FEBS Lett.* 161:230–234.
38. Cantini, M.; Sartore, S.; and Schiaffino, S. 1980. Myosin types in cultured muscle cells. *J. Cell. Biol.* 85:903–909.
39. Samuel, J.L.; Rappaport, L.; Mercadier, J.J.; Lompre, A.M.; Sartore, S.; Triban, C.; Schiaffino, S.; and Schwartz, K. 1983. Distribution of myosin isozymes within single cardiac cells: an immunohistochemical study. *Circ. Res.* 52:200–209.
40. Forsgren, S.; Strehler, E.; and Thornell, L.E. 1982. Differentiation of Purkinje fibers and ordinary ventricular and atrial myocytes in the bovine heart: an immuno- and enzyme histochemical study. *Histochem. J.* 14:929–942.
41. Lompré, A.M.; Schwartz, K.; d'Albis, A.; Lacombe, G.; Thiem, N.V.; and Swynghedauw, B. 1979. Myosin isoenzymes redistribution in chronic heart overload. *Nature* 282:105–107.
42. Alpert, N.R., ed. 1983. *Perspectives in cardiovascular research,* Vol. 7, *Myocardial hypertrophy and failure.* New York: Raven Press.
43. Banerjee, S.K. 1983. Comparative studies of atrial and ventricular myosin from normal, thyrotoxic, and thyroidectomized rabbits. *Circ. Res.* 52:131–136.
44. Cummins, P. 1982. Transitions in human atrial and ventricular myosin light-chain isoenzymes in response to cardiac-pressure-overload-induced hypertrophy. *Biochem. J.* 205:195–204.
45. Schiaffino, S.; Gorza, L.; Saggin, L.; Valfré, C.; and Sartore, S. In press. Myosin changes in hypertrophied human atrial and ventricular myocardium: a correlated immunofluorescence and quantitative immunochemical study on serial cryosections. *Eur. J. Cardiol.*
46. Rumyantsev, P.P., and Kassem, A.M. 1976. Cumulative indices of DNA synthesizing myocytes in different compartments of the working myocardium and conductive system of the rat's heart muscle following extensive left ventricle infarction. *Virchows Arch. B. Cell. Path.* 20:329–342.
47. Oberpriller, J.O.; Ferrans, V.J.; and Carroll, R.J. 1983. Changes in DNA content, number of nuclei and cellular dimensions of young rat atrial myocytes in response to left coronary artery ligation. *J. Mol. Cell Cardiol.* 15:31–42.

10. THE ROLE OF PROTEIN SYNTHESIS AND DEGRADATION IN CARDIAC GROWTH

RADOVAN ZAK

Growth of the heart has attracted the attention of physicians and biologists for many years. To the physician it is important to elucidate how the heart copes with an increased hemodynamic load as well as the characteristics of the enlarged, yet failing, heart. To the biologists, the heart is an excellent model for studies of orderly temporal regulation of genes which takes place during myogenesis and growth. The advantages of the heart for such studies are several. The morphological aspects of cardiac growth have been elucidated quite well both at the gross and ultrastructural levels. The proteins of the heart, especially those involving its contractile apparatus, are abundant, well-characterized, and, in some cases, their genes have been partially defined. Finally, changes in the heart size as well as in the complement of its components can be readily induced in a reversible manner.

AN OVERVIEW OF GENE EXPRESSION IN NUCLEATED CELLS

Control of gene expression takes place both in the nucleus and in the cytoplasm. The steps which take place in the nucleus include transcription of genes, processing of nascent mRNA, and transport of mature mRNA into cytoplasm. In the cytoplasm, the output of protein-synthesizing machinery can be controlled by inactivation (or masking) and degradation of mRNA, by modification of the rate by which mRNA is translated on polysomes, and finally by the degradation of the translated gene product.

M.J. Legato (ed.) THE DEVELOPING HEART, pp. 191–203.

Transcription

Compared to procaryotes, the control of gene expression in nucleated cells is very complex and not completely clarified. It is clear, however, that most of the DNA within a nucleus is transcriptionally inactive at any given time. The conversion of silent to transcribed genes involves a conformational change of the chromatin complex and appears to be governed by its protein components as well as the regulatory sequences of DNA.

The actual process that leads to the opening of repressed genes for their transcription by RNA polymerase is largely obscure. From our general knowledge of gene organization, however, one can postulate the following levels of control: (1) the conformation of chromatin; (2) role of regulatory sequences of the DNA molecule; and (3) the activity and/or amount of RNA polymerase.

As far as the conformational change in chromatin is concerned, it is believed that some uncoiling of the DNA fiber within chromatin is necessary for the transcription to take place. It is well known that the chromosome displays two regions after staining with basic dyes. The non-staining region is referred to as euchromatin, while the staining region is called heterochromatin. Euchromatin corresponds to active genes; heterochromatin responds to inactive genes. This observation, known for many years, has recently been extended to the discovery that DNase I cuts some regions of the DNA molecule more readily than others. The segment of DNA associated with active genes (euchromatin) is hypersensitive to DNAse I action and apparently corresponds to a more open structure of DNA.

It is currently believed that histones act as the blocking agents and inhibitors of DNA transcription. The high degree of homology between histones originating from various types of cells, however, suggests that the histones are the blocking agents but are not involved in the expression of specific genes. The nonhistones of chromatin are more likely candidates for this role for they are highly heterogeneous in molecular weight, cell type specific, and preferentially stimulate transcription of homologous DNA and chromatin.

As far as the role of regulatory nucleotide sequences on the DNA molecule are concerned, considerable progress has been made recently in their analysis. The current view of gene structure is shown in figure 10–1. The gene consists of a series of expressed nucleotide sequences (exons) interdispersed by silent, non-coding intervening sequences (introns). The series of exons and introns is bound on either side by transcribed, but non-coding, regions that demarcate the point where transcription by RNA polymerase is initiated and terminated, respectively. Genes are flanked on either side by non-transcribed regions. Of special interest is the DNase I hypersensitive region which is localized in front (i.e., upstream) of the active gene preceeding the initiation site. Moreover, nucleotide sequences of unusual nucleotide composition were found also to precede the initiation site. These are regions rich in thymidine and adenine (TATA box) and thymidine, cytosine, and adenine (CAAT box). Currently, some evidence points toward the TATA box determining the exact starting point of transcription while the CAAT box (and most likely some additional and more remote segments of DNA) determining the

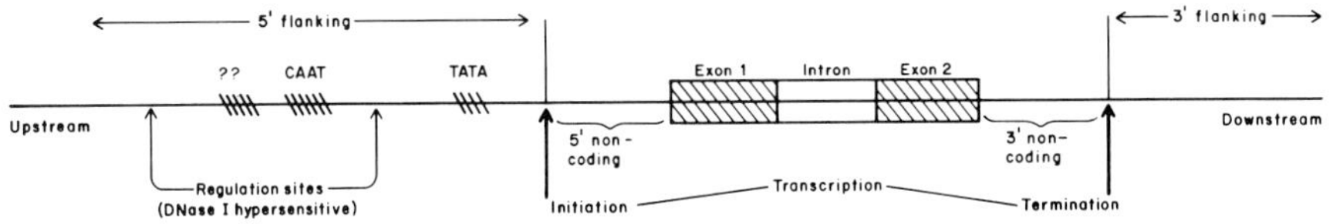

Figure 10–1. Scheme of gene. A typical gene of high organisms is split, i.e., it consists of a series of translated (exons) and non-translated (introns) nucleotide sequences. A gene is demarcated by sites where its transcription by RNA polymerase is initiated and terminated, respectively. Preceding the initiation site (i.e., upstream of DNA) are the nucleotide sequences that are involved in the regulation of transcription. (From [3] by permission.)

efficiency of transcription. The role of downstream flanking sequences is not yet known.

Although the regulatory sequences have been clearly identified, their exact role in the association of RNA polymerase with the DNA template remains unexplained. It is likely though that just as with procaryotes these DNA sequences are targets of molecules which initiate transcription. The attachment of such molecules (e.g., hormones and other growth-regulating substances) somehow results in conformational changes in DNA which allow transcription by RNA polymerase to take place.

In addition to DNA accessibility, control of transcription could also involve modulation of RNA polymerase activity or its amount. Of the three known classes of this enzyme, only one (type II) participates in synthesis of messenger RNA. The remaining two types synthesize ribosomal, transfer and 5S RNA. It is of great interest that polymerase II has been found to be associated with the nucleosome (i.e., the segment of DNA bound to histone octamere). Upon the binding of polymerase to nucleosomes, reversible conformational change follows. Such a structural change is consistent with a loosening in the chromatin structure and the consequence gene opening.

The recent progress in our understanding of gene structure and function which has been very briefly outlined above is quite remarkable. Nevertheless, we still do not know how specific genes are opened in the precise temporal order typical of normal or induced growth.

Post-transcriptional control of gene expression

After the DNA template has been transcribed, the next level of control (referred to as post-transcriptional control) includes: (1) modification of mRNA nucleotides; (2) excission of introns and splicing of the remaining exons; and (3) transport of processed (i.e., mature mRNA) from the nucleus into the cytoplasm.

The nascent mRNA (referred to as primary transcript) is processed in two steps. First, its 5′-end becomes methylated by the addition of 7-methylguanosine (a processed referred to as capping) and the 3′ end receives a "tail" consisting of about 200 adenyl residues (polyadenlyation). The exact role of these two modifications of mRNA is not known. Some data, however, suggest their having a stabilizing effect on the mature mRNA.

The second level of processing follows capping and polyadenylation. This includes the concerted action of endonuclease to excise the introns and ligase which joins the remaining segments of the transcript. Very little is known about the nucleo-cytoplasmic transport of mRNA, but participation of a special class of proteins has been suggested.

Once the mRNA reaches the cytoplasmic compartment, the rate at which it can be translated by the ribosome machinery can be modulated by three mechanisms: (1) rate of mRNA degradation; (2) storage or masking of mRNA; and (3) modulation of protein output by the mRNA-ribosomes complex.

Only a very little is known about the first two mechanisms. Some examples obtained in studies of the heart will be discussed in a later section. As far as the regulation of protein output on polysomes is concerned, multiple control points are likely to exist. The pathway of protein synthesis is a highly complex process consisting of initiation, elongation, and termination steps. In addition to the availability of mRNA discussed in the preceding paragraphs, the following factors might determine protein output: the activity of several enzymes (e.g., peptidyl transferase and amino acyl-tRNa synthetase); the availability of high-energy phosphate (formation of one peptide bond requires hydrolysis of four high energy phosphate, one molecule of ATP needed for amino acid activation, and three molecules of GTP for translation itself); the functional state of ribosomal subunits, including changes in their conformation and phosphorylation. The latter process involves the initiation factor responsible for placing of the first amino acid on the ribosome, one protein component of ribosomes and aminoacyl-tRNA synthetase.

From this brief description of gene regulation it is clear that we are dealing with an extremely complex process which is being unravelled only recently. More detailed description of this rapidly advancing field of molecular biology can be found in recently published review articles [1, 2, 3].

CONTROL OF GENE EXPRESSION IN THE HEART

Studies of gene regulation in the heart are just at their beginning. The main reason why such studies were not pursued in the past is that the heart consists of many types of cells. Consequently, most of its DNA belong to nuclei of non-muscle cells which outnumber myocytes 3 to 1. The second reason is that we do not mutants of myocytes; the third, that myocytes derived from adult myocardium are difficult to grow in culture, and no successful transformation to proliferating cells has been achieved so far. More recently we have learned that even the "typical" muscle proteins, such as myosin, exist in multiple molecular forms, some of which are present in non-muscle cells. As a result, very little is known about the controls of gene expression in the heart.

Transcription

Increased synthesis of RNA is the most striking change occurring in an enlarging myocardium. This was demonstrated both by measurements of RNA levels, pulse labeling of RNA with its precursors [4] and by determinations of DNA-dependent RNA polymerase in nuclei isolated from muscle and non-muscle cells of the heart [5]. In the hyperthyroid heart the activity of polymerase II (enzyme synthesis of messenger RNA) and III (synthesis of 5S RNA) increased in nuclei of both types of cells, while the activity of polymerase I (synthesis of ribosomal RNA) occurred only in the muscle cell nuclei. Determination of total RNA polymerase (i.e., activity which is independent of the regulatory action of chromatin) indicated that the observed changes represent an increase either in the activity or the amount of respective RNA polymerase. Of interest, however, is that the elevation of RNA polymerase activity lags behind the increase in RNA synthesis. The increased synthesis of RNA which occurs in the absence of any change in RNA polymerase activity is consistent with alteration of chromatin conformation and with a consequent increase in the amount of the template available for transcription. That such an increase in template activity does indeed take place is suggested by experiments in which the activity of chromatin isolated from nuclei obtained from hypertrophic hearts has been measured with heterologous RNA polymerase [6].

Although there is no doubt that the increase in RNA synthesis is one of the earliest events taking place after induction of growth, conclusive demonstration of preferential synthesis of mRNA is still elusive. Very shortly after aortic constriction, however, the activity of polymerase-synthesizing mRNA increased somewhat more than the activity of other polymerases [7]. Application of recombinant DNA technology to studies of gene regulation undoubtably will provide a more definitive answer to this important question since the technique allows measurement of the levels of mRNA coding for specific, rather than total, proteins. Thus far, cloned cDNA probes containing incomplete sequences for rat [8] α-myosin heavy chains (HC) and rabbit [9] α- and β-myosin HC have been constructed. The two rabbit probes are very useful for the analysis of differential transcription of the two ventricular myosin messages since both correspond to the amino acid sequence localized in the same portion of the myosin rod. The probes are about 300 nucleotides long and code for amino acid residues 200 to 304 (using the numbering system suggested by Karn and his associates for nematode C. elegance myosin [10]). In addition, cDNA probes are also available for avian [11, 12, 13] and rat [14, 15] skeletal myosin heavy chains (HC).

Analysis of gene structure in the heart has also been initiated, but the data are still very incomplete. Genomic clones for nematode [10], *Drosophila* [16, 17]), chicken [18], rat [8, 19], rabbit [20], and human [21] myosin HC's have been isolated. Today, however, only one heavy-chain gene has been completely reconstructed: the one corresponding to nematode myosin [22]. Data concerning partial characterization of genomic clones specific for rabbit α- and β-ventricular myosin HC's are, however, available [20]. Although the myosin HC is similar in size to that of the nematode, its gene is more than two and one-half times larger. This difference is due primarily to the number of introns. While the nematode gene contains only five intervening sequences, that segment of the rabbit gene so far analyzed that corresponds to 80% of the myosin HC mRNA is interdispersed by at least 28 introns. At present it appears that the introns do not delineate any functional or structural domains of the myosin molecule. The positions of the introns within α- and β-myosin HC genes suggest that they arose by duplication of a common ancestor. So far only limited data are available regarding the chromosomal location of muscle genes. The genes coding for the three myofibrillar proteins — namely, myosin heavy-, myosin light-chain 2, and actin — have been localized on three different chromosomes [23].

Post-transcriptional Events

In addition to DNA transcription, the expression of a gene can be regulated at multiple steps which include all the events from message-processing to the release of completed polypeptides from polysomes. One way to assess whether post-transcriptional events or the transcription itself controls the expression of a particular gene is to compare the rate of protein synthesis with the amounts of its mRNA. We have applied this approach to the study of myosin HCα and HCβ in rabbit ventricle. The rates of heavy-chain synthesis were measured after their separation with HCα and HCβ specific monoclonal antibodies [24] The level of mRNA was determined using cloned cDNA probes described in the previous section.

The correlation between the various rates of protein synthesis was measured after administration of a thyrotoxic dose of thyroid hormone. The ratio of V1 and V3 isomyosin composed of HCα and HCβ, respectively, in the rabbit ventricle depends upon the thyroid state of the animal. While the high levels of T_4 and triiodothyronine (T_3) favor the V1 isomyosin, the hypothyroid state induces transition towards the low ATPase isomyosin in V3. In our experiments the relative amounts of mRNA corresponding to these two classes of heavy chains were determined by S1 nuclease mapping. The principle of this technique is as follows: the total RNA is extracted from a heart and is hybridized with a mRNA specific cDNA probe. Homology between the mRNA and its corresponding cDNA result in the formation of a heteroduplex which is protected against the splitting action of S1 nuclease. Consequently, all of the probe can be recovered at its original size when analyzed by electrophoresis following S1 treatment. In contrast, hybridization of cDNA with heterologous mRNA will be incomplete, and the S1 nuclease will split the probe at the point of nucleotide mismatch. The cDNA will then be recovered as a fragment of its original size. When a mixture of homologous and heterologous mRNA is analyzed with radiolabeled cDNA probes, their relative amounts can be estimated from the ratio of radioactivity associated with the full size and fragmented cDNA after electrophoretic separation.

By the use of this approach we have determined that T_3 induces synthesis of the message for HCα four hours after its administration to propylthiouracyl (PTU)-treated rabbits. The PTU treatment was used to suppress any HCα messenger RNA prior to its induction. A more precise measure of the relative changes in the levels of HCα and HCβ messages was obtained from the measurement of P^{82} radioactivity associated with protected and fragmented cDNA by liquid scintillation counting. The data have shown that by 24 hours after T_3 treatment, half of the transcribed message corresponds to HCα. By 72 hours after treatment the contribution of HCα mRNA to the synthesis of total myosin mRNA rose to 90%.

The correlation between the amount of transcribed mRNA and the synthesis of corresponding myosin HC's was next examined in the normal young rabbit. Both HCα and HCβ are synthesized at about the same fractional rate of $0.14\ d^{-1}$. At one and two days after administration of 200 μg of T_4 per kg of body weight, however, the fractional synthesis rate of HCα increased three-fold, whereas that of HCβ decreased by a similar magnitude. In order to have a meaningful comparison with the total amount of HC messages, we have calculated myosin synthesis rates from the product of the fractional synthesis rate and the pool size of each of the heavy chains. The resulting unit of synthesis rate becomes mg HC synthesized by one gram of heart per day. The synthesis rates of each myosin were then compared with the relative levels of corresponding messages. Both within the euthyroid group of animals, as well as in animals 12 hours after T_4 treatment, the relative synthesis rates and mRNA levels for the two heavy chains were found to be the same. Such a close correlation between protein synthesis and mRNA levels strongly indicates that thyroid hormone regulates the expression of cardiac myosin at one of the steps which precedes translation. This could include gene transcription, processing of nascent mRNA, nucleocytoplasmic transport or storage masking, and degradation of cytoplasmic mRNA. Since our assay of mRNA included all forms of processed transcripts, i.e., nuclear mRNA as well as active and putative inactive cytoplasmic mRNA, the close correlation between the rate of protein synthesis and the total mRNA indicates that neither a change in nucleocytoplasmic transport nor mobilization of "inactive" mRNA plays a significant role in the control of myosin expression.

If such a mobilization took place, the increased synthesis would be accompanied merely by a transfer of mRNA between "active" and "inactive" pools and would not be correlated to the level of total mRNA. The lack of any stored mRNA, however, might not apply to all growth situations. For example, in differentiating myoblasts the rate of protein synthesis has been found to be less than predicted based on the amount of mRNA [25]. These data indicate that not all the newly synthesized mRNA is made available for translation. In support of this conclusion it has also been reported that presumptive myoblasts (i.e., myogenic cells which have not yet initiated synthesis of muscle specific proteins) transcribe, but do not translate, myosin mRNA. Moreover, when myosin HC mRNA isolated from myoblasts is tested for translation in a cell-free system, it has been found to be inactive [26]. Such an inactive mRNA is postulated to be associated with a protein to form discrete protein particles.

Regulation of gene expression at the level of message utilization is also suggested by isolation of ribonucleoprotein particles containing low-molecular-weight RNA (which is distinct from mRNA and is referred to as inhibitory RNA). These particles when isolated from homogenates of chicken embryonic muscle have been found to be potent inhibitors of in vitro protein synthesis [27].

These particles present another possible mechanism for the regulation of mRNA translation.

The control of gene expression at the level of mRNA mobilization, however, appears to be limited only to some systems. There are numerous examples in which the emergence of mRNA coincides with the induction of protein synthesis similarly as in the thyroid-treated hearts. As far as muscles are concerned, the close correlation between message and rate of protein synthesis has been reported for avian myogenesis studied both in vivo [28] and in vitro [29].

Although our data have eliminated the possibility of mRNA mobilization, they do not allow us to estimate whether the level of mRNA, as determined in our experiments, is a function of increased transcription or of differential message degradation. That the degradation rate can, indeed, contribute to the level of mRNA and gene expression is suggested by experiments in which mRNA half-life changed after induction of protein synthesis. For example, in primary cultures of skeletal muscle myoblasts, the half-life of myosin HC mRNA increased substantially after fusion [26]. It is at the time of fusion that the synthesis rate of myosin increases dramatically.

Another level at which the expression of a gene can be controlled following transcription of its message is during regulation of protein output by the ribosomes-mRNA complex. This is referred to as translational control which is a highly complex process involving polypeptide initiation, elongation, and termination. In addition to the amount of mRNA discussed previously, the protein output depends upon the activity of several enzymes (e.g., amino acyl-tRNA transferase), the availability of high-energy phosphates (which are needed for amino acid activation and polypeptide elongation), and the properties of ribosomal sub-units which include their phosphorylation.

Regulation of protein output at the level of a precursor is suggested by the experiments in which the activity of amino acyl-tRNA synthetase fluctuated according to the rates of protein synthesis [30]. In developing cardiac hypertrophy, the activity of the synthetase, however, remained unchanged [31].

The second level of translational control involves the ribosomal cycle which reflects the distribution of ribosomes between their various functional states, i.e., polysomes (in which each ribosome is attached to mRNA and bears a growing peptide chain which is undergoing elongation), monoribosomes (which are single ribosomes), and ribosomal sub-units (which originate from ribosomes which have run off the mRNA and are ready to re-enter the protein-synthesis cycle). The ratio of ribosomes between these functional states reflects the state and activity of protein-synthesizing machinery. For example, the decline in the rate of protein synthesis which is accompanied by the accumulation of ribosomal sub-units is interpreted as a block to the re-entry into the ribosome cycle (i.e., protein initiation). Alternatively, if decreased synthesis is accompanied by a decrease in the number of sub-units and an increase in the number of polysomes, blocking of polypeptide elongation can be anticipated. For example, the anabolic hormones, including insulin [32], enhance the initiation step, while the elongation step is the primary target of energy depletion [33]. In perfused hearts an increased pressure load was found to be accompanied by accelerated peptide-chain initiation [34] and elongation. On the other hand, in the heart enlarged as a consequence of aortic constriction, the proportion of ribosomal sub-units in rats operated upon was similar to that of the control animals [35]. The rate of protein synthesis in the enlarged hearts correlated well with the amount of ribosomal RNA. These data are consistent with the view that increased content of ribosomes, rather than a faster turnover rate of the ribosomal cycle due to enhanced polypeptide initiation and elongation, took place.

Although the regulation of the steps in translation might vary between different models of muscle growth as presented above, it is well documented that the labeling of proteins with radioactive amino acids is increased in all situations when the heart undergoes compensatory growth [36].

Intracellular Turnover of Muscle Proteins

The final steps in which the expression of a gene can be regulated include the degradation of the product of that particular gene. The concept that all intracellular proteins are in a dynamic state and undergoing continuous degradation and resynthesis was originally suggested by Borsook in 1935 and Schoenheimer in 1939 (see [37]). This concept received full support through recent studies. Although the qualitative demonstration of protein turnover is convincing, the actual magnitude of protein half-life is still a matter of controversy. Nevertheless, it appears that the half-life of individual proteins differs substantially, ranging from minutes to several hours or even days.

The reasons for the scarcity of accurate values of protein half-lives are numerous. Some are related to technical difficulties, and others are related to experimental design. As far as the technical problems are concerned, the measurement of protein turnover by the use of radiolabeled tracers is hampered by the fact that non-reutilizable tracers are very rare. Consequently, pulse-labeling kinetics of proteins is difficult to apply. This is because the pool of radioactive-free amino acid is continuously replenished by degradation of labeled proteins which take place during their turnover. Since measurement of the decay in protein radioactivity after its pulse labeling is impossible, one has to evaluate the entire precursor-product relationship in studies of intact animals. Here technical problems arise from the fact that the size of the aminoacyl-tRNA poll is very small, and it is therefore difficult to assess its radioactivity.

An alternative approach to the measurement of protein turnover is to evaluate the rate of protein elimination after its resynthesis has been blocked. Here conceptual problems arise since no specific blocking agents exist and one has to use inhibitors of overall protein synthesis such as cycloheximide. Consequently, this approach has several hazards. First, cycloheximide has been shown in some instances not only to inhibit protein synthesis but also to have an effect on the rate of protein degradation. Secondly, when the synthesis of all proteins is inhibited, it leads to the elimination of proteins which turn over rapidly. Because many key regulatory enzymes appear to be short-lived, they will be eliminated during the experiments and this, of course, will lead to an abnormal situation.

Despite technical difficulties, half-lives of some cardiac proteins have been determined with a high degree of accuracy [38]. According to their half-lives, myofibrillar proteins in rat ventricle fell into two groups: those with a half-life of about five days (myosin heavy chain and α-actinin) and those with half-lives ranging between six to eight days (tropomyosin, troponin, the two myosin light chains, and actin, which appears to be the slowest turning over of compared proteins). These data clearly show that not only individual proteins but also their sub-units such as those of myosin and troponin turn over asynchronously.

The original concept of protein dynamism was later expanded when it was applied to growing systems. The early studies have been confined to the steady state, i.e., conditions under which the amount of studied protein remained constant. On the basis of independently estimated rates of catalase synthesis and destruction, Rechcigl and his associates [39] were the first to demonstrate that both rates, when controlled independently, can lead to a change in the level of an intracellular protein. Although it became obvious that a reduced rate of protein degradation coupled with an increased rate of its synthesis would be the most efficient growth response, the evidence for this concept obtained in subsequent years was contradictory [40]. At present it is difficult to produce a generalization that could predict under which conditions, and in what direction, the rates of synthesis and degradation will change.

Despite contradictory data, however, it appears that liver enlargement during postnatal growth [41], during renegeneration [41], and during nutritional shifts [42, 43, 44] in mammals is accompanied by a decreased rate of degradation and increased rate of synthesis determined for total proteins.

In contrast to studies of liver, measurements of synthesis and degradation rates of total muscle proteins gave conflicting results. For example, growth induced by refeeding of protein-depleted animals in one study led to an almost complete cessation of protein degradation [44], while in another study an increase in degradation was found [45]. In skeletal muscles undergoing a net loss of tissue due to a variety of factors such as diabetes, hypophysectomy, or feeding of low calories or a low-protein diet, both the rates of synthesis and degradation were found to have decreased [46].

Conflicting results have also been reported in investigation of stretch-induced muscle growth. In one group of studies both the rate of protein synthesis and degradation was reduced by stretch applied in vivo [47] as well as in vitro [48]. Similarly, elimination of natural muscle tonus by incubation of perpendicularly sectioned m. soleus resulted in a decreased degradation rate [49]. In contrast, in another model of in vitro stretch, the rate of protein degradation was reduced by stretch [50]. For example, when the muscle was incubated at a slightly increased length (110% of resting length), the release of tyrosine from stretched muscle was reduced. The complexities of the effect of stretch on protein turnover is indicated by the fact that when stretch was applied in the presence of calcium inophore, the synthesis of total proteins was found increased and their degradation reduced. No effect of stretch on either process, however, was noticed in the absence of ionophore.

From this brief outline of studies performed with muscle so far, it is clear that we know very little and face serious technical problems especially when it comes to the evaluation of protein degradation. It thus appears to be premature to make any generalization about the role of protein turnover in determining the rate of muscle growth.

CONCLUSION

At the molecular level, growth can be viewed as an orderly sequential expression of genes. Gene expression is a highly complex process consisting of three major events: transcription of mRNA on the DNA template, translation of the message by protein-synthesizing machinery, and finally selective processing and/or degradation of gene products.

The pattern of gene expression changes from the early stages of cell differentiation through various periods of growth, eventually giving rise to a mature phenotype. There are two gene expression programs in skeletal and cardiac muscles. One program is time-dependent (intrinsic developmental program), and the other is governed by contractile activity and humoral factors. The second program (extrinsic developmental program) includes: type and extend of the functional load, adrenergic input, and the effects of hormones and nutrients. Not all the extrinsic (sometimes called epigenetic) factors have been discovered nor has their interplay been firmly established. Nevertheless, it is clear that upon demand even the mature phenotype undergoes rapid and sometimes profound alteration. So far, this has been clearly demonstrated by the relative contribution of cellular organelles to the total cell mass and for the relative proportions of cellular proteins. Such dynamism of phenotype arises from the fact that the extrinsic developmental program is fully reversible in muscles and that cellular constituents are continuously degraded and resynthesized during their turnover. Consequently, genes can be selectively switched on and off. While switched off, products of the gene will be eliminated by degradation which is part of the turnover process.

The highly complex pathways of gene expression suggest the presence of multiple control sites. At the present time we have just begun to unravel how the activity of the gene is regulated. Understandably the experimental data are often contradictory since growth induced by different stimuli is undoubtedly accompanied by activation of different regulatory sites. Our limited knowledge of the nature of growth stimuli and of their interplay with the multiple immediate steps of gene expression makes it premature to formulate a unified model for the control of cardiac growth.

REFERENCES

1. Brown, D. 1981. Gene expression in eukaryotes. *Science* 211:667–674.
2. Darnell, J.E. 1982. Variety in the level of gene control in eukaryotic cells. *Nature* 297:365–371.

3. Zak, R. 1983. Overview of the growth process. In *Growth of the heart in health and disease,* ed. R. Zak, pp. 1–24. New York: Raven Press.
4. Koide, T., and Rabinowitz, M. 1969. Biochemical correlates of cardiac hypertrophy. II. Increased rate of RNA synthesis in experimental cardiac hypertrophy in the rat. *Circ. Res.* 24:9–18.
5. Cutilletta, A.F.; Rudnik, M.; and Zak, R. 1978. Muscle and non-muscle cell RNA polymerase activity during the development of myocardial hypertrophy. *J. Mol. Cell. Cardiol.* 10:677–687.
6. Florini, J.R., and Dankberg, F.L. 1971. Changes in RNA and protein synthesis during induced cardiac hypertrophy. *Biochem.* 10:530–535.
7. Kako, K.J.; Varnai, K.; and Beznak, M. 1972. RNA synthesis and RNA content of nuclei prepared from hearts during hypertrophy. *Cardiovas. Res.* 6:57–66.
8. Mahdavi, V.; Periasamy; M.; and Nadal-Ginard, B. 1982. Molecular characterization of two myosin heavy chain genes expressed in the adult heart. *Nature* 297:659–664.
9. Sinha, A.M.; Umeda, P.K.; Kavinsky, C.J.; Rajamanickam, C.; Hsu, H.-J.; Jakovcic, S.; and Rabinowitz, M. 1982. Molecular cloning of mRNA sequences for cardiac α- and β-form myosin heavy chains: expression in ventricles of normal, hypothyroid and thyrotoxic rabbits. *Proc. Natl. Acad. Sci.,* 79:5847–5851.
10. McLachlan, A.D., and Karn, J. 1982. Periodic change distributions in the myosin rod amino acid sequence match cross-bridge spacing in muscle. *Nature* 299:226–231.
11. Umeda, P.K.; Sinha, A.M.; Jakovcic, S.; Merten, S.; Hsu, H.-J.; Subramanian K.N.; Zak, R.; and Rabinowitz, M. 1981. Molecular cloning of two fast myosin heavy chain cDNAs from chick embryo skeletal muscle. *Proc. Natl. Acad. Sci.* 78:2843–2847.
12. Hastings, K.E.M., and Emerson, C.P. 1982. cDNA clone analysis of six co-regulated mRNAs encoding skeletal muscle contractile proteins. *Proc. Natl. Acad. Sci.* 79:1553–1557.
13. Freyer, G.A., and Robbins, J. 1983. The analysis of a chicken myosin heavy chain cDNA clone. *J. Biol. Chem.* 258:7149–7154.
14. Nudel, U.; Katcoff, D.; Carmon, Y.; Zevin-Sonkin, D.; Levi, Z.; Shaul, Y.; Shani, M.; and Yaffe, D. 1980. Identification of recombinant phages containing sequences from different rat myosin heavy chain genes. *Nucleic Acids Res.* 8:2133–2144.
15. Nguyen, H.T.; Gubits, R.M.; Wydro, R.M.; and Nadal-Ginard, B. 1982. Sarcomeric myosin heavy chain is coded by a highly conserved multigene family. *Proc. Natl. Acad. Sci.,* 79:5230–5234.
16. Rozek, C.E., and Davidson, N. 1983. *Drosophila* has one myosin heavy chain gene with three developmentally regulated transcripts. *Cell* 32:23–34.
17. Bernstein, S.I.; Mogami, K.; Donady, J.J.; and Emerson, C.P. 1983. *Drosophila* muscle myosin heavy chain encoded by a single gene in a cluster of muscle mutations. *Nature* 302:393–397.
18. Robbins, J.; Freyer, G.A.; Chisholm, D.; and Gilliam, T.C. 1982. Isolation of multiple genomic sequences coding for chicken myosin heavy chain protein. *J. Biol. Chem.* 257:549–556.
19. Wydro, R.M.; Nguyen, H.T.; Gubits, R.M.; and Nadal-Ginard, B. 1983. Characterization of sarcomeric myosin heavy chain genes. *J. Biol. Chem.* 258:670–678.
20. Friedman, D.J.; Umeda, P.K.; Sinha, A.M.; Hsu, H.-J.; Jakovcic, S.; and Rabinowitz, M. 1983. Characterization of genomic clones specifying rabbit α and β ventricular myosin heavy chains. *Circ. Res.* (abstr.) Suppl. III, 68:6.
21. Leinwand, L.A.; Saez, L.; McNally, E.; and Nadal-Ginard, B. 1983. Isolation and characterization of human myosin heavy chain genes. *Proc. Natl. Acad. Sci.* 80:3716–3720.
22. Karn, J.; Brenner, S.; and Barnett, L. 1983. Protein structural domains in the *Caenorhabditis elegans* unc-54 myosin heavy chain gene are not separated by introns. *Proc. Natl. Acad. Sci.* 80:6253–6257.
23. Czosnek, H.; Nudel, V.; Shani, M.; Barker, P.E.; Pravtcheva, D.; Ruddle, F.H.; and Yaffe, D. 1982. The genes coding for the muscle contractile proteins, myosin heavy chain, myosin light chain 2, and skeletal muscle actin are located on three different mouse chromosomes. *EMBO J.* 1:1299–1305.
24. Everett, A.; Sinha, A.M.; Umeda, P.K.; Jakovcic, S.; Rabinowitz, M.; and Zak, R. 1984. Regulation of myosin synthesis by thyroid hormone: relative change in the α- and β-myosin heavy chain mRNA levels in rabbit heart. *Biochem.* 23:1596–1599.
25. Robbins, J., and Heywood, S.M. 1978. Quantification of myosin heavy chain mRNA during myogenesis. *Europ. J. Biochem.* 82:601–608.
26. Buckingham, M.E.; Caput, N.; Cohen, A.; Whalen, R.G.; and Gros, F. 1974. The synthesis and stability of cytoplasmic messenger RNA during myoblast differentiation in culture. *Proc. Natl. Acad. Sci.* 71:1466–1470.

27. Sarkar, S.; Mukherjee, A.K.; and Guha, Ch. 1981. A ribonuclease-resistant cytoplasmic 10 S ribonucleoprotein of chick embryonic muscle. *J. Biol. Chem.* 256:5077–5086.
28. Paterson, B.M., and Bishop, J.O. 1977. Changes in the mRNA population of chick myoblasts during myogenesis *in vitro*. *Cell* 12:451–465.
29. Saidapet, Ch.R.; Munro, H.N.; Valgeirsdottir, K.; and Sarkar, S. 1982. Quantitation of muscle-specific mRNAs using cDNA probes during chicken embryonic muscle development in ovo. *Proc. Natl. Acad. Sci.* 79:3087–3091.
30. Pain, V.M. 1973. Influence of streptozotocin diabetes on the ability of muscle cell sap to support protein synthesis by ribosomes in cell free systems. *Biochim. Biophys. Acta* 308:180–187.
31. Gibson, K., and Harris, P. 1973. Aminoacyl-tRNA synthetase activities specific to twenty amino acids in rat, rabbit and human myocardium. *J. Mol. Cell. Cardiol.* 5:419–425.
32. Morgan, H.E.; Jefferson, L.S.; Wolpert, E.B.; and Rannels, D.E. 1971. Regulation of protein synthesis in the heart muscle. II. Effect of amino acid and insulin on ribosomal aggregation. *J. Biol. Chem.* 246:2163–2170.
33. Jefferson, L.S.; Wolpert, E.B.; Giber, K.E.; and Morgan, H.E. 1971. Regulation of protein synthesis in heart muscle. III. Effect of anoxia on protein synthesis. *J. Biol. Chem.* 246:2171–2178.
34. Kira, Y.; Kochel, P.J.; Gordon, E.E.; and Morgan, H.E. 1984. Aortic perfusion pressure as a determinant of cardiac protein synthesis. *Am. J. Physiol.* 246:C247–258.
35. Siehl, D.; Chua, B.H.L.; and Morgan, H.E. 1984. More ribosomes account for faster protein synthesis in hypertrophying rat hearts. *Am. J. Physiol.*, in press.
36. Zak, R., and Rabinowitz, M. 1979. Molecular aspects of cardiac hypertrophy. *Ann. Rev. Physiol.* 41:539–552.
37. Zak, R.; Martin, A.; and Blough, R. 1979. Assessment of protein turnover by use of radioisotopic tracers. *Physiol. Rev.* 59:407447.
38. Zak, R., and Galhotra, S.S. 1983. Contractile and regulatory proteins. In *Cardiac metabolism*, ed. A.J. Drake-Holland and M.I.M. Noble, pp. 339–363. New York: Wiley and Sons.
39. Price, V.E.; Sterling, W.R.; Tarantola, V.A.; Hartley, R.W.; and Rechcigl, M. 1962. The kinetics of catalase synthesis and destruction *in vivo*. *J. Biol. Chem.* 237:3468–3475.
40. Goldberg, A.L., and St. John, A.C. 1976. Intracellular protein degradation in mammalian and bacterial cells, Part 2. *Ann. Rev. Biochem.* 45:747–803.
41. Swick, R.W., and Ip, M.M. 1974. Measurement of protein turnover in rat liver with ^{14}C-carbonate. *J. Biol. Chem.* 249:6836–6841.
42. Dunaway, G., and Weber, G. 1974. Effects of hormonal and nutritional changes on rates of synthesis and degradation of hepatic phosphofructokinase isozymes. *Arch. Biochm. Biophys.* 162:629–637.
43. Conde, R.D., and Scornik, O.A. 1976. Role of protein degradation in the growth of livers after a nutritional shift. *Biochem. J.* 158:385–390.
44. Young, V.R.; Stothers, S.C.; and Vilaire, G. 1971. Synthesis and degradation of mixed proteins and composition changes in skeletal muscle of malnourished and refed rats. *J. Nutr.* 101:1379–1390.
45. Funabiki, R.; Watanabe, Y.; Nishizawa, N.; and Hareyama, S. 1976. Quantitative aspects of the myofibrillar protein turnover in transient state on dietary protein depletion and repletion revealed by urinary excretion of N-methylhistidine. *Biochim. Biophys. Acta* 451:143–150.
46. Millward, D.J.; Garlick, P.J.; Nnanyelugo, D.O.; and Waterlow, J.E. 1976. The relative importance of muscle protein synthesis and breakdown in the regulation of muscle mass. *Biochem. J.* 156:185–188.
47. Laurent, G.J.; Sparrow, M.P.; and Millward, D.J. 1978. Turnover of muscle protein in the fowl: changes in rates of protein synthesis and breakdown during hypertrophy of the anterior and posterior latissimus dorsi muscles. *Biochem. J.* 176:407–417.
48. Goldspink, D.F. 1977. The influence of immobilization and stretch on protein turnover of rat skeletal muscle. *J. Physiol.* 264:267–282.
49. Seider, M.J.; Kapp, R.; Chen, Ch. P.; and Booth, F.W. 1980. The effects of cutting or of stretching skeletal muscle *in vitro* on the rates of protein synthesis and degradation. *Biochem. J.* 188:247–254.
50. Etlinger, J.D.; Kameyama, T.; Toner, K.; van der Westhuyzen, D.; and Matsumoto, K. 1980. Calcium and stretch-dependent regulation of protein turnover and myofibrillar disassembly in muscle. In *Plasticity of muscle*, ed D. Pette, pp. 541–557. Berlin: Walter de Gruyter Co.

11. ONTOGENY OF PERIPHERAL BLOOD VESSELS

KENT HERMSMEYER

There are numerous indications that the vasculature changes considerably in both form and function during ontogenic development, strongly affecting cardiovascular responsiveness and regulation. One of the most important landmarks in growth of blood vessels may be the time of functional adrenergic innervation. Interaction between sympathetic neurons and vascular muscle cells is important, not only for the immediate regulation of vascular tone but also for the long-term interactions regulating sensitivity to vasoactive agents, membrane excitability, and growth. This chapter will review the development of vasoconstrictor function by the sympathetic nervous system and also the development of trophic interactions, but it will not address the embryonic origin of blood vessels covered in embryology textbooks and also summarized in the *Handbook of Physiology* [1, 2].

NERVE-MUSCLE ASSOCIATIONS

In the neonatal stages, blood vessels are rapidly growing to serve the ever-increasing mass of organs [3]. Research on mechanisms controlling the process of formation of blood vessels suggests that growth of the sympathetic nervous system, which can be demonstrated anatomically from early embryonic stages, is closely associated with growth of the blood vessels [1]. There is now growing evidence that the growth of the vascular muscle cells is under the control of sympathetic innervation. In studies where the innervation to a blood vessel was removed, there

This research was supported by Grants HL 14388 and HL 16328 from the National Institutes of Health.

M.J. Legato (ed.) THE DEVELOPING HEART, pp. 205–210. *Martinus Nijhoff Publishing. Boston/The Hague/Dordrecht/Lancaster.*

was a lack of normal growth of new muscle cells [4]. The sympathetic nervous system must not be necessary for all growth, however, because blood vessels develop normally when neonatal animals are deprived of their sympathetic innervation by treatment with 6-hydroxydopamine or anti-nerve growth factor [5]. It is not clear whether growth characteristics of blood vessels in which innervation was prevented are determined by the brief exposure to growing nerves before destruction of nerve endings, or whether the nervous system control of growth has redundant control systems, a subject which needs considerable additional research.

NERVE-MEDIATED VASOCONSTRICTION

Whether the sympathetic nervous system has complete or partial control of vascular muscle growth, it certainly has immediate instantaneous control of vasoconstriction. Although one can demonstrate growth of nerve fibers in association with blood vessels even at embryonic stages, it is only in the neonatal animal that the sympathetic neurons become functional. Electrical and pharmacological stimulation suggest that only in postnatal rats does sympathetically controlled contraction of vessels begin to occur [6, 7]. This vasoconstriction may occur even without demonstrable function of the ganglionic synapse in the sympathetic nervous system [7]. Of course, the timecourse of functional adrenergic innervation will depend greatly on the exact circulation being studied. Recent studies suggest that brain, heart, and kidney vessels might be controlled by innervation first, with the blood flow to less vital organs not under sympathetic control until later stages [8]. There is, however, a striking lack of available data on the development of functional sympathetic control in different circulations.

NERVE-MEDIATED SENSITIVITY MODULATION

In the few available systematic studies, there appear to be responses to exogenous adrenergic stimulation at the earliest stages that blood vessels can be recognized, suggesting that receptors are inherently programmed in the muscle cells. However, there is clearly modulation of the sensitivity to adrenergic agents, which occurs especially at the time of sympathetic innervation and is likely to be due to a trophic influence of the sympathetic nervous system [9, 10, 11]. While it is likely that norepinephrine release by itself plays a role in modulating norepinephrine sensitivity in blood vessels [12], there are probably other causes as well.

Recent experiments on isolated single cells from neonatal animals have shown an extremely high inherent sensitivity to catecholamines (subnanomolar ED_{50} concentrations), which could be profoundly decreased on first exposure to norepinephrine [13]. The direct correlation between the sensitivity to norepinephrine and the degree to which the sensitivity was decreased on first exposure to norepinephrine led to postulation of the theory of physiological desensitization. Control of norepinephrine sensitivity may be one of the important functions of the sympathetic nervous system during development. It is possible, although this has not been systematically explored, that the sympathetic nervous system releases small amounts of norepinephrine which gradually increase during development. This exposure to catecholamines could decrease the inherent high sensitivity of the unstimulated cell, resulting in the reduced sensitivity appropriate to the adult animal. Certainly much additional research is needed in the area of modulation of norepinephrine sensitivity in blood vessels, as this is a likely pathophysiological mechanism that would not be detected in short-term experiments.

ALTERED BLOOD VESSEL FUNCTION IN HYPERTENSION

One example of the possible importance of sympathetic trophic function on development of blood vessels is provided by the studies of cross-innervation of vascular muscle in the spontaneously hypertensive rat (SHR) and its genetic counterpart (WKY). We have recently described experiments in which caudal arteries (a densely innervated resistance artery) were transplanted into anterior eye chambers of rats to allow innervation by sympathetic neurons that normally serve the iris of the host animal; because the anterior eye chamber is an immunologically privileged site, there is no problem with tissue rejection, and the blood vessels retain the appropriate morphology and develop innervation patterns similar to those found in situ [14].

We carried out the four possible combinations of donor and host animals between SHR and WKY, and studied the transplanted arteries six weeks later. We found that when the donor artery was from an animal two weeks old, there was conversion of membrane properties to those appropriate for the host animal in the cross-innervation experiments, i.e., the artery from a hypertensive animal had the norepinephrine sensitivity and membrane potential electrogenesis characteristics of its normotensive host, and the artery of a normotensive animal developed the norepinephrine sensitivity and membrane potential electrogenesis characteristics of the hypertensive host [15, 16, 17]. However, when exactly the same experiment was done using donor tissue from animals 12 weeks of age, there was no crossover of membrane properties [14]. Therefore, the conversion by transplantation only occurred if the blood vessels were at a correct age to be "triggered" by the hypothetical sympathetic trophic substance.

To determine whether the crossover of membrane properties was due to the sympathetic innervation specifically or to humoral factors present in the host animal, we repeated the experiment with the prevention of sympathetic innervation in half of the animals by previous superior cervical ganglionectomy and found that denervation prevented the crossover of membrane properties [10]. A summary of data from those experiments is shown in figure 11–1 where the sympathetic trophic influence is evident in the reinnervated set but not in the ganglionectomized (denervated) set.

These experiments appear specifically to demonstrate a trophic influence of the sympathetic nervous system as the factor that triggers abnormal responsiveness in hypertension. Furthermore, studies of the increase in blood pressure in SHR and WKY have shown that the first significant difference in blood pressure is measured at age three weeks [18], a time when functional innervation to the peripheral blood vessels is likely to begin. These facts appear to suggest that one of the etiologic

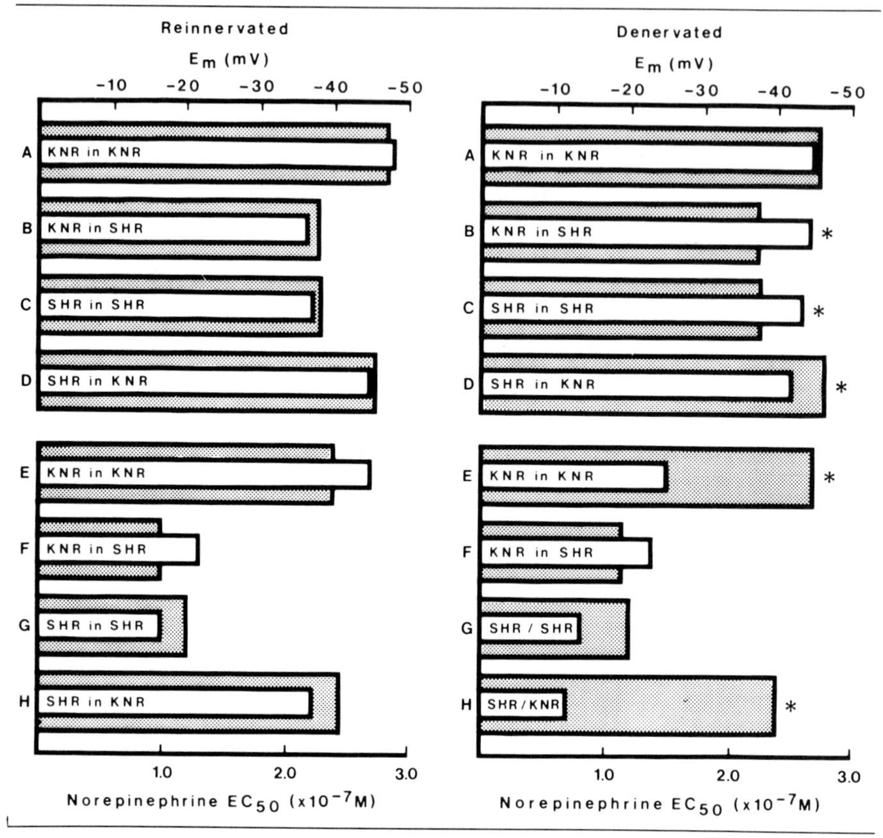

Figure 11–1. Summary of cross-innervation experiments, in which reinnervation was established or was prevented by denervation, demonstrates the trophic influence of the sympathetic nervous system in the expression of the altered basis for membrane potential that contributes to increased norepinephrine sensitivity in the spontaneously hypertensive rat. On the left side (reinnervated), the transplant (inner open bars) of Kyoto-Wistar normotensive rat (KNR) arteries into host spontaneously hypertensive rats (SHR) caused conversion of membrane potentials (measured at 16°C to inhibit electrogenic ion transport) to be similar (bar B) to those found in the caudal artery of the host animal (outer stippled bar). Similarly, SHR transplants into the KNRs resulted in the arteries having properties expected for normotensive rats (bar D). These membrane potential data in bars A through D were reflected in the norepinephrine sensitivity shown in bars E through H, where there was also conversion of the KNR normally lower sensitivity to the higher norepinephrine sensitivity (lower ED_{50}) found in the SHR (F). Norepinephrine EC_{50} represent mid-point values on a concentration-contraction curve. Similarly, SHR transplanted into KNR behaved like KNR (H). The data set on the right (denervated) show that prevention of reinnervation (ganglionectomy) prevents the six-week transplantation procedure from causing crossover of membrane potential properties and tends to cause the same result in concentration of norepinephrine to cause half-maximal contraction. The complication is that norepinephrine sensitivity is also increased by the denervation (post-junctional supersensitivity), making differences in sensitivity more difficult to determine. However, there also appears to be some tendency for SHR to show increased sensitivity (H), suggesting that an influence of the sympathetic neurons even prior to age two weeks may have begun a tendency toward increased norepinephrine sensitivity in SHR. (Redrawn with permission from the American Heart Association; Abel and Hermsmeyer, 1981).

factors in certain forms of hypertension may be altered membrane characteristics of vascular muscle caused by abnormal trophic function of the sympathetic nervous system.

NEED FOR ADDITIONAL RESEARCH

There are relatively few areas with such important consequences that are being so little studied as the ontogenic development of the vasculature. A number of very important questions remain to be explored which are mentioned here with the hope of encouraging additional investigators to join the search. One important question is that of how much blood vessel growth might be due to hyperplasia as opposed to hypertrophy. Studies by Owens and Schwartz [19] have suggested that stimuli for development of additional sets of DNA are altered in situations that might include altered sympathetic function, such as may occur in hypertension. Furthermore, both in normal blood vessels and during pathological processes there is development of increased cell ploidy (DNA content) which has only recently been recognized as an ongoing process in blood vessels [19]. This may be important in the regulation of blood vessel contractility and in the development of the increased stiffness of blood vessels that occurs with aging. Recent studies have demonstrated increased myocardial cell division after sympathectomy in newborn rats [20]. These processes, which determine the extent and timing of the preparation for cell division, are certainly worthy of additional research.

Still another important unanswered question is that of the influence of dual innervation, or possibly multiple transmitters released by a single nerve ending. It has been recognized that release of ATP, along with norepinephrine, can have important effects on blood vessel function [21]. Instead of or in addition to multiple release, one lab has suggested that there may be changeover of nerve endings from release of one transmitter to another during normal development in the sympathetic nervous system [22].

Several important questions relate to the possibility of multiple transmitter release from a neuron. If there is multiple transmitter release, does the sequence of release of the transmitter, especially during ontogeny, regulate the development of the vascular muscle cells? This is an experiment that can be carried out in isolated cells in culture, where transmitter substances and, indeed, identified neurons with known release characteristics, can be used to explore the effect of different transmitters and different sequences of transmitter exposure on function of the muscle cells. If the transmitter released by a neuron changes during development, does the change cause altered muscle cell function? Are the changes in transmitter release caused by a trophic factor released by the muscle cells? Is timing important?

Because the function of blood vessels is modulated by so many factors involving neurons, circulating humoral substances, and local influences, there is a great need for additional study of development of the peripheral circulation. All of the disciplines that can contribute to the understanding of the function and pharmacology of control of the peripheral circulation are greatly needed to contribute knowledge and allow for a more complete understanding of this aspect of the circulation.

REFERENCES

1. Van Mierop, L.H.S. 1979. Morphological development of the heart. In *Handbook of physiology,* Section 2: The cardiovascular system, ed. R. M. Berne, N. Sperelakis, and S. R. Geiger, pp. 1–28. Bethesda, MD: American Physiological Society.
2. Manasek, F.J. 1979. Organization, interactions, and environment of heart cells during myocardial ontogeny. In *Handbook of physiology,* Section 2: The cardiovascular system, ed. R.M. Berne, N. Sperelakis, and S.R. Geiger, pp. 29–42. Bethesda, MD: American Physiological Society.
3. Legato, M.J. 1979. Cellular mechanisms of normal growth in the mammalian heart. I. Qualitative and quantitative features of ventricular architecture in the dog from birth to five months of age. *Circ. Res.* 44:250–262.
4. Bevan, R.D. 1975. Effect of sympathetic denervation on smooth muscle cell proliferation in the growing rabbit ear artery. *Circ. Res.* 37:14–19.
5. Levi-Montalcini, R. 1975. NGF: an uncharted route. In *Neurosciences: Paths of discovery,* ed. F.G. Worden, J.P. Swazey, and G. Adelman, pp. 245–265. Cambridge & London: MIT.
6. Boatman, D.L.; Schaffer, R.A.; Dixon, R.L.; and Brody, M.J. 1965. Function of vascular smooth muscle and its sympathetic innervation in the newborn dog. *J. Clin. Invest.* 44:241–246.
7. Mills, E., and Smith, P.G. 1983. Functional development of the cervical sympathetic pathway in the neonatal rat. *Fed. Proc.* 42:1639–1642.
8. Buckley, N.M.; Brazeau, P.; and Gootman, P.M. 1983. Maturation of circulatory responses to adrenergic stimuli. *Fed. Proc.* 42:1643–1647.
9. Aprigliano, O. 1983. Neural influences and norepinephrine sensitivity in the rat portal vein. *Fed. Proc.* 42:257–262.
10. Abel, P.W., and Hermsmeyer, K. 1981. Sympathetic cross-innervation of SHR and genetic controls suggests a trophic influence on vascular muscle membranes. *Circ. Res.* 49:1311–1318.
11. Fronek, K. 1983. Trophic effect of the sympathetic nervous system on the vascular smooth muscle. *Annals of Biomedical Engineering* 2:36.
12. Abel, P.W.; Trapani, A.; Aprigliano, O.; and Hermsmeyer, K. 1980. Trophic effect of norepinephrine on the rat portal vein in organ culture. *Circ. Res.* 47:770–775.
13. Hermsmeyer, K., and Mason, R. 1982. Norepinephrine sensitivity and desensitization of cultured single vascular muscle cells. *Circ. Res.* 50:627–632.
14. Campbell, G.R.; Chamley-Campbell, J.; Short, N.; Robinson, R.; and Hermsmeyer, K. 1981. Effect of cross-transplantation on normotensive and spontaneously hypertensive rat arterial muscle membrane. *Hypertension* 3:534–543.
15. Hermsmeyer, K. 1976. Electrogenesis of increased norepinephrine sensitivity of arterial vascular muscle in hypertension. *Circ. Res.* 38:362–367.
16. Hermsmeyer, K. 1981. Membrane potential mechanisms in experimental hypertension. In *New trends in arterial hypertension, INSERM Symposium No. 17,* ed. M. Worcel et al., pp. 175–187. Amsterdam: Elsevier/North Holland Biomedical Press.
17. Hermsmeyer, K. 1982. Electrogenic ion pumps and other determinants of membrane potential in vascular muscle (the 1982 Henry Pickering Bowditch Lecture). *Physiologist* 25:454–465.
18. Lais, L.T.; Rios, L.L.; Boutelle, S.; DiBona, G.F.; and Brody, M.J. 1977. Arterial pressure development in neonatal and young spontaneously hypertensive rats. *Blood Vessels* 14:277–284.
19. Owens, G.K., and Schwartz, S.M. 1982. Alterations in vascular smooth muscle mass in the spontaneously hypertensive rat. *Circ. Res.* 51:280–289.
20. Kugler, J.D., Gillette, P.C.; Graham, S.P.; Garson, A., Jr.; Goldstein, M.A.; and Thompson, H.K., Jr. 1980. Effect of chemical sympathectomy on myocardial cell division in the newborn rat. *Pediatr. Res.* 14:881–884.
21. Worcel, M., Saiag, B.; and Chevillard, C. 1980. An unexpected mode of action for hydralazine (HYD). *Trends in Pharmacol. Sci.* 1:136–138.
22. Potter, D.D.; Furshpan, E.J.; and Landis, S.C. 1983. Transmitter status in cultured rat sympathetic neurons: plasticity and multiple function. *Fed. Proc.* 42:1626–1632.

12. HYPERTENSION IN THE YOUNG

SUZANNE OPARIL

Detection and management of hypertension in children and adolescents are made difficult by the lack of generally agreed-upon diagnostic criteria for abnormally elevated blood pressure, by the lability of blood pressure in this age group, and by our lack of experience with both pharmacologic and non-pharmacologic management of hypertension in the young. Despite the widespread view that primary hypertension may begin in childhood, it is frequently not possible to determine in any given case whether blood pressure increments that occur with growth are physiological or represent the origins of hypertension that will develop fully in adulthood. During normal childhood and adolescence, blood pressure increases as a result of growth and maturation [1, 2] and is so heavily dependent on age that a nomogram of blood pressure versus age must be used to define the normal range.

Figure 12–1 is a chart of selected percentiles of blood pressure for children ages 2 to 18 developed by the Task Force on Blood Pressure Control in Children of the National Heart, Lung and Blood Institute from studies conducted in Muscatine, Iowa; Rochester, Minnesota; and Miami, Florida [3]. Sustained blood pressure levels (obtained on at least three occasions) that are above the 95th percentile were considered abnormal with recognition that any cutoff point represents an arbitrary decision at any age. These criteria for the normal range of blood pressure in the pediatric population have been criticized on the basis of the racial composition of the study groups, the methods that were used in measuring blood pressure, sample size, and failure to take into account weight, height, and maturation of the study subjects. Since it has been shown that degree of adiposity and chronological

M.J. Legato (ed.) THE DEVELOPING HEART, pp. 211–243.

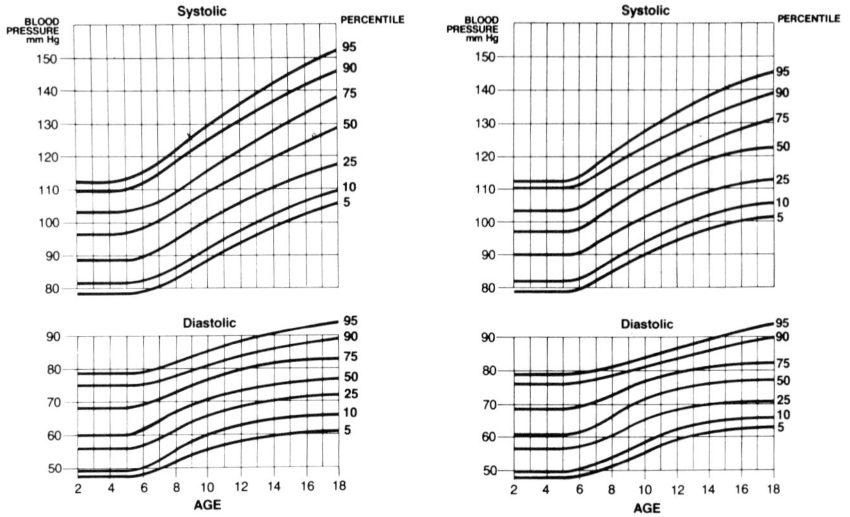

Figure 12–1. (*Left*) Percentiles of blood pressure measurement in boys (right arm, seated). (*Right*) Percentiles of blood pressure measurement in girls (right arm, seated). (Reprinted from the Report of the Task Force on Blood Pressure Control in Children. 1979. *Pediatrics* 59 (suppl 2): 802–803.)

and skeletal age are related to blood pressure in childhood and adolescence [1, 2], these factors must be considered when developing accurate reference standards for blood pressure in children. As a practical general guideline for use in clinical practice, however, only blood pressure and age are considered. Table 12–1 gives such a rough estimate of the upper limit of normal diastolic pressure.

The Task Force stressed that the charts and tables are not intended for use in assessment of an individual child's blood pressure at a single point in time, but rather for the plotting of pressures during growth and maturation. The Task Force recommended that blood pressure measurements be obtained and plotted at least once a year as part of a child's general evaluation. Because there is considerable variation in most children's pressures, a single elevated measurement in an apparently healthy child does not necessarily reflect disease, so it is necessary to repeat these measurements over time.

MEASUREMENT OF BLOOD PRESSURE IN CHILDREN

The Task Force on Blood Pressure Control in Children [3] recommended that children three years of age and older have their blood pressures measured annually as part of their continuing health care. In addition, blood pressure should be measured in younger children and infants who are ill. Methods of measuring blood pressure in children are the same as in adults, but accurate measurements are more difficult to obtain for the following reasons: (1) various arm sizes require the availability and selection of a cuff of appropriate size; (2) readings are difficult to obtain in anxious or restless children; (3) blood pressure is more labile in children and adolescents than in adults.

The Task Force recommended that the largest cuff that will fit snugly and comfortably around the child's arm be selected. The inflatable bladder contained within the cuff should completely encircle the arm without overlapping and without obstructing the antecubital fossa. Use of cuffs that are too narrow or short for the patient will, as in adults, cause artificially elevated readings to be obtained.

Prior to measuring the blood pressure, every effort should be made to reassure the child. If possible, the child should be sitting quietly with the right arm fully exposed and supported at heart level. Infants should be placed in the supine position. The cuff should be wrapped snugly around the upper arm and the diaphragm of the stethoscope placed lightly in the antecubital fossa. Because normal blood pressures in children are so low, major errors can be caused by heavy pressure of the stethoscope on the brachial artery. To obtain accurate blood pressure determinations, the cuff should be inflated to approximately 30 mmHg above the point at which the radial pulse disappears, then the pressure in the cuff released slowly (2 to 3 mmHg/sec). Muffling (phase IV) of the Korotkoff sounds is used as the indicator of diastolic pressure in infants and children because phase V (disappearance of Korotkoff sounds) may not occur. For infants in whom accurate blood pressure measurements cannot be made by auscultation, an electronic device which employs the Doppler technique should be used. Because of the lability of blood pressure in children, measurements that are elevated or in the high normal range should be repeated several times during a given visit at a time when every effort is made to assure a quiet resting state. Further, measurements should be repeated at least three times at different visits before diagnostic evaluation or treatment is undertaken.

The extreme lability of blood pressure in the young was emphasized by the results of two recent studies of large numbers of adolescents. In one study of 650 high school students, elevated blood pressure values were found in 7.7%, 11.0%, and 15.5% of freshman, sophomore, and junior students on initial screening. When re-examined 1 and 12 months later, hypertension persisted in 3.4%, 4.7%, 5.8%, and 1.8%, 4.2%, and 7.9% of the three groups, respectively [4]. Approximately one-quarter of the group was overweight, and there was a family history of hypertension in 28%, 33%, and 55% of each group, respectively. In another study of approximately 950 black urban adolescents, the prevalence of hypertension on first, second, and third visits was 10.2%, 3.9%, and 1.8%, respectively [5].

Diagnostic Evaluation of the Asymptomatic Hypertensive Child

Asymptomatic children and adolescents who have blood pressure levels above the 95th percentile (figure 12–1) on three or more occasions should be evaluated. The medical history should include questioning about drugs (oral contraceptives and corticosteroids) and foodstuffs (licorice, excess calories, and perhaps excess salt) that predispose to hypertension. Evidence of other cardiovascular risk factors (diabetes, cigarette smoking, obesity, and hyperlipidemia) and symptoms suggestive of secondary causes of hypertension should be sought. The family history should be reviewed with emphasis on the familial occurrence of hypertension and premature coronary artery disease and of target organ damage (history of stroke, heart attack, or renal failure of unknown origin) in parents or grandparents who are less than 50 years old.

The physical examination should detect evidence of target organ damage and exclude secondary causes of hypertension (table 12–2) by funduscopic and cardiovascular examination wherever possible. The peripheral pulses should be palpated carefully, and palpation and auscultation of the abdomen carried out to search for masses and bruits. Basic laboratory testing recommended for this stage of evaluation is similar to that in adult hypertensives. It should include urinalysis, urine culture, hemoglobin/hematocrit, serum electrolytes, blood urea nitrogen, serum creatinine, and a chest x-ray. In addition, because of the extremely high incidence of renal causes of hypertension in children (table 12–3), a renal imaging study such as abdominal ultrasound, intravenous pyelography, digital subtraction angiography, or abdominal CT scan is indicated in the asymptomatic child with hypertension. Any clues detected in the history or physical examination may direct further studies. As a rule of thumb, more extensive diagnostic evaluation is indicated in children with more severe hypertension.

DIFFERENTIAL DIAGNOSIS OF PERSISTENT HYPERTENSION IN CHILDREN

As shown in table 12–3, secondary hypertension appears to be more common in children than in adults, among whom it accounts for only 2% to 5% of patients. It should be emphasized that the numbers in table 12–3 are based on three series of patients treated in major referral centers [6, 7, 8]. Secondary hypertension may be more common among these children than among those who present in a pediatrician's office without prior selection [9]. Further study is needed to determine the absolute incidence of secondary causes of hypertension in ambulatory pediatric populations. Because of the risk of future cardiovascular morbidity and mortality in childhood hypertensives, the difficulty of obtaining compliance with medical antihypertensive treatment, and the hazard of cumulative toxicity of antihypertensive drugs, a more aggressive search for secondary causes of hypertension is indicated in these patients than in adult hypertensives.

Renal Disorders

Renal and renovascular diseases are the most common causes of secondary hypertension in children and adolescents, accounting for 60% to 80% of cases in most large series (table 12–3). Table 12–4 shows the relative frequency of the various forms of renal hypertension in children based on a large, recently published series [10]. As is shown, hypertensive children often present with small scarred kidneys. Such children are frequently asymptomatic but may have the signs or symptoms of prior urinary tract infections or of renal insufficiency. The diagnosis, clinical course, and treatment of the various forms of renal parenchymal disease which are associated with hypertension in children are discussed elsewhere [10, 11].

All children who present with persistent hypertension should be screened for renal disease. Initial laboratory evaluation should include a urinalysis, urine culture, blood urea nitrogen, serum creatinine, and electrolytes. The finding of proteinuria suggests glomerular disease, but can also be seen with pyelonephritis or obstructive uropathy. If proteinuria is discovered on routine urinalysis, quantitative determination should be obtained with a 24-hour urine collection. Under basal conditions protein excretion should be less than 60 mg/m^2/day. Examination of the urine sediment may provide further clues to the presence of renal parenchymal disease. Hyaline, granular, and waxy casts can be seen in chronic glomerulonephritis; white cell and granular casts are found in chronic pyelonephritis. Documentation of urinary tract infection by urine culture raises the possibility of pyelonephritis, but sterile urine does not rule out this diagnosis. Elevations of blood urea nitrogen and serum creatinine are found in patients with chronic renal parenchymal disease of any etiology.

The rapid sequence intravenous pyelogram provides assessment of renal size and configuration and of the configuration of the upper portion of the collecting system and thus may give evidence of intrinsic renal disease or of obstructive uropathy. Symmetrical decreases in kidney size are characteristic of chronic glomerulonephritis, although frequently a late finding. Dilatation of the collecting system is seen in obstructive uropathy. Cortical scarring suggests chronic pyelonephritis. In addition, certain congenital disorders such as polycystic disease or the Ask-Upmark kidney (segmental hypoplasia associated with severe hypertension) have characteristic radiographic appearances.

The usefulness of the rapid sequence intravenous pyelogram as a screening procedure for renovascular disease remains a matter of debate, as a high percentage of both false positives and false negatives have been reported. However, certain findings are suggestive of renal artery stenosis and may aid in a decision regarding angiography (figure 12–2). Delayed appearance of contrast in the calyceal system of the affected kidney is an important sign of unilateral renal arterial disease. This delay is usually seen in films taken at one, two, and three minutes following injection of contrast. Other findings suggestive of renal artery stenosis include a difference in the size of the two kidneys of greater than 1 cm, hyperconcentration of contrast in the pelvis and ureter of the affected kidney, and ureteral notching secondary to collateral vessels. Any of these abnormalities raises the suspicion of a stenotic lesion, but their absence does not exclude renovascular disease.

The radioisotope renogram is also used in the initial evaluation and followup of children with suspected renovascular hypertension. The principle of the scan is that ^{125}I or ^{131}I-hippuran is both filtered at the glomerulus and secreted by the renal tubule, so that its clearance rate approximates that of the effective renal plasma flow. Levels of radioactivity contained within the renal parenchyma and collecting system are recorded continuously and plotted against time. Figure 12–3 is a radioisotope renogram from a patient with tight stenosis in the right renal artery and

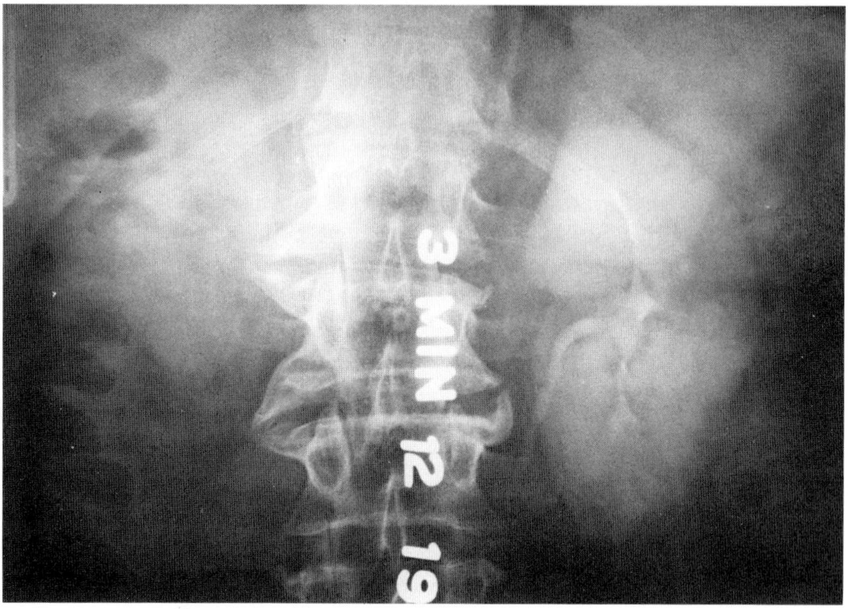

Figure 12–2. Rapid sequence intravenous pyelogram from a patient with tight stenosis in the right renal artery of several years' duration and a normal left kidney. The three-minute film shows a small right kidney without contrast in the calyceal system.

a normal left kidney. The normal curve (solid line) has an early rapid rise, representing delivery of the isotope to the renal parenchyma by blood flow. This rise is terminated by a peak followed by a gradual decline, approaching the baseline asymptomatically 20 to 30 minutes after the initial injection. The peak and decline occur as the isotope is cleared from the renal parenchyma and collecting system, and passes down the ureters to the bladder. The curve described by the kidney distal to the stenotic renal artery (hashed line) shows a delay and a decrease in height of the early peak and delayed return of the excretory phase of the slope toward the baseline, reflecting a delayed excretory phase. These changes are reversible with relief of the stenotic lesion.

Findings on excretory urogram or renal scintigraphy suggestive of a renovascular lesion, the presence of an epigastric bruit, and resistance to medical antihypertensive therapy are considered indications for selective renal angiography. This procedure should be considered in any child with a persistent diastolic blood pressure greater than 100 mmHg who does not have an obvious nonrenovascular cause for hypertension. As a rule, the more severe the hypertension and the younger the patient, the more likely there are to be angiographic abnormalities. Angiographic criteria for anatomically significant renal artery stenosis are summarized in table 12–5 and illustrated in figure 12–4. Although in experienced hands the complication rate of renal angiography is low, the procedure is not without risk. In a large comparative study involving 2,719 patients of all ages, the mor-

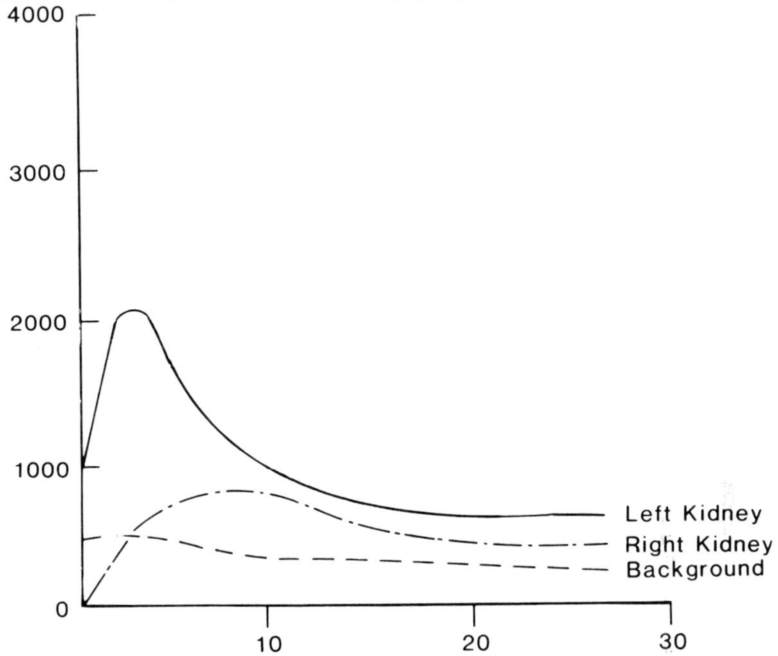

Figure 12–3. Radioisotope renogram from a patient with tight stenosis in the right renal artery and a normal left kidney. Renal blood flow is markedly impaired on the right.

tality rage was 0.11% [12]. Major complications occurred in 1.2% and minor complications in 2.7% of patients studied. These included hemorrhage or thrombosis at the site of entry of the catheter and renal infarction. In addition, many centers use general anesthesia in younger patients, adding to the risk of the procedure. All of these factors should be taken into consideration when a decision is made regarding angiography in hypertensive children.

Anatomic demonstration of renal artery stenosis in a hypertensive child does not establish a renal vascular etiology for the hypertension, since renal artery stenosis can occur in normotensive people. Documentation of the functional significance of an anatomic lesion requires the demonstration of increased renin production by the involved kidney. This is done by differential renal vein renin measurement. When renin activity in the venous effluent from the involved kidney is 1.5 or more times that of the uninvolved side and when the uninvolved side can be shown not to produce renin, the probability of improvement in blood pressure after surgery is high [13]. To demonstrate that the uninvolved kidney produces no renin it is necessary to show either that there is no arterial-venous difference in renin activity across the renal vascular bed or no stepup in renin activity from vena cava to renal vein blood. Stimulation of renin release with maneuvers such as tilting, salt deprivation, or administration of vasodilators, diuretics,

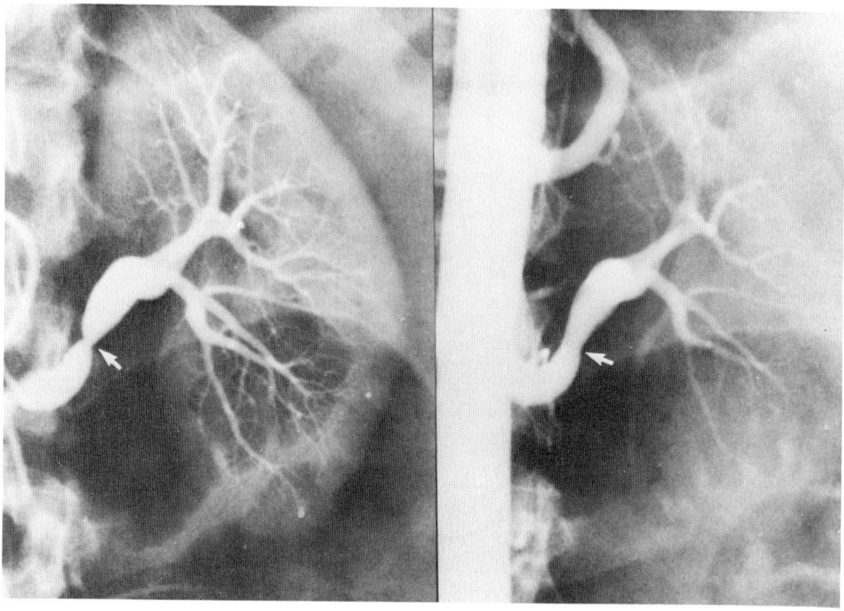

Figure 12–4. *Left Panel* — Arrow indicates site of high-grade stenosis in the renal artery of an adolescent with fibromuscular disease. *Right Panel* — The same artery following successful percutaneous dilatation. (Photograph courtesy of Frederick Keller, M.D., University of Alabama in Birmingham, and Joseph Rosch, M.D., Oregon health Sciences Center, Portland.)

or converting enzyme inhibitors has been shown to produce selective enhancement of renin release from the affected kidney [14, 15]. The diagnosis of surgically remediable unilateral renal artery stenosis can be missed when renal vein sampling is performed in the unstimulated state [15] or when renin-suppressing antihypertensive drugs have not been discontinued prior to study. When renal artery stenosis is segmental, increased renin release is restricted to that portion of the kidney distal to the lesion. Renin is suppressed in the remaining portion of the affected kidney, so the diagnosis of surgically remediable disease can be missed when samples are obtained from the main renal veins only [16]. Mixing of blood from the suppressed normal portion of the kidney with blood from the affected area can obscure the source of elevated renin. Thus segmental renal vein sampling is used to define indications for partial nephrectomy in hypertensive patients with branch renal artery stenoses and other localized lesions such as renal infarction, renal cysts, or renin secreting tumors [16].

Digital subtraction angiography is a promising new technique shown to be useful in the diagnosis of renal artery stenosis [17, 18]. Following rapid injection of a bolus of contrast medium through a large central venous catheter, images of the renal arteries and renal parenchyma are recorded directly on an image intensifier, digitized, and stored in a computer that electronically subtracts, reconstructs, enhances, and magnifies them. Images are displayed on a video screen. The extrarenal vessels are well visualized in 70% to 90% of cases, and an IVP can be obtained from the same dose of contrast medium if the study is carried out for a longer period. Overall accuracy of digital subtraction angiography compared to standard renal angiography was 71% in a recent study [18]. Because of its low cost and safety and because its performance does not require hospitalization, digital subtraction angiography is being looked to as the technique of the future in screening for and perhaps even in definitive diagnosis of renal artery stenosis. This method could potentially be adapted to videodensitometry, permitting comparisons of blood flow to the two kidneys [17].

The treatment of choice for renovascular hypertension in children is revascularization via either percutaneous transluminal angioplasty or surgery. Medical therapy is reserved for patients who are not candidates for surgery or angioplasty due to the anatomy of their lesions or their underlying medical condition. Most patients with renovascular hypertension require a complex three- or four-drug medical regimen to achieve adequate control of blood pressure. Such regimens are expensive, inconvenient, and produce a high incidence of adverse effects and are thus of limited efficacy in children with renovascular hypertension. The principles governing the choice of antihypertensive drugs in hypertensive children are discussed later in this chapter.

Percutaneous transluminal angioplasty of the renal artery is useful in the treatment of selected patients with renovascular hypertension [19, 20, 21, 22, 23, 24, 25, 26]. The procedure involves introduction of a selective catheter by the Seldinger technique via the femoral artery into the stenotic renal artery, then withdrawal of the selective catheter with a guide wire in place and replacement with a Gruntzig dilating catheter which has a balloon at the tip [23]. The dilating catheter is advanced beyond the stenosis and the balloon inflated several times with diluted contrast material at a pressure of four to eight atmospheres. Pressure gradients are measured before and after dilatation, and flush aortagrams are performed to document the morphologic effect of the intervention. Since the procedure causes trauma to the internal layer of the renal artery at the dilatation site, prevention of thrombosis during and after angioplasty is important. Accordingly, heparin is usually administered during the procedure, and aspirin/dipyridamole treatment is initiated two to three days prior to dilatation and continued for at least three months thereafter. Antispasmodics and vasodilators are also administered during the procedure.

Complications of percutaneous transluminal angioplasty are uncommon and generally not fatal. Occlusion of the renal artery secondary to intimal dissection or thrombosis, renal artery perforation, retroperitoneal bleeding, and balloon rupture have been reported. Since many of these complications require emergency surgery, transluminal renal artery dilatation should be done only with the backup of a vascular surgeon. Hypotension in the immediate post-angioplasty period can be a problem, particularly in patients who have had severe hypertension or who have been maintained on large doses of antihypertensive drugs up to the time of angioplasty. Accordingly, special attention should be given to keeping these patients well hydrated.

Dilatation of renal arterial lesions is technically successful (residual stenosis <30% of the luminal diameter and >20% reduction of the aorta-renal pressure gradient) in 90% of adult patients with fibromuscular disease [22, 23]. Although use of the procedure in the pediatric age group is too limited to permit assessment of its success rate, early experience has been promising [25]. Further, fibromuscular lesions, the most common cause of renal artery stenosis in children [27], have been shown to be susceptible to dilatation in the adult experience. Additional study is needed to determine whether the small size of the renal arteries in children will present a significant barrier to attempts at percutaneous dilatation.

Long-term results of transluminal dilatation of renal artery stenosis, as assessed by vessel patency and effect on blood pressure, are encouraging in the adult age group [20, 21, 23, 26]. Results in patients with fibromuscular disease are particularly favorable. In the largest reported long-term followup study, in which 89 patients, 51 with atherosclerotic disease and 31 with fibromuscular disease, were followed for a mean of 16 months (range 4 to 40 months), technically successful dilatation was accomplished in 87% of the fibromuscular stenoses patients, and 93% of those patients with technically successful angioplasties showed sustained improvement in blood pressure control [26]. All patients with fibromuscular disease were either cured or improved. Similar results have been reported in other studies of adult populations [20, 21, 23]. When renal artery stenosis and hypertension recur after successful dilatation, repeat dilatation is often effective [21].

Thus, percutaneous transluminal dilatation of the renal artery offers an attractive new method of revascularizing the kidney in patients with renovascular hypertension. Its advantages include much lower cost and lower morbidity and mortality than renal vascular surgery, greater ease of documenting results by recording pressures across the stenosis before and after dilatation, and ease of repetition if restenosis occurs. Disadvantages include technical failure in 10% to 20% of cases due to inability to pass through the stenosis or inability to dilate very firm lesions. Although extensive experience with this procedure is restricted to the adult age group and although the procedure is too new to permit adequate assessment of its long-term success rate at present, early experience in children with renovascular hypertension has been promising [25].

Surgical treatment of renovascular hypertension in children has recently been shown to be as effective in lowering blood pressure as comparable surgery in adults with this syndrome [28, 29]. In a series of 27 children with renovascular hypertension, all of whom had fibromuscular disease of the renal artery(ies), hypertension was cured in 59% and improved in 19% [28]. Surgery was unsuccessful in only 22% (six patients). Of the six patients in whom surgery was unsuccessful, four had bilateral disease. For patients with unilateral disease, the cure rate was 70% and the improvement rate 20 percent. There was only one death in this series, and post-operative morbidity was restricted to two patients who developed postoperative retroperitoneal hematomas. The surgical procedure of choice in children with renovascular hypertension is aortorenal bypass with autogenous hypogastric artery. Saphenous vein grafts have proven unsatisfactory due to a high incidence of aneurysmal dilatation, and splenorenal shunts have a high incidence of occlusion, presumably due to their small caliber. If, as is frequently the case, the diseased kidney is atrophic or small intrarenal vessels are extensively involved, ablative surgery may be necessary. Nephrectomy should be avoided and functioning renal parenchyma preserved wherever possible, however, since fibromuscular renal artery disease in children is frequently bilateral and progressive, ultimately leading to renal atrophy [28].

Renin-secreting tumors represent another cause of surgically curable hypertension in children. Several kinds of neoplasms, including Wilms tumors, intra- and extra-renal hemangiopericytomas, and renal carcinomas, can produce hypertension by secreting large amounts of renin. These tumors are frequently small, ranging in size from the dimensions of a "rice kernel" to 4 cm in diameter, and frequently cannot be visualized on angiography [30]. Differential renal vein renin measurements, coupled with the finding of elevations in peripheral vein plasma renin activity, are useful in making the diagnosis of renin-secreting tumor. Segmental sampling is needed for tumor localization [16]. Excision of the tumor removes the source of renin and cures the hypertension.

Coarctation

Coarctation of the aorta is a common form of congenital heart disease, accounting for 13% to 15% of hypertension in children [11]. Coarctation is the most common cause of hypertension in neonates and infants. Infants in whom coarctation is discovered usually present with symptoms of congestive heart failure and/or insufficient perfusion of the lower extremities. In symptomatic infants, the coarctation is usually placed before the insertion of the ligamentum ductus and is referred to as the preductal or infantile type of coarctation. This form of coarctation is frequently associated with other congenital abnormalities, such as patent ductus arteriosus, ventricular septal defect, bicuspid aortic valve, subaortic stenosis; and transposition of the great vessels. In contrast, the majority of children with juxtaductal coarctation are asymptomatic and therefore present later in life. Diagnosis is usually made upon finding hypertension in early childhood or adolescence associated with absent, significantly diminished, or delayed femoral pulses and decreased or unobtainable blood pressure in the lower extremities. A systolic gradient between upper and lower extremities of 10 mmHg is considered hemodynamically significant. A midsystolic murmur is usually heard over the anterior chest and commonly radiates to the back. Other systolic or continuous murmurs may be heard over the lateral chest and are thought to be secondary to flow through large dilated collaterals.

Findings compatible with coarctation on chest x-ray include the "3" sign formed by indentation of the aorta at the site of the coarctation with pre- and post-stenotic dilation and notching of the inferior border of the ribs due to erosion by

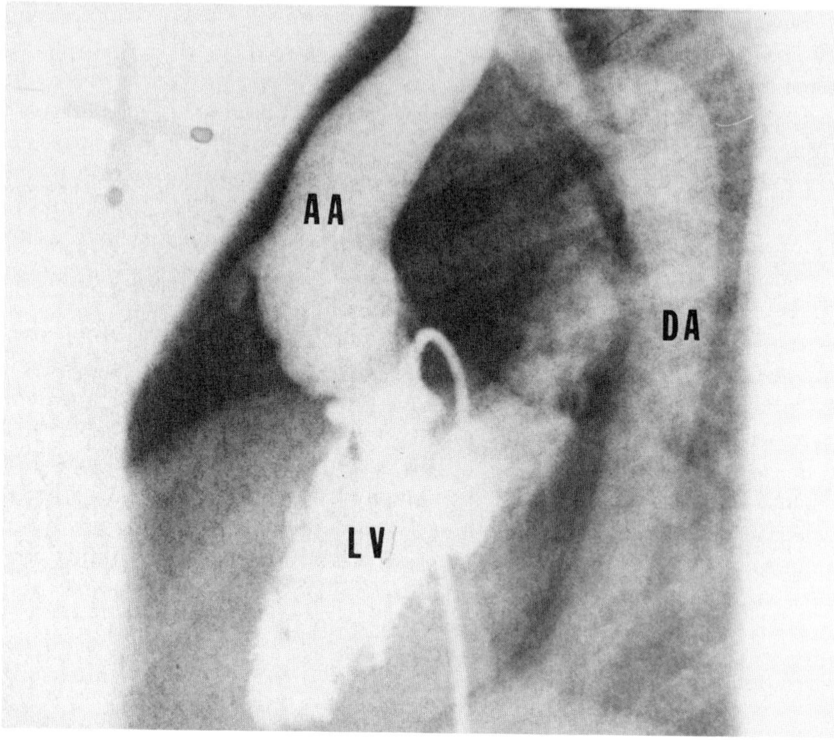

Figure 12–5. Lateral view of a left ventriculogram demonstrating normal ascending aorta (AA), coarctation (arrow) which appears to be in the region of the ductus, although the ductus is not clearly visualized, and postductal dilatation in the descending aorta (DA). (Photograph courtesy of Benigno Soto, M.D., University of Alabama in Birmingham.)

dilated collaterals. Electrocardiogram and echocardiogram usually show left ventricular hypertrophy. A bicuspid aortic valve has been reported to occur in up to 85% of patients with coarctation and may be demonstrated by echocardiography. Measurement of systolic blood pressure before and after treadmill exercise has been shown to be a useful noninvasive index of the severity of coarctation [31]. If the increase in systolic blood pressure with exercise is equal to or greater than 35 mmHg, there is a low probability that the pressure gradient at catheterization will be greater than 50 mmHg. Measurement of blood-pressure changes following treadmill exercise has also been used in the evaluation of persistent or recurrent hypertension following surgical repair of coarctation [32, 33]. An arm-to-leg systolic pressure gradient of more than 35 mmHg following exercise [32] and a post-exercise systolic pressure of more than 200 mmHg in the arms [33] have been taken as indications for recatheterization. Left heart catheterization with aortography is indicated to establish the severity of the aortic gradient, to localize and determine the length of the coarctation (figure 12–5), and to identify associated cardiac abnormalities prior to surgical correction.

The treatment of choice for coarctation is surgery [34]. The timing of the operation appears to have a major influence on outcome. In a recent retrospective review of 234 hypertensive children and adolescents who underwent surgical repair of coarctation, 21% of those who were operated in the first year of life had residual hypertension due to persistent coarctation, while only 6% of those who were operated between ages one and five had residual hypertension and persistent coarctation [35]. When surgery was performed after the age of six, the incidence of residual hypertension increased to 30% and was frequently unassociated with a persistent aortic gradient. Similar findings have been reported by others [36, 37, 38] and have been attributed to persistent changes in the resistance vessels of patients who underwent late repair of coarctation. These findings, coupled with the increased mortality in neonates and infants who undergo surgical repair of coarctation, indicate that surgery should ideally be performed between the ages of one and five years.

As indicated previously, surgical repair of aortic coarctation, even when anatomically successful, frequently does not cure the associated hypertension. Further, patients with coarctation appear to have an increased risk of premature death from cardiovascular disease regardless of surgical intervention and whether or not hypertension persists. The increased risk in these patients has been shown to correlate with the duration of preoperative hypertension [39]. All children should be followed regularly after surgery and, if they remain hypertensive, should be reevaluated for recurrence of the coarctation. Most of these children will not have demonstrable persistent of recurrent coarctation and will require medical treatment for their hypertension [40].

ESSENTIAL HYPERTENSION

Diagnosis

Essential hypertension, or hypertension of unknown etiology, accounts for 1% to 16% of cases of sustained blood pressure elevation in children (table 12–3). The diagnosis of essential hypertension is made in children, as in adults, by exclusion of secondary causes. Children with essential hypertension are likely to be obese [41] and have a family history of hypertension [42, 43]. A positive family history of hypertension has been reported in 30% to 88% of hypertensive children [11, 42, 43].

Pathogenesis

Heredity is a major factor in determining blood pressure [44] and the mechanisms by which blood pressure is genetically controlled are both diverse and interrelated. Studies in normotensive first-degree relatives of essential hypertensives have demonstrated important differences in electrolyte excretion and circulating renin levels that may make this group susceptible to the development of hypertension [45, 46]. Following sodium loading, relatives of hypertensives had higher blood pressures, higher plasma renin values, and more sluggish natriuretic responses than did controls. Thus, inherited abnormalities in the renin-angiotensin-aldosterone system, including perhaps alterations in sympathetic nervous drive to renin release, and/or an intrinsic defect in the ability of the kidney to handle volume overload may contribute to the pathogenesis of essential hypertension.

Increased activation of the sympathetic nervous system and an exaggerated pressor response to mental stress have been described in normotensive adolescent female offspring of hypertensive parents in comparison to age and sex matched offspring of normotensive parents [47]. Sympathetic neural and pressor responses to mental stress were further enhanced in the former group when a high salt diet was consumed, suggesting that the increased sympathetic nervous system activity seen in hypertensive kindreds is accompanied by an enhanced pressor response to sodium loading. Exposure to competitive mental tasks has also been shown to be accompanied by significant reductions in urine volume and urinary sodium excretion in normotensive adolescent male offspring of hypertensive parents and in young men with borderline hypertension [48]. The amount of sodium retained was directly related to the magnitude of the heart rate increase during exposure to stress in these subjects, suggesting that both responses were mediated via the sympathetic nervous system. This interpretation is reinforced by the results of studies in the dog [49, 50] and in the genetically hypertensive rat [51] which demonstrate that stress-induced sodium retention is abolished by infusion of propranolol or by surgical ablation of the renal sympathetic nerves. Further, long-term increases in blood pressure have been shown to develop in dogs exposed to daily shock avoidance combined with saline infusion, although neither the stress nor the saline alone altered baseline blood pressure [52]. Taken together, these findings suggest that psychological stress induces changes in renal excretory function that may play an important role in long-term blood pressure regulation. In addition, the human data suggest that the tendency to retain sodium during stress may be familial and may reflect a predisposition to develop hypertension.

An association between increased sympathetic activity, increased dietary sodium chloride, and hypertension has been demonstrated in the spontaneously hypertensive rat of the Okamoto strain (SHR) [53, 54, 55]. The SHR is a genetic model in which sympathetic neural activity is increased during the developing phase of hypertension. Bilateral renal denervation, a maneuver which reduces renal sympathetic tone and prevents sodium retention in the developing SHR, delays the development of hypertension in young SHR but does not alter renal sodium handling or blood pressure in the mature animal with fixed hypertension [53]. Further, ingestion of a high salt diet in young SHR enhances the severity of the hypertension, increases peripheral sympathetic activity, and alters the norepinephrine content of hypothalamic nuclei known to be involved in the regulation of blood pressure [54]. In contrast, no response to oral salt loading is seen in mature SHR. Taken together, these findings suggest that in the developmental phase of genetic hypertension, sympathetic nervous system activity is increased and acts as a trigger to set off a chain of events that lead to the development of fixed systemic hypertension. In the presence of increased sodium intake, there is further enhancement of sympathetic activity, and the hypertension is earlier in onset and more severe. These mechanisms are operational in the developing individual only and disappear once stable hypertension becomes established. Further study is needed to document these phenomena in man and to examine possible mechanisms by which sodium can enhance sympathetic activity.

Mechanisms which have been postulated to link sodium, the sympathetic nervous system, and the pathogenesis of systemic hypertension include alterations in: synthesis, storage and turnover of biogenic amines in central and peripheral neurons [56], baroreflex sensitivity [57], vascular reactivity to alpha adrenergic agonists and central and peripheral sympathetic stimulation, sensitivity of pre- and postsynaptic mechanisms that govern the release and reuptake of biogenic amines [58], renal sodium handling [59], and membrane sodium transport [60, 61].

Abnormalities in cation transport by red blood cell membranes have been reported in patients with essential hypertension and in their first degree relatives, but not in patients with secondary hypertension [60, 61, 62]. Ouabain-insensitive maximum red-cell sodium-lithium countertransport has been shown to be increased two- to three-fold in patients with essential hypertension compared with normotensive controls [60]. Further, normotensive teenage sons of parents who both have essential hypertension have higher rates of sodium-lithium counter-transport than age-matched sons of normotensive parents (figure 12–6) [62]. In contrast, the maximum rate of ouabain-insensitive sodium-potassium co-transport is reduced in red cells of patients with essential hypertension [61]. It is not clear whether counter-transport and co-transport are carried out by different transport systems or represent different aspects of the same system [62]. Further, an etiologic relationship between the transport abnormalities and the pathogenesis of hypertension has not been established. It has been postulated that the transport abnormalities may also involve arteriolar smooth muscle cells and/or endocrine cells which secrete vasoactive substances, either of which could increase peripheral vascular resistance and thereby cause hypertension [62].

Evidence has been adduced that sodium ion may produce effects on nerve terminals and vascular smooth muscle by altering calcium handling by the cell. Calcium is important for the uptake and binding of norepinephrine in nerve endings, for the release of catecholamines from nerve endings and the adrenal medulla [63], and for vascular smooth muscle contraction [64]. It has been postulated that there is a direct sodium-calcium counter-exchange system across biological membranes in which sodium concentration gradients influence the movement of calcium across cell membranes [65]. In this transport system, an increase in intracellular sodium decreases the transport of calcium out of the cell, thereby increasing the concentration of active calcium within the cell. In the vascular smooth muscle cell a decrease in the binding of calcium to the plasma membrane with a consequent increase in intracellular concentration of ionized calcium could make the cell membrane less stable and increase the sensitivity of the cell to vasoactive stimuli. In the neuron a similar change in calcium handling would decrease retention, storage, and reuptake of norepinephrine, making more of the transmitter available to receptors.

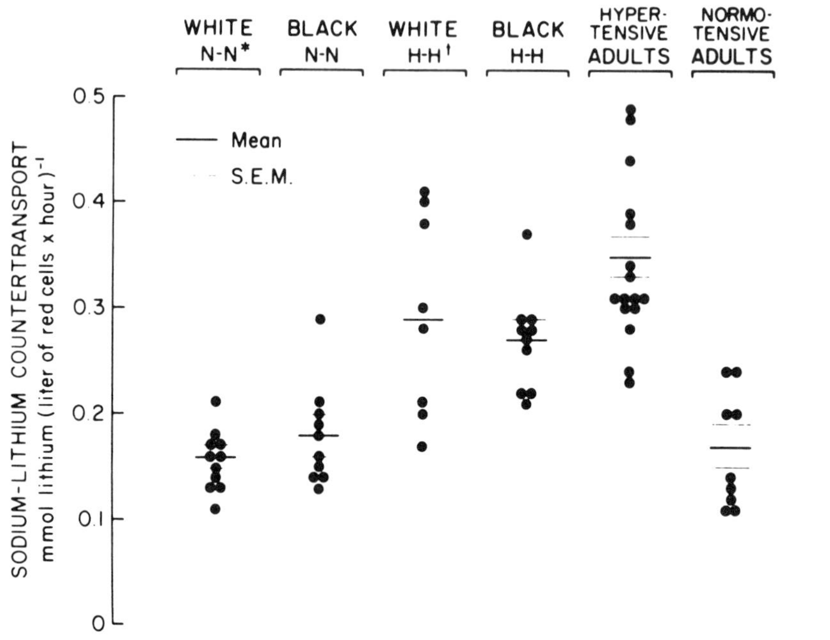

Figure 12–6. Maximal rate of sodium-lithium efflux in red cells of sons of normotensive and hypertensive parents and in the cells of two comparison groups of adults. *NN denotes that both parents were normotensive and had family histories negative for essential hypertension. †HH denotes that both parents had confirmed essential hypertension. (Reprinted from Woods, J.W.; Falk, R.J.; Pittman, A.W.; Klemmer, P.J.; Watson, B.S.; and Namboodiri, K. 1982. Increased red-cell sodium-lithium countertransport in normotensive sons of hypertensive parents. *N. Engl. J. Med.* 306:593–595.)

Sodium-transport abnormalities in hypertensive subjects have been ascribed to the actions of circulating inhibitor(s) of membrane pumping mechanisms rather than to intrinsic defects in the membranes themselves [66, 67]. According to this interpretation, chronic hypertension is due to the presence of a circulating sodium-transport inhibitor, or natriuretic factor, which is released in response to increased extracellular fluid volume. The latter may, in turn, be related to a genetic abnormality in the kidney. Further study is needed to elucidate the role(s) of these sodium-transport abnormalities in the pathogenesis of hypertension and to establish or refute their usefulness as genetic markers for hypertension. If sodium-transport abnormalities turn out to be predictive of the development of high blood pressure, they could become useful diagnostic tests for the "pre-hypertensive state" in the pediatric age group.

Elevations in serum uric acid [68] and increases in left ventricular mass [69, 70] have been described in children and adolescents with modest blood pressure elevations. Thus serum uric acid measurements and echocardiography have been suggested as diagnostic procedures for detecting a hypertensive diathesis in children whose blood pressures are not yet definitely abnormal. It is uncertain whether the increases in circulating uric acid reflect subtle abnormalities in renal hemodynamics or impairment in renal tubular transport mechanisms. The cardiac abnormalities may be secondary to increased sympathetic nervous system activity or related to the hyperdynamic circulatory state which is considered by some investigators to represent the earliest phase in the development of systemic hypertension [71]. Further study is needed to assess the utility of these procedures in making an early diagnosis of hypertension in children and adolescents and in monitoring natural history and responses to therapy.

Treatment

Non-pharmacologic means of lowering blood pressure or of preventing hypertension from developing should be employed in children with marginal blood pressure elevations and those at high risk for developing essential hypertension, i.e., children who are obese or have strong family histories of hypertension or cardiovascular disease. Weight control, restriction of dietary sodium, exercise, and avoidance of smoking are recommended for such children. These recommendations are made in the absence of firm evidence based on controlled trials that non-pharmacologic interventions are effective in preventing or ameliorating hypertension in children and adolescents. For children in whom non-pharmacologic measures are ineffective and those in whom hypertension is sustained at levels high enough to constitute a risk for the development of target organ damage, pharmacologic therapy should be instituted (for guidelines see the section, "Medical Treatment of Hypertension in Children and Adolescents."

ENDOCRINE DISORDERS

As shown in table 12–3, endocrine disorders are responsible for 6% to 9% of childhood hypertension. Table 12–2 lists the various endocrine causes of hypertension in children.

Pheochromocytoma

Pheochromocytoma represents a rare but remedial cause of systemic hypertension in children. It occurs in 0.04% to 0.1% of hypertensive patients. Less than 5% of pheochromocytomas occur in children. Patients with pheochromocytoma are almost invariably symptomatic, so it is legitimate to use symptoms and physical signs in screening patients for pheochromocytoma and to perform biochemical tests only on patients who have clinical evidence of the disorder. The most frequent symptoms of pheochromocytoma are headache and excessive sweating. Weight loss and palpitations are also common. Sustained hypertension is present in 88% of children with this disorder [72]. Postural hypotension may occur, particularly in children whose tumors secrete epinephrine predominately. Diabetes mellitus is common. Von Recklinghausen's neurofibromatosis is associated with pheochromocytoma, so the physical examination should include inspection for cafe au lait spots and neurofibromas and the family history should include specific questions about the syndrome.

Diagnosis of pheochromocytoma requires the demonstration of increased concentrations of catecholamines and/or their metabolites in plasma or urine and the localization of tumor via CT scan, radionuclide techniques, or angiography. Measurements of 24-hour urinary excretion of VMA (3-methoxy, 4-hydroxymadelic acid) and of metanephrine and normatanephrine are the most commonly used screening tests for pheochromocytoma because of their low cost and simplicity [73]. Free catecholamines in urine can also be measured. Assay of plasma catecholamines offers advantages over measurement of urinary catecholamines and/or metabolites in that it eliminates the unrealiability inherent in 24-hour urine collections and artifacts secondary to interference from drugs and angiographic contrast media. Measurement of plasma catecholamines is more reliable than 24-hour urinary metanephrines and/or VMA in screening for pheochromocytoma as long as appropriate precautions are taken in obtaining plasma samples [74]. Patients must be resting and the stress of venipuncture must be avoided by insertion of an indwelling catheter 20 to 30 minutes before sampling. Plasma catecholamine levels are usually enormously elevated in patients with pheochromocytoma and tend to remain elevated even at times when blood pressure is normal [74], suggesting that circulating catecholamines released from the tumor may not be entirely responsible for the hypertension seen in the syndrome.

The specificity of plasma catecholamine measurements in the diagnosis of pheochromocytoma can be enhanced by repeating the measurements following administration of clonidine [75] or pentolinium [76]. In normal subjects or patients with essential hypertension, stimulation of central-adrenergic receptors with clonidine or administration of a ganglion blocker causes suppression of plasma norepinephrine by blunting central sympathetic outflow. In patients with pheochromocytoma in whom circulating norepinephrine levels depend on autonomous release from the tumor, these agents do not affect plasma norepinephrine levels. Interestingly, both groups show similar reductions in supine blood pressure and heart rate following administration of these agents, suggesting that catecholamines secreted by the tumor are not responsible for the maintenance of hypertension in patients with pheochromocytoma.

Once a presumptive diagnosis of pheochromocytoma has been made by demonstrating elevations in plasma catecholamines or a positive clonidine or phentolamine suppression test, localization studies are indicated. In many cases the tumor can be localized by CT scanning [76] (figure 12–7). For pheochromocytomas that are smaller than the resolving power of the CT scan, particularly if they are located outside the adrenal, other means of localization may be required. This is frequently the case in children, in whom the incidence of multiple tumors, many of which are extra adrenal, is much higher (25% to 51%) than in adults (10%) [72, 77]. The guanethidine analog [^{131}I] meta-iodobenzylguanide (^{131}-I MIBG) is useful in imaging pheochromocytomas that are undetectable by CT scan [78]. Tumors detected by ^{131}I-MIBG scanning include both benign and malignant and intra- and extra-

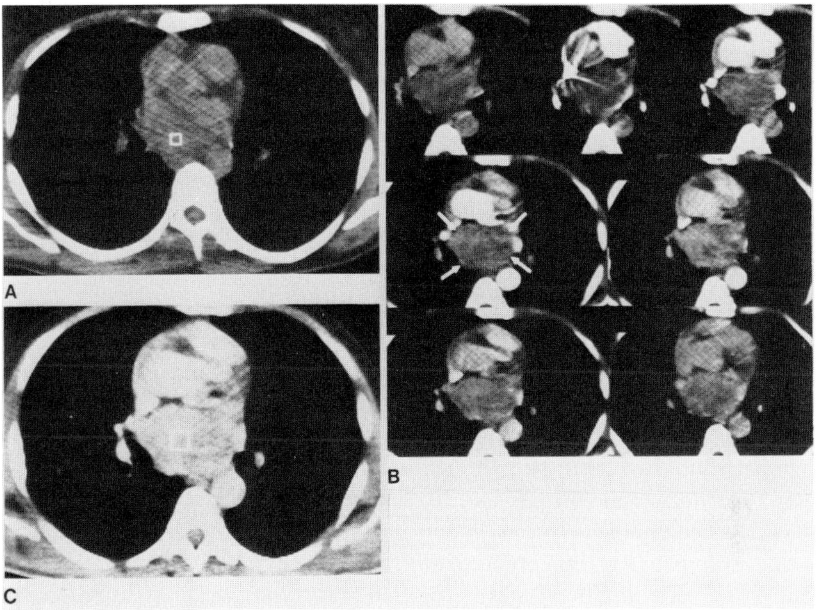

Figure 12–7. Intrathoracic pheochromocytoma which presented as a distortion in the mediastinal contour. On precontrast scan the mass (cursor — Panel A), is isodense with adjacent structures. Dynamic scanning after bolus intravenous injection of iodinated contrast material in a single tomographic plane shows a clear distinction of hypervascular tumor (arrows, Panel B) from adjacent cardiovascular structures. In Panel C, one minute after bolus injection, the tumor (cursor) is nearly isodense with adjacent cardiovascular structures. (Reprinted from Francis, I.R.; Glazer, G.M.; Shapiro, B.; Sisson, J.C.; and Gross, B.H. 1983. Complementary roles of CT and ^{131}I-MIBG scintigraphy in diagnosing pheochromocytoma. *AJR* 141:719–725.)

adrenal pheochromocytomas which have various patterns of hormone secretion. Recently use of a combination of ^{131}I-MIBG and CT scanning has been shown to be particularly useful in detecting extra-adrenal pheochromocytomas [79]. False negative results have been reported with both CT and MIBG scans, however. In such cases selective arteriography and/or differential venous catheterization with measurement of regional plasma catecholamines are helpful in localizing the tumor. Arteriography and venography are indicated only if scanning procedures are negative, as they are associated with risk to the patient.

Treatment of pheochromocytoma is surgical unless the tumor is metastatic or the patient is inoperable for other reasons. Patients should be prepared with α-adrenergic blocking agents (phenoxybenzamine or prazosin) for one to two weeks prior to surgery in order to block the pressor effects of circulating norepinephrine and expand the intravascular volume [80, 81]. In selected cases β-adrenergic blockers should be added later [80]. Because of the high incidence of multiple and extra-adrenal tumors in children, it is recommended that the tumor be removed through a midline abdominal incision and that examination of both adrenals as well as a careful search for extra-adrenal tumors be made at the time of operation [72, 78].

If surgery cannot be performed, patients can be managed medically with α- and β-adrenergic blockers or the new combined α- and β-adrenergic antagonist labetalol [80, 81, 82]. Alpha methyl para tyrosine (Demser [R]) inhibits tyrosine hydroxylase, the rate-limiting enzyme in catecholamine biosynthesis, thus reducing catecholamine production by the tumor and secondarily lowering blood pressure [83]. It is the drug of choice in the management of inoperative pheochromocytoma, but its many adverse effects, including sedation, diarrhea, galactorrhea, anxiety, tremulousness, and crystalluria, limit its clinical usefulness.

Congenital Adrenal Hyperplasia

The rare syndromes 11β- and 17α-hydroxylase deficiency result in the hypertensive forms of congenital adrenal hyperplasia [84]. In 11β-hydroxylase deficiency a decrease in cortisol synthesis results in stimulation of ACTH secretion and overproduction of steroid precursors. Plasma levels of androgen are increased, producing virilization. Hypertension is thought to be secondary to increased production of desoxycorticosterone. In 17α-hydroxylase deficiency there is overproduction of mineralocorticoids and decreased production of glucocorticoids and sex steroids, resulting in hypogonadism. Hypertension in this syndrome is also thought to be secondary to desoxycorticosterone excess. In both syndromes, treatment with replacement doses of glucocorticoid hormones corrects both the hypertension and the humoral and electrolyte abnormalities.

Hyperaldosteronism

Hyperaldosteronism is a syndrome of hypertension associated with the excessive secretion of aldosterone and secondary hypokalemia and suppressed plasma renin activity. The pathogenesis of the hypertension remains poorly understood, although increases in plasma volume have been described in some cases, and increased peripheral vascular resistance is a constant finding.

Glucocorticoid-remediable hyperaldosteronism is a rare disorder that usually presents at an early age and occurs in kindreds, apparently being transmitted by an autosomal dominant gene not linked to HLA [84, 85]. Aldosterone production in patients with untreated glucocorticoid-remediable hyperaldosteronism is under the control of ACTH and may occur in the zona fasciculata rather than the glomerulosa. The clinical and biochemical features of the disease are similar to those of primary hyperaldosteronism secondary to adrenal cortical adenoma or bilateral adrenal hyperplasia except that aldosterone secretion is completely suppressed within 48 hours of dexamethasone (1 mg) administration. This response is so constant that a positive dexamethasone suppression test is diagnostic. In children with glucocorticoid-remediable hyperaldosteronism, the hypertension also responds to glucocorticoid treatment; in adults, blood pressure may not respond despite adequate suppression of aldosterone [84]. Thus early diagnosis is critical.

In primary hyperaldosteronism, there is autonomous or relatively autonomous overproduction of aldosterone by the adrenal gland. In children the syndrome is rare and is usually secondary to bilateral adrenal hyperplasia; adrenal cortical adenoma and carcinoma are less common causes. A presumptive diagnosis of primary hyperaldosteronism can be made on the basis of measurements of increased plasma and/or urinary aldosterone, suppressed renin, and potassium wasting in the sodium-repleted state. In adults the best single test for identifying patients with primary hyperaldosteronism is the measurement of aldosterone excretion after three days of salt loading [86]. Urinary aldosterone measurements made in patients on an ad lib diet are not adequate to make a diagnosis of primary hyperaldosteronism because of the substantial overlap with essential hypertension and secondary aldosteronism. One reason for this phenomenon is that hypokalemia inhibits aldosterone production, so that plasma and urinary aldosterone in severely hypokalemic subjects with untreated primary hyperaldosteronism may be in the normal range for potassium-replete normal subjects. Primary hyperaldosteronism is distinguishable from glucocorticoid-remediable hyperaldosteronism by failure of aldosterone to be suppressed to undetectable levels with dexamethasone [84].

Once a presumptive diagnosis of aldosterone excess has been made, additional procedures are indicated to differentiate between bilateral adrenal hyperplasia and adrenal adenoma or carcinoma as sources for the hormone. The distinction is important in guiding therapy, since patients with adrenal adenomas frequently have both their hyperaldosteronism and their hypertension completely corrected by unilateral adrenalectomy, whereas those who have bilateral hyperplasia often remain hypertensive even after both adrenals are removed [87].

Since the clinical and biochemical characteristics of primary hyperaldosteronism secondary to bilateral adrenal hyperplasia and adrenal cortical adenoma or carcinoma are so similar, the differential diagnosis can only be made by localizing the tumor. A number of techniques are available for localizing aldosterone producing adrenal tumors. Direct adrenal vein sampling by percutaneous catheterization and measurement of adrenal venous aldosterone levels is the most accurate method of localizing the site of the tumor [88]. In experienced hands, this technique is better than 90% accurate in tumor localization. Simultaneous cortisol determinations are used to confirm that adrenal vein sampling has been achieved. Computed tomographic (CT) scanning can resolve adrenal tumors of > 1.0 cm in diameter [89]. In patients with suspected bilateral adrenal cortical hyperplasia, CT scan reveals either bilateral adrenal enlargement or normal-appearing glands. CT scans have largely replaced ^{125}I-19-iodocholesterol scintiscans in the diagnostic evaluation of patients with primary hyperaldosteronism [89].

The treatment of choice for hypertension secondary to an aldosterone-producing adenoma is unilateral adrenalectomy. In 68% to 83% of cases, surgery restores both blood pressure and electrolytes to normal [90]. In primary hyperaldosteronism secondary to adrenal carcinoma, the treatment of choice is surgical removal of the neoplasm, particularly if there is no evidence of metastasis. Alternatively, pharmacologic therapy, including aldosterone antagonists (spironolactone), and inhibitors of steroid biosynthesis such as aminoglutethimide (Cytodren^{R}) or Trilostane^{R} or the adrenolytic agent o,p^{1}-DDD can be used [90]. In patients with primary hyperaldosteronism secondary to bilateral adrenal hyperplasia, remission of hypertension following bilateral total adrenalectomy is seen in only 18% to 35% of cases [90]. Chronic spironolactone administration has been shown to be effective in controlling hypertension in these patients for periods of up to nine years. For patients who are intolerant of spironolactone, amiloride may be an appropriate substitute. Alternatively, conventional antihypertensive therapy may be used.

Cushing's Syndrome

Cushing's syndrome is a rare disorder, particularly in children. It is accompanied by hypertension in approximately 85% of affected patients. The etiology of hypertension in Cushing's syndrome is uncertain but is thought to be most likely related to the mineralocorticoid effects of cortisol excess. Excess cortisol also stimulates production of renin substrate, resulting in elevated levels of angiotensin II, and causes increased vascular reactivity to endogenous vasoconstrictors. The most common cause of Cushing's syndrome in childhood is a benign adrenal tumor which produces excess glucocorticoids and androgens. Rarely, adrenal carcinoma causes Cushing's syndrome. Virilization is common in adrenal carcinoma, and its presence should raise the question of adrenal malignancy. In a patient in whom physical findings such as moon facies, hirsutism, truncal obesity, acne, easy bruisibility, and hyperpigmented striae raise the question of Cushing's syndrome, the diagnosis can confirmed by the demonstration of increased plasma levels of cortisol that do not suppress following administration of dexamethasone [91]. Arteriography, venography, and abdominal CT scan are useful in localization of the tumor. Although surgical removal of the tumor usually cures the hypertension, some patients have a persistent increase in blood pressure which is thought to be secondary to fixed vascular damage. Such patients require medical antihypertensive treatment.

HYPERTENSIVE CRISIS IN CHILDHOOD

Hypertensive crisis in children, as in adults, is usually defined as a sudden acute rise in blood pressure accompanied by encephalopathy, congestive heart failure, and/or any other life-threatening cardiovascular complication of the hypertension. Hypertensive encephalopathy has been attributed to (1) cerebral edema which forms as a result of sudden major increases in intracerebral capillary pressure and (2) intense focal spasm of intracerebral arteries with consequent ischemic damage to intracerebral vessels and brain parenchyma. Further study is needed to fully delineate the pathogenesis of hypertensive encephalopathy in human subjects. Congestive heart failure occurs in the setting of accelerated hypertension because of a combination of increased afterload on the heart and increased activity of neurohumoral mechanisms which cause sodium and water retention.

Children with hypertensive encephalopathy usually present with fatigue, lethargy, visual disturbances, vomiting, and sometimes seizures and focal neurological deficits [92]. If the hypertension is associated with congestive heart failure, dyspnea, tachypnea, orthopnea, and cyanosis are usually present. The history and physical examination should include, in addition to those items mentioned under the diagnostic evaluation of the asymptomatic hypertensive child (see the earlier section entitled "Diagnostic Evaluation of the Asymptomatic Hypersensitive Child" in this chapter), careful recording of blood pressure in both upper and lower extremities, palpation of peripheral pulses, and a complete neurological examination.

The most likely etiologies of hypertensive crisis in childhood are listed in table 12–6 [92]. In older children, the most common causes of hypertensive crisis are acute post-streptococcal glomerulonephritis and other nephritides. A recent history of pharyngitis or scarlet fever would suggest the former diagnosis. In infants, the most common cause of hypertensive crisis is coarctation of the aorta; the second most common, renal artery thrombosis secondary to catheterization of the umbilical artery in the neonatal period. In both cases, a presumptive diagnosis can be made by history and physical examination and confirmed by angiography.

The laboratory evaluation of children with hypertensive crisis should include those tests obtained routinely in the diagnostic evaluation of the asymptomatic hypertensive child and, in addition, a throat culture and immunologic evaluation to rule out post-streptococcal glomerulonephritis and other nephritides which have an immunologic basis.

The treatment of hypertensive crisis in children, as in adults, requires hospital admission. The principles of therapy are similar to those used in the treatment of hypertensive crisis in adults, but clinical experience is less extensive [11, 92].

MEDICAL TREATMENT OF HYPERTENSION IN CHILDREN AND ADOLESCENTS

The Task Force on Blood Pressure in Children has recommended medical therapy for children 3 to 12 years old with diastolic pressures >90 mmHg and for children older than 12 years with diastolic pressures >100 mmHg [3]. Pediatric patients with less severe hypertension (diastolic pressures above the 95th percentile but <90 mmHg in the 3-to-12-year age group and above the 95th percentile but <100 mmHg in the 13-to-18-year age group) and who are free of the complications of hypertension should be treated nonmedically. Nonpharmacologic treatment for these children should include weight reduction wherever appropriate, sodium restriction and counseling regarding avoidance of cigarette smoking, the need for careful monitoring if oral contraceptives are used, and the advisability of regular exercise. It should be recognized that these treatment guidelines are arbitrary since the clinical course of hypertension in children has not been clearly delineated and the efficacy of neither pharmacologic nor nonpharmacologic antihypertensive treatment has been established in this age group.

Accordingly, when pharmacologic therapy is instituted, efforts should be made to control blood pressure with the lowest doses and the smallest number of drugs possible. For example, when a "first-step" drug does not control blood pressure adequately, consideration should be given to changing to a different drug rather than adding a second agent. Caution should be exercised in the administration of antihypertensive drugs to children and adolescents for the following reasons: (1)

experience with these agents in children is insufficient to rule out important adverse effects on growth and development; (2) the cumulative adverse effects of prolonged administration of antihypertensive therapy are unknown; (3) many of the antihypertensive agents used in adults have not been evaluated for safety and efficacy in children; and (4) recent studies in adults with mild uncomplicated essential hypertension have called into question the efficacy of antihypertensive treatment in preventing ischemic heart disease and have even suggested that diuretic treatment is associated with an increased incidence of sudden cardiac death in some sub-groups of patients [93]. Since most children with essential hypertension have mild uncomplicated disease, these concerns are particularly pertinent in the pediatric group.

In all cases the behavioral measures mentioned previously should be used as adjuncts to pharmacologic therapy. Patients should be followed carefully to insure satisfactory blood pressure control and encourage adherence to the management program. In cases where a combination of pharmacologic and non-pharmacologic therapy results in both good control of blood pressure and disappearance of risk factors, discontinuation of pharmacologic therapy should be considered. Non-pharmacologic therapy and careful followup should be continued in these instances.

The selection of specific antihypertensive drugs for use in children is somewhat arbitrary and largely based on experience in adults. The clinical pharmacology of antihypertensive drugs in children has been recently reviewed elsewhere [94] and is beyond the scope of this chapter. In the United States a diuretic has been used traditionally as the initial drug in the antihypertensive regimen. The advantages of this approach are that many patients with mild to moderate hypertension can be controlled on a diuretic alone, that many diuretics are effective when administered once daily, and that diuretic administration is associated with few subjective adverse effects. On the other hand, recent evidence suggests that control of mild hypertension with diuretics does not prevent ischemic heart disease [93]. Further, diuretic administration exacerbates a number of coronary disease risk factors — including circulating lipid, glucose, and uric acid levels, causes electrolyte imbalance, and may be associated with an increased incidence of sudden cardiac death in some patient groups [95]. For these reasons, it may be wise to begin the pharmacologic treatment of hypertension in children with another agent, such as a β-adrenergic blocking drug. For patients who cannot be controlled on a single agent, a stepped-care approach comparable to that used in adults (figure 12–8) is recommended with the caveat that both the dose and the number of agents administered should be kept to a minimum. It is unlikely that this approach will compromise the efficacy of pharmacologic therapy in controlling blood pressure in children and adolescents, since it would be expected that young patients with minimal fixed structural changes in resistance vessels would respond readily. It has been demonstrated previously that young compared to older adult hypertensives appear to respond better to antihypertensive drugs [96].

Recommended Drug Regimen for Children with Hypertension, by Age and Severity*

	Age (yr) 0-6	7-12	13-18
	Diastolic Blood Pressure = 90-100 mm Hg		
Step 1	Consultation	Thiazide	Individualize therapy†
Step 2‡		add β blocker§ / add methyldopa	
Step 3‖		add vasodilator / add vasodilator	
Step 4	Consultation with specialist recommended for refractory patients		
	Diastolic Blood Pressure = 100-114 mm Hg		
Step 1	Consultation and hospitalization recommended	Thiazide	
Step 2‡		add β blocker§	add methyldopa
Step 3‖		add vasodilator	add vasodilator
Step 4	Consultation with specialist recommended for refractory patients		
	Diastolic Blood Pressure > 114 mm Hg		
	All ages from 0-18: Consultation and hospitalization recommended		

Figure 12–8. *After appropriate evaluation and nondrug management, drug therapy may be considered indicated. †Drug therapy not always required. Consider weight control and avoidance of salt abuse. Systematic follow-up. Recheck annually at minimum. ‡Reserpine has been omitted from recommended drug regimens because of the many adverse effects it can produce in children, e.g., emotional depression and lability, sedation, stuffy nose. (Reprinted from the Report of the Task Force on Blood Pressure Control in Children. 1979. *Pediatrics* 59 (suppl 2): 802–803.)

A recent study of the effect of pharmacologic treatment on blood pressure in children whose pressures tracked in the upper decile suggests that antihypertensive treatment may be effective in this population but presents some difficulties in interpretation [97]. Children aged 8 to 18 with blood pressure in the highest decile were randomly allocated into treatment and non-treatment groups. Treatment included a combination of dietary guidance, modification of school lunches, parental involvement, and pharmacologic treatment with a combination of low doses of propranolol twice daily and chlorthalidone once or twice daily. Systolic and diastolic blood pressures in the treatment group after six months of observation were significantly lower, 5 and 3 mmHg, respectively, than in the control group. The physiologic significance of this small reduction in blood pressure is uncertain, but it does represent a shift from the upper to the middle deciles of mean blood pressure. Further study with longer periods of followup is needed to adequately evaluate the long-term effects of such interventions on hypertension in children.

Table 12–1. Estimates of Upper Limit of Normal Diastolic Pressure According to Age

Age	Diastolic Pressure
14 to 18 years	> 90 mmHg
10 to 14 years	> 85 mmHg
6 to 10 years	> 80 mmHg
6 years	> 75 mmHg

Table 12–2. Causes of Persistent Hypertension in Children

1. Renal
   a. Acute glomerulonephritis
   b. Hemolytic-uremic syndrome
   c. Bilateral obstructive uropathy
   d. Congenital defects, polycystic kidneys
   e. Unilateral renal disorders
      (1) Renal artery abnormalities
         a. Stenosis
         b. Neurofibromatosis
         c. Thrombosis
         d. Trauma
         e. Fistula
         f. Fibromuscular dysplasia
         g. External compression
      (2) Unilateral parenchymal disease, pyelonephritis, congenital defects, obstructive uropathy, radiation nephritis, infarction
      (3) Perirenal masses
   f. Anaphylactoid purpura (Henock-Schonlein) nephritis
   g. Renal transplantation (rejection, steroid-related)
   h. Acute renal failure
   i. Anephric patients
   j. Blood transfusions in patients with renal disease
   k. Genitourinary surgery
   l. Renal biopsy
   m. Tumors of the kidney (Wilms' tumor, juxtaglomerular cell tumors, tuberous sclerosis)
   n. Collagen disease (periarteritis, lupus, dermatomyositis)
   o. Chronic glomerulonephritis and chronic pyelonephritis
   p. Heavy metal poisoning
   q. Amyloidosis (familial form)
   r. Fabry's disease (angiokeratoma corporis diffusum)
   s. Familial nephritis (Alport's disease, medullary cystic disease)
   t. Renal tubular acidosis with nephrocalcinosis

2. Endocrine
   a. Pheochromocytoma
   b. Congenital adrenal hyperplasia
   c. Hyperthyroidism (systolic only)
   d. 17-Hydroxylase deficiency
   e. Aldosteronism (primary)
   f. Neuroblastoma
   g. Cushing's disease
   h. Liddle's syndrome
   i. Hyperparathyroidism
   j. Ovarian tumors

3. Vascular system
   a. Coarctation of the aorta
   b. Polycythemia
   c. Anemia (systolic only)
   d. Pseudoxanthoma elasticum
   e. Takayasu's arteritis
   f. Radiation aortitis
   g. Patent ductus arteriosus (systolic only)
   h. Arteriovenous fistula (systolic only)
   i. Leukemia
   j. Subacute bacterial endocarditis
   k. Cardiac problems (heart block, aortic insufficiency)

Table 12–2. Continued

4. Metabolic
   a. Diabetes mellitus (renal involvement)
   b. Gouty nephropathy
   c. Acute intermittent porphyria
   d. Hypercalcemia
   e. Hypernatremia

5. Neurological
   a. Dysautonomia (Riley-Day syndrome)
   b. Neurofibromatosis
   c. Increased intracranial pressure (any cause, especially tumors, infection, trauma)
   d. Guillain-Barre syndrome
   e. Poliomyelitis
   f. Anxiety

6. Drug-related
   a. Steroid administration (corticosteroids and DOCA)
   b. Heavy metals (mercury and lead)
   c. Reserpine overdose
   d. Amphetamine overdose
   e. Intravenous alpha-methyldopa
   f. Sympathomimetic drugs (nose drops, cough medicine, cold preparations)
   g. Excessive ingestion of licorice
   h. Ingestion of birth control pills

7. Essential

8. Miscellaneous
   a. Burns
   b. Stevens-Johnson syndrome
   c. Cyclic vomiting with dehydration
   d. Hypertension related to stretching of the femoral nerve (leg traction)

Source: Gruskin, A.G.: Clinical evaluation of hypertension in children. *Primary Care* 1:233–242, 1974.

Table 12–3. Causes of Persistent Hypertension in Children*

	Place of Study					
	Toronto		London		New York	
Condition	No. of Cases	%	No. of Cases	%	No. of Cases	%
Renal Disorders	66	73	78	78	62	61
Coarctation	—	—	15	15	13	13
Essential	15	16	1	1	7	7
Endocrine	8	9	—	—	6	6
Obesity	—	—	1	1	6	6
Miscellaneous	2	2	5	5	7	7
Total	91	100	100	100	101	100

Source: New, M.I.: Report of the Hypertension Task Force: Part A, Pediatrics. U.S. Department of Health, Education and Welfare Public Health Service, National Institutes of Health Publication No. 79-1628, (1979).
*Patients 0–18 years of age, Toronto 1960–1974, London 1970–1975, New York 1970–1977.

Table 12–4. Etiology of 238 Cases of Renal Hypertension in Children

	n	Percent
Scarred small kidneys		32.3
Histologically defined	30	
Without histology	47	
Glomerular diseases	70	29.4
Rapidly progressive glomerulonephritis		
Steroid-resistant primary nephrosis		
Focal and segmental glomerulosclerosis		
Membranous nephropathy		
Minimal glomerular lesions		
Renovascular lesions	47	19.7
Fibromuscular dysplasia		
Renal artery aneurysm		
Renal artery thrombosis		
Takayasu's disease		
Neurofibromatosis		
Pseudoxanthoma elasticum		
Ehlers-Danlos syndrome		
Obstructive uropathy	18	7.5
Hemolytic uremic syndrome	15	6.3
Miscellaneous		4.6
Wilms' and other tumors	4	
Trauma	3	
Polycystic kidneys	3	
Sequelae of renal ischemia	1	

Source: Broyer, M., Bacri, J.L., and Royer, P.: Renal forms of hypertension in children: Report on 238 cases. In *Hypertension in Children and Adolescents*, G. Giovanelli, M.I. New, G. Gorini (eds.) New York, Raven Press, 1981, pp 201–208.

Table 12–5. Arteriographic Findings in Significant Renal-Artery Stenosis

1. Reduction of luminal diameter by < 70%.
2. Presence of collateral circulation
3. Post-stenotic dilatation
4. Diminished velocity of flow across stenosis
5. Decreased kidney size
6. Pressure gradient of < 25% of the systolic pressure across the stenosis

Table 12–6. Etiologies of Hypertensive Crisis in Childhood

Common Causes	Rare Causes

A. Acute poststreptococcal glomerulonephritis
B. Other nephritides
  1. Hemolytic uremic syndrome
  2. Henoch-Schonlein purpura nephritis
  3. Lupus erythematosus
  4. Membranoproliferative glomerulonephritis
C. Chronic pyelonephritis
D. Segmental hypoplasia (Ask-Upmark kidney)
E. Renal vascular disease
F. Chronic renal failure in patients on dialysis
G. Renal transplantation
H. Treatment with adrenocortical steroids
I. Miscellaneous
  1. Drugs
  2. Poisons
  3. Other pathologies
	A. Pheochromocytoma
	B. Essential hypertension
	C. Drugs and poisons
	  1. Adrenocortical steroids
	  2. Hypotensive agents
	    a. Reserpine overdosage
	    b. Methyldopa infusion
	    c. Clonidine withdrawal
	  3. Methamphetamine
	  4. Phencyclidine
	  5. Excess ingestion of licorice
	D. Hypercalcemia
	E. Hypernatremia
	F. Renin-secreting tumors
	G. Pseudoxanthoma elasticum
	H. Inflammatory bowel disease (usually within 3 days after surgery)

Source: Balfe, J.W., Rance C.P.: Recognition and management of hypertensive crises in childhood. *Pediatr. Clin. North Am.* 25:159–174, 1978.

ACKNOWLEDGMENTS

The author would like to thank Drs. Bonita Falkner and Alan Gruskin for their advice in preparation of the chapter and Robin Roberson for her excellent secretarial assistance.

REFERENCES

1. Harlan, W.R.; Cornoni-Huntley, J.; and Leaverton, P.E. 1979. Blood pressure in childhood: The National Health Examination Survey. *Hypertension* 1:559–565.
2. Cornoni-Huntley, J.; Harlan, W.R.; and Leaverton, P.E. 1979. Blood pressures in adolescence: The United States Health Examination Survey. *Hypertension* 1:566–571.
3. Blumenthal, S.; Epps, R.P.; Heavenrich, R.; Laar, R.M.; Leiberman, E.; Mirkin, B.; Mitchell, S.C.; Naito, V.B.; O'Hare, D.; Smith, W.M.; Tarazi, R.; and Upson, D. 1977. Report of the Task Force on Blood Pressure Control in Children. *Pediatrics* 59:797–818.
4. Falkner, B.; Hamstra, B.; Lombardo, R.; and Shirk, J. 1981. Blood pressure variability in adolescent males. *Int. J. Ped. Neph.* 2:177–180.
5. Katz, S.H.; Hediger, M.L.; Schall, J.I.; Bowers, E.J.; Barker, W.F.; Aurand, S.; Eveleth, P.B.; Gruskin, A.B.; and Parks, J.S. 1980. Blood pressure, growth and maturation from childhood through adolescence. *Hypertension* 2:155–169.
6. Rance, C.P.; Arbus, G.S.; Balfe, J.W.; and Kooh, S.W. 1974. Persistent systemic hypertension in infants and children. *Pediatr. Clin. North. Am.* 21:801–824.
7. Gill, D.G.; DaCosta, B.M.; Cameron, J.S.; Joseph, M.C.; and Chantler, C. 1976. Analysis of 100 children with severe and persistent hypertension. *Arch. Dis. Child.* 51:951–956.
8. New, M.I. 1979. *Report of the Hypertension Task Force: Part A, Pediatrics.* U.S. Department of Health, Education and Welfare Public Health Service, NIH Publication No. 79-1628.
9. Londe, S.; Bourgoignie, J.L.; Robson, A.M.; and Goldring, D. 1971. Hypertension in apparently normal children. *J. Pediatr.* 78:569–577.
10. Broyer, M.; Bacri, J.L.; and Royer, P. 1981. Renal forms of hypertension in children: Report on 238 cases. In *Hypertension in children and adolescents,* ed. G. Giovanelli, M.I. New, and G. Gorini, pp. 201–208. New York: Raven Press.

11. Mongeau, J.G., and Fouron, J.C. 1983. Hypertension in children, including coarctation of the aorta. In *Hypertension,* ed. J. Genest, O. Kuchel, P. Hamet, and M. Cantin, pp. 1041–1059. New York: McGraw-Hill.
12. Reiss, M.D.; Brookstein, J.J.; and Bleifer, K.H. 1972. Radiologic aspects of renovascular hypertension. IV. Arteriographic complications. *JAMA* 221:374–378.
13. Dillan, M.J.; Shah, V.; and Barratt, T.M. 1978. Renal vein renin measurements in children with hypertension. *Br. Med.* 22:168–170.
14. Re, R.; Novelline, R.; Escourrour, M.T.; Athanasoulis, C.; Burton, J.; and Haber, E. 1978. Inhibition of angiotensin-converting enzyme for diagnosis of renal-artery stenosis. *N. Engl. J. Med.* 298:582–586.
15. Strong, C.G.; Hunt, J.C.; Sheps, S.G.; Tucker, R.M.; and Bernatz, P.E. 1971. Renal venous renin activity: enhancement of sensitivity of lateralization by sodium depletion. *Am. J. Cardiol.* 27:602–611.
16. Schambelan, M.; Glickman, M.; Stockigt, J.R.; Biglieri, E.G. 1974. Selective renal-vein renin sampling in hypertensive patients with segmental renal lesions. *N. Engl. J. Med.* 290:1153–1157.
17. Hillman, B.J.; Ovitt, T.W.; Capp, P.; Prosnitz, E.H.; Osborne, R.W., Jr.; Goldstone, P.; Zukoski, C.F.; and Malone, J.M. 1982. The potential impact of digital video subtraction angiography on screening for renovascular hypertension. *Radiology* 142:577–579.
18. Novick, A.C.; Buonocore, E.; and Meaney, T.F. 1982. Digital subtraction angiography for postoperative evaluation of renal arterial reconstruction. *J. Urol.* 127:14–17.
19. Cicuto, K.P.; McLean, G.K.; Oleage, J.A.; Freiman, D.B., Grossman, R.A.; and Ring, E.J. 1981. Renal artery stenosis: anatomic classification for percutaneous transluminal angioplasty. *AJR* 137:599–601.
20. Colapinto, R.F.; Stronell, R.D.; Harries-Jones, E.P.; Gildiner, M.; Hobbs, B.B.; Farrow, G.A.; Wilson, D.R.; Morrow, J.D.; Logan, A.G.; Birch, S.J. 1982. Percutaneous transluminal dilatation of the renal artery: follow-up studies on renovascular hypertension. *AJR* 139:727–732.
21. Mahler, F.; Probst, P.; Haertel, M.; Weidmann, P.; and Krneta, A. 1982. Lasting improvement of renovascular hypertension by transluminal dilatation of atherosclerotic and nonatherosclerotic renal artery stenoses. *Circulation* 65:611–617.
22. Sos, T.A.; Saddekni, S.; Sniderman, K.W.; Weiner, M.; Beinart, C.; Pickering, T.G.; Case, D.B.; Vaughan, E.D., Jr.; and Laragh, J.H. 1982. Renal artery angioplasty: techniques and early results. *Urol. Radiol.* 3:223–231.
23. Sos, T.A.; Vaughan, E.D.; Pickering, T.G.; Case, D.; Sniderman, K.W.; Sealey, J.E.; and Laragh, J.H. 1983. Renal vascular hypertension: A. Diagnosis of renovascular hypertension and evaluation of "surgical" curability. In *Transluminal angioplasty,* ed. W.R. Castaneda-Zuniga, pp. 57–61. New York: Thieme-Stratton.
24. Tegtmeyer, C.J.; Brown, J.; Ayers, C.A.; Wellons, H.A.; and Stanton, L.W. 1980. Percutaneous transluminal dilatation of the renal arteries. *Radiology* 135:589–599.
25. McCook, T.A.; Mills, S.R.; Kirks, D.R.; Heaston, D.K.; Seigler, H.F.; Malone, R.B.; and Osofsky, S.G. 1980. Percutaneous transluminal renal artery angioplasty in a 3 1/2 year old hypertensive girl. *J. Pediatr.* 97:958–960.
26. Sos, T.A.; Pickering, T.G.; Sniderman, K.; Saddekni, S.; Case, D.B.; Silane, M.F.; Vaughan, E.D., Jr.; and Laragh, J.H. 1983. Percutaneous transluminal renal angioplasty in renovascular hypertension due to atheroma or fibromuscular dysplasia. *New Engl. J. Med.* 309:274–279.
27. Makker, S.P., and Moorthy, B. 1979. Fibromuscular dysplasia of renal arteries: an important cause of renovascular hypertension in children. *J. Pediatr.* 95:940–945.
28. Novick, A.C.; Straffon, R.A.; Stewart, B.H.; and Benjamin, S. 1978. Surgical treatment of renovascular hypertension in the pediatric patient. *J. Urol.* 119:794–799.
29. Franklin, S.S.; Young, J.D., Jr.; Maxwell, M.H.; Foster, J.H.; Palmer, J.M.; Cerny, J.; and Varady, P.D. 1975. Operative morbidity and mortality in renovascular disease. *JAMA* 231:1148–1153.
30. Conn, J.W. 1974. Primary reninism, a new syndrome. *Verhandlungen der Deutschen Gesellschaft fur Innere Medizin* 80:171–182.
31. Alpert, B.S.; Kartodihardjo, W.; Harp, M.S.; Izukawa, T.; and Strong, W.B. 1981. Exercise blood pressure response — a predictor of severity of aortic stenosis in children. *J. Pediatr.* 98:763–765.

32. Connor, T.M. 1979. Evaluation of persistent coarctation of aorta after surgery with blood pressure measurement and exercise testing. *Am. J. Cardiol.* 43:74–78.
33. Freed, M.D.; Rocchini, A.; Rosenthal, A.; Nadas, A.S.; and Castaneda, A.R. 1979. Exercise-induced hypertension after surgical repair of coarctation of the aorta. *Am. J. Cardiol.* 43:253–258.
34. Waldhausen, J.A., and Nahrwold, D.L. 1966. Repair of coarctation of the aorta with subclavian flap. *J. Thorac. Cardiovasc. Surg.* 51:532–533.
35. Liberthson, R.R.; Pennington, D.G.; Jacobs, M.L.; and Daggett, W.M. 1979. Coarctation of the aorta: review of 234 patients and clarification of management problems. *Am. J. Cardiol.* 43:835–840.
36. Sorland, S.J.; Rostad, H.; Forfang, K.; and Abyhohm, G. 1980. Coarctation of the aorta: a follow-up study after surgical treatment in infancy and childhood. *Acta Paediatr. Scand.* 69:113–118.
37. Beerman, L.B.; Neches, W.H.; Patnode, R.E.; Fricker, F.J.; Matthews, R.A.; and Park, S.C. 1980. Coarctation of the aorta: late results after surgery. *Am. J. Dis. Child.* 134:464–466.
38. Hubbell, M.M., Jr.; O'Brien, R.G.; Krovetz, L.J.; Mauck, H.P.; and Tompkins, D.G. 1979. Status of patients 5 or more years after correction of coarctation of the aorta over age 1 year. *Circulation* 60:74–80.
39. Maron, B.J.; Humphries, O.J.; Rowe, R.D.; and Mellits, E.D. 1973. Prognosis of surgically corrected coarctation of the aorta. *Circ. Res.* 47:119–126.
40. Nanton, M.A., and Olley, D.M. 1976. Residual hypertension after coarctectomy in children. *Am. J. Cardiol.* 37:769–772.
41. Voors, A.W.; Webber, L.S.; Frerichs, R.R.; and Berenson, G.S. 1977. Body height and body mass as determinants of basal blood pressure in children. In: The Bogalusa heart study. *Am. J. Epidemiol.* 106:101–108.
42. Feinleib, M.; Garrison, R.; Borhani, N.; Rosenman, R.; and Christian, J. 1975. Studies of hypertension in twins. In *Epidemiology and control of hypertension,* ed. O. Paul, pp. 3–20. New York: Stratton Intercontinental.
43. Biron, P.; Mongeau, J.G.; and Bertrand, D. 1975. Familial aggregation of blood pressure in adopted and natural children. In *Epidemiology and control of hypertension, ed. O. Paul, pp. 397–405. New York: Stratton Intercontinental.*
44. Platt, R. 1963. Heredity in hypertension. *Lancet* 1:899–904.
45. Grim, C.E.; Luft, F.C.; Miller, J.Z.; Brown, P.L.; Gannon, M.A.; and Weinberger, M.H. 1979. Effects of sodium loading and depletion in normotensive first degree relatives of essential hypertensives. *J. Lab. Clin. Med.* 94:764–771.
46. Grim, C.E.; Luft, F.C.; Miller, J.Z.; Rose, R.J.; Christian, J.C.; and Weinberger, M.H. 1980. An approach to the evaluation of genetic influences on factors that regulate arterial blood pressure in man. *Hypertension* 2:134–142.
47. Falkner, B.; Onesti, G.; and Angelakos, E. 1981. Effect of salt loading on the cardiovascular response to stress in adolescents. *Hypertension* 3:II195–II199.
48. Light, K.C.; Koepke, J.P.; Obrist, P.A.; Willis, P.W., IV. 1983. Psychological stress induces sodium and fluid retension in men at high risk for hypertension. *Science* 220:429–431.
49. Grignolo, A.; Koepke, J.P.; and Obrist, P.A. 1982. Renal function, heart rate, and blood pressure during exercise and avoidance in dogs. *Am. J. Physiol.* 242:R482–R490.
50. Koepke, J.P.; Light, K.C.; and Obrist, P.A. 1983. Neural control of renal excretory function during behavioral stress in conscious dogs. *Am. J. Physiol.* 245:R251–R258.
51. Lundin, S., and Thoren, P. 1982. Renal function and sympathetic activity during mental stress in normotensive and spontaneously hypertensive rats. *Acta Physiol. Scand.* 115:115–124.
52. Anderson, D.E. 1982. Behavioral hypertension mediated by salt intake. In *Circulation, neurobiology and behavior,* ed. O. Smith, R. Galosy, and S. Weiss, pp. 247–257. New York: Elsevier.
53. Winternitz, S.R.; Katholi, R.E.; and Oparil, S. 1980. Role of the renal sympathetic nerves in the development and maintenance of hypertension in the spontaneously hypertensive rat. *J. Clin. Invest.* 66:971–978.
54. Winternitz, S.R.; Wyss, J.M.; Meadows, J.R.; and Oparil, S. 1982. Increased norepinephrine content of hypothalamic nuclei in association with worsening of hypertension following high sodium intake in the young spontaneously hypertensive rat. *Clin. Sci.* 63:339s–343s.
55. Winternitz, S.R., and Oparil, S. 1982. Sodium-neural interactions in the development of spontaneous hypertension. *Clin. Exp. Hypertension* A4(4/5):751–760.

56. De Champlain, J.; Krakoff, L.; and Axelrod, J. 1968. Relationship between sodium intake and norepinephrine storage during development of experimental hypertension. *Circ. Res.* 23:479–491.
57. Berecek, K.H.; Winternitz, S.R.; Wyss, J.M.; and Oparil, S. 1983. Exaggerated cardiovascular responsiveness to central sympathetic and alpha adrenergic stimulation in spontaneously hypertensive rats receiving high sodium intake. *Clin. Res.* 31:487A.
58. Lokhandwala, M.F., and Eikenburg, D.C. 1983. Presynaptic receptors and alterations in norephinephrine release in spontaneously hypertensive rats. *Life Sci.* 33:1527–1542.
59. Bianchi, G.; Cusi, D.; Gatti, M.; Plupi, G.; Ferrari, P.; Barlassina, C.; Picotti, G.P.; Bracchi, G.; Colombo, G.; Gori, D.; Velis, O.; and Mazzei, D. 1979. A renal abnormality as a possible cause of essential hypertension. *Lancet* 1:173–177.
60. Canessa, M.; Adragna, N.; Solomon, H.S.; Connolly, T.M.; and Tosteson, D.C. 1980. Increased sodium-lithium countertransport in red cells of patients with essential hypertension. *N. Engl. J. Med.* 302:772–776.
61. Garay, R.P.; Elghozi, J.L.; Dagher, G.; and Meyer, P. 1980. Laboratory distinction between essential and secondary hypertension by measurement of erythrocyte cation fluxes. *N. Engl. J. Med.* 302:769–771.
62. Woods, J.W.; Falk, R.J.; Pittman, A.W.; Klemmer, P.J.; Watson, B.S.; and Namboodiri, K. 1982. Increased red-cell sodium-lithium countertransport in normotensive sons of hypertensive parents. *N. Engl. J. Med.* 306:593–595.
63. Keen, P.M., and Bogdanski, D.F. 1970. Sodium and calcium ions in uptake and release of norepinephrine by nerve endings. *Am. J. Physiol.* 219:677–682.
64. Bohr, D.F. 1964. Electrolytes and smooth muscle contraction. *Pharmacol. Rev.* 16:85–111.
65. Blaustein, M.P. 1977. Sodium ions, calcium ions, blood pressure regulation, and hypertension: a reassessment and a hypothesis. *Am. J. Physiol.* 232(3):C165–C173.
66. Haddy, F.J., and Overbeck, H.W. 1976. The role of humoral agents in volume expanded hypertension. *Life Sci.* 19:935–948.
67. De Wardner, H.E., and MacGregor, G.A. 1983. The role of a circulating inhibitor of Na+-K+-ATPase in essential hypertension. *Am. J. Nephrol.* 3:88–91.
68. Prebis, J.W.; Gruskin, A.B.; Polinsky, M.S.; and Baluarte, H.J. 1981. Uric acid in childhood essential hypertension. *J. Pediatr.* 98:702–707.
69. Schieken, R.M.; Clarke, W.R.; and Lauer, R.M. 1981. Left ventricular hypertrophy in children in the upper quintile of the blood pressure distribution: the Muscatine study. *Hypertension* 3:669–675.
70. Zahka, K.G.; Neill, C.A.; Kidd, L.; Cutilletta, M.A.; and Cutilletta, A.F. 1981. Cardiac involvement in adolescent hypertension: echocardiographic determination of myocardial hypertrophy. *Hypertension* 3:664–668.
71. Julius, S., and Schork, M.A. 1971. Borderline hypertension: a critical review. *J. Chronic Dis.* 23:723–754.
72. Stackpole, R.H.; Melicow, M.M.; and Uson, A.C. 1963. Pheochromocytoma in children. *J. Pediatr.* 63:315–330.
73. Sheps, S.; Van Heerden, J.; and Sheedy, P., II. 1981. Current approach to the diagnosis of pheochromocytoma. In *Secondary forms of hypertension: current diagnosis and management,* ed. M.D. Blaufox and C. Bianchi C, pp. 11–18. New York: Grune and Stratton.
74. Bravo, E.L.; Tarazi, R.C.; Gifford, R.W.; and Stewart, B.H. 1979. Circulating and urinary catecholamines in pheochromocytoma: diagnostic and pathophysiologic implications. *N. Engl. J. Med.* 301:682–686.
75. Bravo, E.L.; Tarazi, R.C.; Fouad, F.M.; Vidt, D.G.; and Gifford, R.W. 1981. Clonidine-suppression test: a useful aid in the diagnosis of pheochromocytoma. *N. Engl. J. Med.* 305:623–626.
76. Stewart, G.H.; Brown, E.L.; Haaga, J.; Meaney, T.F.; and Tarazi, R. 1978. Localization of pheochromocytoma by computed tomography. *N. Engl. J. Med.* 299:460–461.
77. Greenberg, R.E. 1975. Pheochromocytomas. In *Endocrine and genetic diseases of childhood and adolescence,* ed. L.I. Gardner, pp. 898–907. New York: W.B. Saunders.
78. Sisson, J.C.; Frager, M.S.; Valk, T.W.; Gross, M.D.; Swanson, D.P.; Wieland, D.M.; Tobes, M.C.; Beierwaltes, W.H.; and Thompson, N.W. 1981. Scintigraphic localization of pheochromocytoma. *N. Eng. J. Med.* 305:12–17.
79. Francis, I.R.; Glazer, G.M.; Shapiro, B.; Sisson, J.C.; and Gross, B.H. 1983. Complementary roles of CT and ^{131}I-MIBG scintigraphy in diagnosing pheochromocytoma. *AJR* 141:719–725.

80. Manger, W.M., and Gifford, R.W., Jr. 1977. *Pheochromocytoma*. Berlin: Springer-Verlag.
81. Wallace, J.M., and Gill, D.P. 1978. Prazosin in the diagnosis and treatment of pheochromocytoma. *JAMA* 240:2752–2753.
82. Rosei, E.A.; Brown, J.J.; Lever, A.F.; Robertson, A.S.; Robertson, J.I.S.; and Trust, P.M. 1976. Treatment of pheochromocytoma and of clonidine withdrawal hypertension with labetalol. *Br. J. Clin. Pharmacol.* Suppl 3, 3(4):809–815.
83. Sjoerdsma, A.; Engelman, K.; Spector, S.; and Udenfriend, S. Inhibition of catecholamine synthesis in man with alpha-methyl-tyrosine, an inhibitor of tyrosine hydroxylase. *Lancet* 2:1092–1094.
84. New, M.I.; Oberfield, S.E.; and Levine, L.S. 1983. Hypertension in children. In *Hypertension,* ed. J. Genest, O. Kuchel, P. Hamet, and M. Cantin, pp. 853–889. New York: McGraw-Hill.
85. New, M.I.; Oberfield, S.E.; Levine, L.S.; Dupont, B.; Pollack, M.; Gill, J.R.; and Bartter, F.C. 1980. Demonstration of autosomal dominant transmission and absence of HLA linkage in dexamethasone suppressible hyperaldosteronism. *Lancet* 1:550–551.
86. Bravo, E.L.; Tarazi, R.C.; Dustan, H.P.; Fouad, F.M.; Textor, S.C.; Gifford, R.W.; and Vidt, D.G. 1983. The changing clinical spectrum of primary aldosteronism. *Am. J. Med.* 74:641–651.
87. Biglieri, E.G. 1976. A perspective on aldosterone abnormalities. *Clin. Endocrinol.* 5:399–410.
88. Ganguly, A.; Bergstein, J.; Grim, C.E.; Yum, M.N.; Weinberger, M.H. 1980. Childhood primary aldosteronism due to an adrenal adenoma: preoperative localization by adrenal vein catheterization. *Pediatrics* 65:605–609.
89. White, E.A.; Schambelan, M.; Rost, C.R.; Biglieri, E.G. Moss, A.A.; Korobkin, M. 1980. Use of computed tomography in diagnosing the cause of primary aldosteronism. *N. Engl. J. Med.* 303:1503–1507.
90. Weinberger, M.H. 1983. Primary aldosteronism. In *Hypertension,* ed. J. Genest, O. Kuchel, P. Hamet, and M. Cantin, pp. 922–939. New York: McGraw-Hill.
91. Liddle, G.W. 1960. Tests of pituitary-adrenal suppressibility in the diagnosis of Cushing's syndrome. *J. Clin. Endocrinol.* 12:1539–1560.
92. Balfe, J.W., and Rance, C.P. 1978. Recognition and management of hypertensive crises in childhood. *Pediatr. Clin. North. Am.* 25:159–174.
93. Multiple Risk Factor Intervention Trial Research Group. 1982. Multiple risk factor intervention trial: risk factor changes and mortality results. *JAMA* 248(12):1465–1477.
94. Mirkin, B.L., and Sinaiko, A.R. 1978. Childhood hypertension: new therapeutic agents and management concepts. In *Clinical pharmacology and therapeutics: a pediatric perspective,* ed. B.L. Mirkin, pp. 251–272. Chicago: Yearbook Medical Publishers.
95. Oliver, M.F. 1982. Risks of correcting the risks of coronary disease and stroke with drugs. *N. Engl. J. Med.* 306:297–298.
96. Buhler, F.R.; Burkart, F.; Lutold, B.E.; et al. 1975. Antihypertensive beta blocking action as related to renin and age: A pharmacologic tool to identify pathogenic mechanisms in essential hypertension. *Am. J. Cardiol.* 36:653–669.
97. Berenson, G.S.; Voors, A.W.; Webber, L.S., et al. 1975. A model of intervention for prevention of early essential hypertension in the 1980's. *Hypertension* 5:41–54.

13. CORONARY ATHEROSCLEROSIS — EXTENSION OF DEVELOPMENTAL PROCESSES BEGINNING IN INFANCY

ADAM SCHNEEWEISS AND HENRY N. NEUFELD

The clinical presentation of coronary atherosclerosis in adults is known today to be only the tip of an iceberg. As with several other organs in the body, structural changes which may be termed pre-atherosclerotic changes begin early in infancy or even before birth.

Development of the coronary arteries does not stop at birth but continues during infancy and childhood. Examination of these changes in human embryoes, infants, and children has revealed a correlation between various genetically determined structural changes in the coronary arteries in infancy and childhood, and the development of coronary atherosclerosis in adults. Moreover, studies on "natural models" such as coarctation of the aorta and the Progeria syndrome have enabled investigators to evaluate their developmental process in childrens' coronary arteries as it affects the development of coronary artery disease in adults, and to determine the role of various risk factors present in the childhood in modifying the pattern of these structural changes.

This chapter addresses two aspects of the subject. The first part presents data suggesting that coronary atherosclerosis is a pediatric problem related to developmental changes. The second part focuses on the normal development of the coronary circulation with emphasize on changes possibly related to adulthood atherosclerosis.

IS CORONARY ATHEROSCLEROSIS A PEDIATRIC PROBLEM?

The theory that atherosclerosis is a pediatric problem has been confirmed during more than two decades of research. This conclusion is based on the findings of four main findings:

M.J. Legato (ed.) THE DEVELOPING HEART, pp. 245–251.

1. The presence of atherosclerotic and other obstructive lesions in the coronary arteries of children and young adults.
2. The presence of known risk factors for coronary atherosclerosis in infancy and childhood.
3. The finding that progressive structural changes in coronary arteries in infancy and childhood are related to coronary atherosclerosis in adult life.
4. The finding that known risk factors for atherosclerosis can produce premature structural coronary changes in infancy and childhood, leading to coronary obstructive lesions in children or young adults.

The Actual Finding of Coronary Atherosclerotic and Pre-atherosclerotic Changes in Infants, Children, and Young Adults

Research in this direction gained impetus when autopsy examinations of young U.S. soldiers in Korea revealed coronary atherosclerosis which in some cases was quite advanced [1]. Further evidence was found in autopsy studies during the International Atherosclerosis Project [2] and in young, apparently healthy men in whom the disease process must have begun in childhood.

The Hutchinson-Gilford Progeria syndrome may be considered a natural model of accelerated developmental changes and premature aging. Most patients with this rare syndrome die in the first decade of life from complications of severe coronary atherosclerosis.

Coronary obstructive lesions of various types have also been found in several inherited disorders of metabolism such as Sandhoff's disease [3], Fabry's disease [4], alkaptonuria [5], and other conditions such as pseudoxanthoma elasticum.

The Presence of Known Risk Factors for Coronary Atherosclerosis in Infancy and Childhood

Several of the known risk factors for coronary atherosclerosis may be present in infancy. Hypertension in children has come to be recognized in recent years as a public health problem, with incidence ranging between 1% and 4% [6]. Hypertension is an established risk factor for coronary atherosclerosis even in the absence of other risk factors [7, 8, 9]. Hypertension has been suggested to aggravate atherosclerosis by stretching the arterial wall, which increases its premeability to cholesterol and lipoproteins. The not yet fully developed coronary arteries of an infant are at least as vulnerable to the deleterious effect of hypertension as are the coronary arteries of an adult. This subject is discussed further in the next section of this chapter.

Hyperlipidemia may be associated with an increased risk for coronary atherosclerosis. The only type for hyperlipidemia that presents a major risk factor in early childhood is type II hyperlipidemia (according to the classification of Fredrickson et al. [10]), which presents as hypercholesterolemia with xanthomatosis. It is a genetically determined dominant trait with an incidence of 0.5% to 2%. It was shown that youthful hypercholesteremics are about 30 times more susceptible to premature myocardial infarction than their age-matched normocholesteremics

[11]. These premature infarctions result from a prolonged abnormal developmental process during childhood, while patients in the homozygote state are especially affected. Those at the heterozygote state are also at increased risk.

Tobacco smoking is an important risk factor for development of coronary atherosclerosis. It is usually not found in the pediatric age group under consideration in this chapter, but it may pose a problem at the two extremes of the group: in older children where tobacco smoking may be found, and among pregnant women whose smoking may damage the coronary arteries of the fetus. A recent electronmicroscopic study revealed a pronounced thickening of the basement membrane of the umbilical artery of fetuses of smoking mothers [12]. These changes were not found in fetuses of non-smoking mothers, and they are very similar to changes described in the coronary arteries of infants from a coronary atherosclerosis population [13].

Diabetes mellitus is a major risk factor for coronary atherosclerosis in adults. The effect of diabetes on the development of atherosclerosis may be due to the hyperglycemia which has been proven to be a mild risk factor in itself, or by its deleterious influence of serum lipids. The prevalence of juvenile onset diabetes as 1/1000 children and is characterized by the same lipid profile as adult onset diabetes [5, 14]. Males are also considered to be at major risk for coronary atherosclerosis although the exact mechanism for this preference of the disease has not been determined. The greater incidence of coronary atherosclerosis in males is evident especially at young ages where male infants and children show more pronounced preatherosclerotic changes than females. This point is discussed further in the next section.

It may be summarized that most risk factors for atherosclerosis are present in infancy and childhood, and their interference with the normal development of the coronary arteries may result in premature atherosclerosis.

Structural Changes in the Coronary Arteries in Infants and Children Related to Coronary Atherosclerosis in Adults

The coronary arteries undergo continuous structural changes in infancy and childhood, including formation of a new layer of the arterial wall — the musculoelastic layer — and proliferation changes in the intima and media, leading to asymmetric narrowing of the arterial lumen (described in detail in the next section of this chapter). They are evident in the first weeks of life. Several studies performed in the United States, Finland, Israel, and other countries showed them to be more common in population groups whose adults have a higher incidence of coronary atherosclerosis.

Studies from our Institute [15] demonstrated that histologic changes in the intima and media in infancy and childhood were more common in males than females. They were also more pronounced in infants of Ashkenazi Jews, a population group which has a high incidence of coronary atherosclerosis in adulthood, and were found to be almost absent in infants and children of Yemenite Jews and Bedouins, populations in Israel with a very low incidence of coronary artherosclerósis [13]. A study in Finland showed that structural changes in coronary artery walls were prominent in infants from the eastern part of the country and minimal in infants from the western part, a finding that correlates well with the distribution of the incidence of coronary artherosclerosis in the adult population. [16]. Similar findings were obtained in studies of various ethnic groups in the United States [2, 17] and Haiti [17]. Moreover, we found in Israel that the difference in coronary artery structural changes between the sexes (greater in males) were pronounced in infants and children of the population groups with high incidence of atherosclerosis such as Ashkenazi Jews, and minimal in infants and children of those populations with a low incidence of atherosclerosis, such as Yemenite Jews and Bedouins.

Known Risk Factors for Coronary Atherosclerosis Can Produce Premature and Accelerated Structural Changes in the Coronary Arteries

The coronary arteries and coronary microcirculation of infants and children with most risk factors for coronary atherosclerosis are not readily accessible for histologic examination. There are, however, several natural models on which one can study histologically the effect of hypertension on the coronary circulation; these are cases of coronary hypertension induced by congenital lesions such as supravalvular aortic stenosis and coarctation of the aorta.

As early as 1962, Neufeld et al. [18] reported the necropsy findings in a two-year-old boy with tubular narrowing of the ascending aorta (hypoplasia of the ascending aorta), a rare form of supravalvular aortic stenosis. The media of the coronary arteries was thickened and contained coarse elastic fibers. The authors attributed these changes to the increased systolic pressure in the coronary arteries and the ascending aorta, proximal to the stenotic segment. In 1968, Vlodaver and Neufeld [19] histologically studied the coronary arteries of 15 patients with coarctation of the aorta and observed significant histological changes. The intima was thickened due to proliferative and degenerative changes of the elastic fibers and proliferation of the collagenous tissue. The media was markedly thickened with proliferation of the elastic fibers. The intimal changes were limited to the large coronary arteries, while the medial changes extended also to the small coronary arteries. In addition, young men with coarctation of the aorta exhibited premature coronary atherosclerosis as well as larger than normal coronary capacity.

In a recent study from our Institute, Schneeweiss et al. [20] reported structural changes also in the coronary microcirculation of patients with coarctation of the aorta. Eleven patients ranging in age from 4 to 20 years were studied. Small pieces of myocardium from the left ventricle were obtained during cardiac surgery and examined by electron microscope. Marked changes were found in the arterioles, while more distal segments of the coronary microcirculation — the precapillary sphincters and the capillaries — appeared normal. The walls in the arterioles were thickened. The muscular layer appeared fragmented, and smooth muscle cells had lost their spindle-shaped form and contained abnormal vacuoles and swollen mitochondria. In severely affected areas, the smooth muscle cells were completely atrophic or replaced by collagenous agglomerations. In advanced cases, the intima was also affected with vacuolization of endothelial cells and penetration of collagen. Two patterns of change were observed: in children the changes exhibited a patchy pattern with islets of structural changes interspersed among areas of normal tissue; in young adults, the changes were more diffuse, occupying in some cases the whole circumference of the arteriolar wall.

These findings confirm that coronary changes are produced by coronary hypertension and are related to the duration of this effect. The coronary arterial and microcirculatory changes may at account at least partially for the increased mortality rate in patients with coarctation of the aorta undergoing surgical correction at a relatively late age.

The effect of tobacco smoking on the circulation was demonstrated in Asmussen's studies of the umbilical arteries. Fetuses of smoking mothers had pronounced thickening of the basement membrane of the umbilical arteries. The rough endoplasmic reticulum was dilated, the endothelial cells were endematous, and there were contractile-type myocytes beneath the intima. These changes were not found in fetuses of non-smoking mothers, and although not proven to be directly related to coronary atherosclerosis, they bear a close similarity to the changes described in infants from groups with a high incidence of coronary artery disease.

DEVELOPMENTAL PROCESSES AND PRE-ATHEROSCLEROTIC CHANGES IN THE CORONARY ARTERIES IN INFANCY AND CHILDHOOD

The coronary arterial walls undergo continuous changes throughout infancy and childhood. Since some of these changes were shown to be related to atherosclerosis, only coronary arteries of fetuses may be considered completely normal. A histologic study reported by Neufeld [21] of coronary arteries of fetuses up to the age of 34 weeks was based on this assumption which does not, however, take into account the influence of risk factors via the material circulation.

The normal fetal coronary arterial wall consists of: (1) the intima — a single layer of endothelial cells; (2) the internal elastic membrane — a homogenous elastic material surrounding the intima as a continuous tube with longitudinal corrugations; (3) the media — a few layers of smooth muscle cells and some elastic fibers; and (4) the adventitis — an external layer of loose connective tissue.

At 34 weeks gestation, the simple structure of the coronary arterial wall becomes more complex and a musculoelastic layer develops between the media and the intima. This layer is formed by a series of degenerative and proliferative changes including splitting of the internal elastic membrane, proliferation of smooth muscle cells within the split areas, and an increase in the amount of elastic fibers in the internal layers of the media. The musculoelastic layer may occupy most of the media, leaving only two to three rows of smooth muscle cells in the outer layer. Structural changes appear also in the intima which thickens due to formation of cushions composed of fibroblasts, elastic fibers, and droplets of acid mucopolysaccharides [15]. These changes are common to many ethnic groups.

They are usually seen a few days after birth, but may in a few instances exist before birth. The changes are completed at the age of one year by which time the arterial wall has assumed a new shape. The most common site of occurrence of these changes is the proximal part of the left anterior descending coronary artery.

Studies in our institute and other centers have shown these intimal and medial changes to be more extensive in males than in females. The sex difference was more prominent in population groups with a high incidence of coronary atherosclerosis in adult life. It is therefore concluded that at least some of these changes may be considered to be pre-atherosclerotic.

In 1967 Vlodaver and Neufeld [22] developed a hemodynamic concept to explain the physiologic role of the structural changes. After birth, the pressure in the coronary arteries rises and causes increased tension in the arterial wall, bringing about initial degenerative changes. Later, during the adaptive phase, a musculoelastic layer is formed which combines the elastic and active tensions, enabling it to withstand the elevated pressure and prevent deleterious deformation of the arterial wall.

SUMMARY

Continuous structural changes, which are considered to be adaptive in nature, occur in the walls of the coronary arteries throughout the first year of life. They were found to be related to known risk factors for atherosclerosis and were more common in groups with a high incidence of coronary atherosclerosis in adult life. It is therefore concluded that developmental processes in infancy may form preatherosclerotic changes in the coronary arteries in the first years of life.

REFERENCES

1. Enos, W.F.; Holmes, R.H.; and Beyer, J. 1953. Coronary disease among United States soldiers killed in action in Korea. Preliminary report. *JAMA* 152:1090.
2. McGill, H.C. 1968. *The geographic pathology of atherosclerosis*. Baltimore: Williams & Wilkins.
3. Blieden, L.C.; Desnick, R.; and Carter, J. 1974. Cardiac involvement in Sandhoff's disease; an inborn error of glyco-sphingolipid metabolism. *Am. J. Cardiol.* 34:83.
4. Blieden, L.C., and Molle, J.H. 1974. Cardiac involvement in inherited disorders of metabolism. *Prog. Cardiovasc. Dis.* 10:615.
5. Strong, W.B. 1978. Is atherosclerosis a pediatric problem? In *Atherosclerosis: its pediatric aspects*, ed. W.B. Strong, p. 1. New York: Grune & Stratton.
6. Fixler, D.E. 1978. Epidemiology of childhood hypertension. In *Atherosclerosis: its pediatric aspects*, ed. W.B. Strong, p. 177. New York: Grune & Stratton.
7. Wissler, R.W., and Geer, J.C. 1972. *The pathogenesis of atherosclerosis*. Baltimore: Williams & Wilkins.
8. Frees, E.D. 1969. Hypertension and atherosclerosis. *Am. J. Med.* 46:735.
9. Kannel, W.B.; Schwartz, M.J.; and McNamara, P.M. 1969. Blood pressure and risk of coronary heart disease: the Framingham study. *Dis. Chest.* 56:43.
10. Frederickson, D.S.; Levy, R.J.; and Lees, R.S. 1967. Fat transport in lipoproteins — an integrated approach to mechanisms and disorders. *N. Engl. J. Med.* 276:24.
11. Thomas, C.B.; Ross, D.C.; and Duszynski, K.R. 1978. Youthful hypercholesterolemia: its associated characteristics and role in premature myocardial infarction. In *Atherosclerosis: its pediatric aspects*, ed. W.B. Strong, p. 131. New York: Grune & Stratton.
12. Asmussen, I. 1979. Fetal cardiovascular system as influenced by maternal smoking. *Clin. Cardiol.* 2:246.

13. Vlodaver, Z.; Kahn, H.A.; and Neufeld, H.N. 1964. The coronary arteries in early life in three different ethnic groups. *Circulation* 29:541.
14. Kannel, W.B.; Dawber, T.R.; and McNamara, P.M. 1966. Detection of the coronary prone adult: the Framingham study. *J. Iowa. Med. Soc.* 56:26.
15. Neufeld, H.N., and Vlodaver, Z. 1977. Structural changes of coronary arteries in young age groups. In *International cardiology*. Vol. 2 of Cardiovascular Clinics series, ed. A.N. Brest and P.D. White, p. 55. Philadelphia: F.A. Davis.
16. Personen, E.; Norio, R.; and Sarna, S. 1975. Thickenings in the coronary arteries in infancy as an indication of genetic factors in coronary heart disease. *Circulation* 51 and 52:218.
17. Groon, D.; Mckee, E.E.; Adkins, W.; Pean, V.; and Hudicourt, E. 1964. Developmental patterns of coronary and aortic atherosclerosis in young Negroes of Haiti and the United States. *Ann. Intern. Med.* 61:900.
18. Neufeld, H.N.; Wagenvoort, C.A.; Ongley, P.A.; and Edwards, J.E. 1962. Hypoplasia of ascending aorta: an unusual form of supravalvular aortic stenosis with special reference to localized coronary arterial hypertension. *Am. J. Cardiol.* 10:746.
19. Vlodaver, Z., and Neufeld, H.N. 1968. The coronary arteries in coarctation of aorta. *Circulation* 37:449.
20. Schneeweiss, A.; Sherf, L.; Lehrer, E.; Lieberman, Y.; and Neufeld, H.N. 1982. Segmental study of the terminal coronary vessels in coarctation of the aorta. *Am. J. Cardiol.* 49:1996.
21. Neufeld, H.N.; Wagenvoort, C.A.; Edwards, J.E. 1962. Coronary arteries in fetuses, infants and young adults. *Lab. Invest.* 11:837.
22. Vlodaver, Z., and Neufeld, H.N. 1967. The musculo-elastic layer in the coronary arteries: a histological and hemodynamic concept. *Vasc. Dis.* 4:136.

INDEX